Trail Guide to the Body

"Order and simplification are the first steps toward the mastery of a subject - the actual enemy is the unknown."

Thomas Mann, *The Magic Mountain*

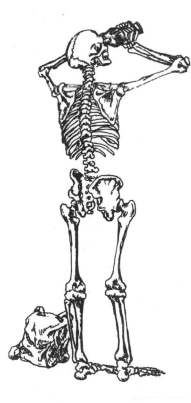

Trail Guide to the Body

How to locate muscles, bones and more

Andrew Biel, LMP

Licensed Massage Practitioner

Illustrations by Robin Dorn, LMP

Licensed Massage Practitioner

Published by Andrew Biel, LMP
PO Box 6107, Boulder, CO 80302 USA
(800) 775-9227

Associate Editors
Kate Bromley, MA, LMP
Lauriann Greene, LMP
Clint Chandler, LMP

Graphic Design by Jackie A. Phillips

Printed by Consolidated Press, Seattle, WA

**Library of Congress
Cataloging-in-Publication Data**

Biel, Andrew R.
Trail Guide to the Body:
How to locate muscles, bones and more

Includes bibliographical references.

Includes index.

ISBN: 0-9658534-0-3
Library of Congress Catalog Number: 97-093767

8

Grateful acknowledgment is
made to reprint an excerpt from:

The Magic Mountain by Thomas Mann
Copyright © 1927. Used by permission
of Random House, a division of Alfred Knopf, Inc.

Four Quartets by T.S. Eliot
Copyright © 1943. Used by permission
of Harcourt Brace & Company

Disclaimer
The purpose of this book is to provide information for hands-on therapists on the subject of palpatory anatomy. This book does not offer medical advice to the reader, and is not intended as a replacement for appropriate health care and treatment. For such advice, readers should consult a licensed physician.

Table of Contents

Six - Pelvis and Thigh 191

Seven - Leg and Foot 241

Preface

"We shall not cease from exploration.
And the end of all our exploring
Will be to arrive where we started
And know the place for the first time."

T.S. Eliot, *Four Quartets*

Many years ago, as a skinny ten-year old, I remember pinching the flesh under my armpit only to accidently locate a muscle. When I moved my arm a certain way, the flesh hardened and slipped between my fingers. "Wow," I remember saying, "I didn't think I *had* any muscles!"

I told my parents of my new discovery and they suggested I check the encyclopedia to see which muscle I had found. The Latin names only confused me, but for months I showed everyone I met my one and only muscle.

I continued to be fascinated with the parts and pieces of the body, and how they all seemed to work together to create movement, breath - life. During my bodywork training I learned that the mystery muscle of my arm pit was the latissimus dorsi. I also discovered how to palpate other muscles, as well as tendons, bones and other tissues throughout the body. I realized the importance of palpation for assessing tissues and performing safe and effective manual techniques.

As an instructor of bodywork, palpatory anatomy and kinesiology, I knew of many books which described and illustrated the anatomy of the body. I found few, however, that demonstrated how to manually locate and explore its structures. *Trail Guide to the Body* is designed to do just that: to teach you to navigate, map and "gain your bearings" on the human body.

In preparation for any journey, it helps to know the lay of the land you will be traveling. As a manual therapist, the territory of exploration is arguably the most fascinating and challenging of all terrains - the human body. All health care providers need a thorough understanding of the location, texture and interrelationship of the body's structures. Yet, the "hands-on" practitioner cannot take the guided bus tour of the body, viewing it from afar and only hearing of its amazing qualities. Instead, his profession lies in deep exploration of an environment that is never the same on any two individuals. He must roll up his sleeves and discover the native terrain with both hands.

So welcome to the human body! You are about to embark on the journey of a lifetime, and this book will serve as your trusty guide.

Acknowledgements

While creating this book, I was blessed to work with many helpful, inspiring and loving people. First of all, my many thanks to Robin Dorn for the incredible dedication, joy and artistry she has brought to this book.

I am deeply grateful to Jackie Phillips and Diana Thompson for their wonderful guidance, suggestions, patience and confidence.

Many thanks to Kate Bromley, Lauriann Greene, Clint Chandler, Claire Gipson and Marty Ryan for their editing and insight.

This book could not have been completed without the assistance and patience of Roger Williams, Leslie Grounds, Kirk Butler, Dave Oder, Sylvia Burns, Damon Williams, Brian Weyand, Chad Herrin, Pantelis Zafiriou and Obie Roe during the modeling and photography.

Thank you to Vicky Stolsen, Jay Hilwig and Sandy Johns of Consolidated Printing for answering so many questions. Thanks to Sandy Merrell for her proofreading and Damon Williams for his help with the translations. Thanks also to Steve Goldstein, Meta Gunzenhauser, Alison Kim, Paige Reed and Dawn Schmidt.

Thanks to all the students, faculty and staff of Seattle Massage School including Jamie Alagna, Coleen Renee, Cynthia Wold, Alexis Brereton, Deb Brockman, Mary Bryan, Patrick Bufi, Sean Castor, Jessica Elliott, Alyce Green-Davis, Joanne Guidici, Laura Goularte, Diana Kincaid, Sean McDaniel, Steve Miller, Vicky Panzeri, Paula Pelletier, Anita Quinton, Zdenka Vargas and Tonya Yuricich.

Thank you to the faculty, students and staff of the Boulder College of Massage Therapy including Gaye Franklin, Mary Ellen O'Malley and Christopher Quinn.

I am very grateful to the following people for their expertise, research and encouragement: Leon Chaitow, Sandy Fritz, Darlene Hertling, Al Schwartz, John White, Sharon Babcock, Cynthia Christy, Ann Ekes, Barb Frye, Daniel Gebo, Jim Holland, George C. Kent, Don Kelley, Lee Haines, Darlene Hertling, Mary Marzke, Susan Parke, Annie Thoe, Jeannie Waschow and John Zurhourek.

Thank you to the many people who assisted with the editing process including Adam Bailey, Nancy Benerofe, Vicky Fosie, Dawn Fosse, Joanne Fowler, Petra Guyer, Debra Harrison, Llysa Holland, Ian Hubner, Erica King, Elinore Knutson, Dave Lawrence, Andrew Litzky, Kate McConnell, Becky Masters, Micheal Max, Audra Meador, Chris Meier, Debra Nelli, Eric Newberg, Jillian Orton, Dee Reeder, Penny Rosen, Janice Schwartz, Gerald Sexton, Joy Shaw and Danny Tseng.

Special thanks to my family for their endless support and encouragement. Last, thanks to Tiger and Kenya - if it weren't for you, this would have been finished a lot sooner.

This book is dedicated to my mother and father.

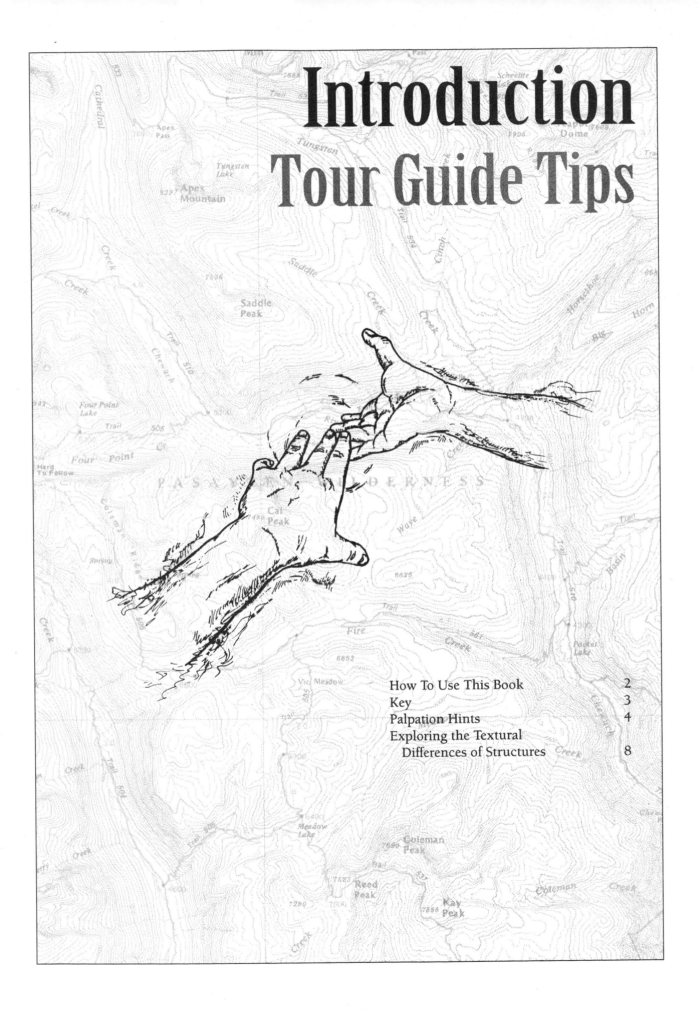

Introduction
Tour Guide Tips

How To Use This Book

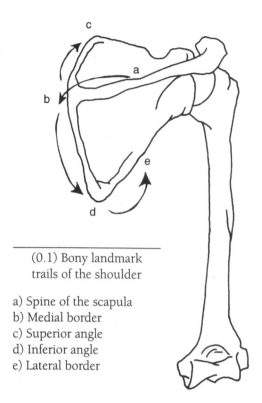

(0.1) Bony landmark
trails of the shoulder

a) Spine of the scapula
b) Medial border
c) Superior angle
d) Inferior angle
e) Lateral border

Trail Guide to the Body includes six chapters, each focusing on a region of the body. Outlined first are the topographical contours which can be seen on the surface of the skin. This is followed by the bones and bony landmarks (the bone's hills, dips and ridges). The bony landmarks can be thought of as your "trail markers." They are utilized as stepping off points to locate muscles and tendons. Finally, other structures such as ligaments, nerves, arteries and lymph nodes are located.

When possible, a region's bony landmarks have been strung together to form a trail. (0.1) These trails are designed to help you understand the connection between structures. Without a path to follow, you, the traveler, could be lost in a jungle of flesh and bones with no idea of your "base camp" location. You and your travel partner will find the journey more enjoyable and valuable if you have a trail leading to your destination point.

Since bodies come in a variety of sizes and shapes, it may seem unrealistic that one trail guide would apply to all bodies. If the terrain is never the same, what is the use of a map? Even though the topography, shape and proportion of each person is unique, the body's composition and structures are virtually identical on all individuals. The difference is simply qualitative: it is easy to find many structures on a person with a slender build and more challenging on a physique with bulky muscles or a large amount of adipose (fatty) tissue.

Trail Guide to the Body is designed around the following scenario: You follow along with the text and palpate on a partner (friend or classmate) who is on a bodywork table or seated in a chair. If you are a student, you are advised to proceed step-by-step, repeat certain methods when necessary and explore the body along the way. If you are a more experienced practitioner, you may want to pick and choose your destinations.

The procedures outlined in *Trail Guide to the Body* are gentle and rarely uncomfortable, yet it is best to practice on an individual who has no serious health conditions. Your partner may either wear loose, thin clothing or be undressed and draped under a sheet to enable you to palpate specifically.

Sometimes your partner will be asked to passively lie or sit on the table. At other times, he may be asked to move a limb, bend a joint or contract a group of muscles. These movements should be done smoothly and specifically to the text's instructions to allow you to explore the region thoroughly.

Talk to your partner before palpating, so he will understand his role. Also, clarify beforehand which areas of the body you would like to palpate and explore, so he will know what to expect.

Key

Name of structure

Introduction describing a structure's function, depth and relationship to other structures

Action,
Origin and Insertion sites,
Nerve innervation of the muscle

Step-by-step instructions how to palpate a structure

Alternative palpatory routes

"Check It" questions will confirm your location. They may ask you about your location in relationship to a nearby structure or ask you or your partner to create a movement. Unless otherwise indicated, the answers to the questions should be "Yes!"

After you have become familiar with a muscle, look for the compass to give you the essential location and bony landmarks to palpate a muscle. The action refers to a movement your partner can perform to feel the muscle contract.

Sternocleidomastoid

The sternocleidomastoid (SCM) is located on the lateral and anterior sides of the neck. It has a large belly which is composed of two heads: a flat, clavicular head and a slender, sternal head. (5.1) Both heads merge together to attach at the mastoid process behind the ear. The SCM is superficial, completely accessible and often visible when the head is turned to the side.

A -
Unilateral:
Laterally flex the head to the same side
Rotate the head to the opposite side
Bilateral:
Flex the neck
Assist in inspiration

O - Top of manubrium, medial clavicle

I - Mastoid process of temporal bone, lateral superior nuchal line of occiput

N - Spinal accessory

SCM

(5.1)

1) Supine, with practitioner at head of the table. Locate the mastoid process of the temporal bone and medial clavicle.
2) Draw a line between these landmarks to delineate the location of the SCM.
3) Ask your partner to raise her head ever-so-slightly off the table as you palpate the SCM. It will usually protrude visibly. (5.2)

With your partner relaxed, can you grasp the SCM between your fingers and outline its thickness and shape? How much space is between the clavicular attachments of the SCM and trapezius?

(5.2)

Location - Superficial, anterior neck
Bony Landmarks - Mastoid process, clavicle
Action - "Flex your head"

sternocleidomastoid   **ster**-no-**kli**-do-**mas**-toyd

Translations and pronunciations

Look for Mr. Bones sharing cautionary advice

Check out the grey boxes for palpation tips, comparitive anatomy and other curiosities

The techniques described in *Trail Guide to the Body* should be viewed as helpful tour guides. When first palpating, it is best to follow the specific instructions. After you have located a structure, it is recommended that you adapt and explore other methods to discover approaches which work best for you.

When possible, an optional method for locating a structure has been included. Like any worthwhile roadtrip, veering off course to explore other areas often leads to wonderful discoveries. Please feel free to veer.

Palpation Hints

Palpation means "to examine or explore by touching (an organ or area of the body), usually as a diagnostic aid." It is an art and a skill which involves (1) locating a structure, (2) becoming aware of its characteristics and (3) assessing its quality or condition so you can determine how to treat it.

The first two aspects of palpation - locating and being attentive to the body's structures - require a thorough knowledge of functional anatomy and experience by mindful, hands-on practice. This is the focus of *Trail Guide to the Body*. Assessment - the third aspect of palpation - is a vast subject requiring a book of its own.

An experience of all senses, palpation includes hands and fingers, open eyes, listening ears, calm breath and a quiet mind. As you explore the terrain and texture of the body, be sure to bring along all of your sensing tools.

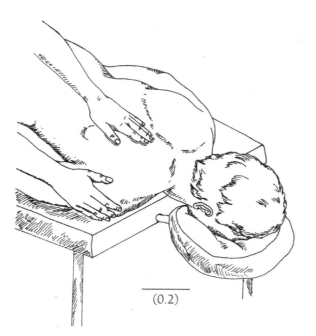

(0.2)

Making Contact

Allow your hands and fingers to be responsive and sensitive. Relaxed, patient hands will allow the body's contour, temperature and structures to come more easily into your awareness. (0.2)

While palpating, you may find it helpful to periodically close your eyes and focus your awareness into your hands. For greater sensitivity and stability, try laying one hand upon the other, using the top hand to create the necessary pressure, while the bottom hand remains relaxed. This will allow the bottom hand to stay receptive, while the top hand directs movement and depth.

Smaller structures can be located by using one or two fingertips. Many structures are best palpated by using your whole hand. Ample contact with the region you are palpating will make for more specific palpation and greater understanding of the inter-relationship of structures. Using a full hand also helps to define a structure's complete shape by sculpting out all of its sides and edges.

Working Hard versus Working Smart

Often in the excitement of trying to locate something (whether it be a muscle or a set of car keys), you search so earnestly that your mental and physical awareness can begin to diminish. You begin to "work hard." Frustration appears, your breath stalls and your hands ultimately become insensitive. Instead of working hard, you can work smart by reading the information about the structure before you palpate. As you palpate, try to visualize what you are accessing and verbalize to your partner what you are feeling.

"Work smart" by locating these structures on your own body before you palpate on your partner.

palpate   L. *palpare*, to touch

Self-palpation will improve your own kinesthetic understanding of what you are looking for on your partner and improve your own self awareness. Also, read the information aloud. Hearing the language as you are reading the text will assist in your retention and understanding of the information.

Last, be patient with your learning process. Allow yourself to "make a wrong turn and get lost" on the body. Chances are you are in close proximity to what you are seeking. By letting your senses recognize the body's trail signs, you will get to where you want to be.

Here is a simple exercise to increase your tactile sensitivity and palpatory skills. You will need a phone book and a human hair. Lay the hair beneath a single page of the phone book. Close your eyes, palpate through the page and try to locate the hair. When you find it, reposition the hair and add another page. Continue to add pages until you can no longer locate the hair. How many pages can you palpate through? 5? 10? 15?!

Less Is More

As you begin exploring the body, you may not feel things as readily as you wish. A common response is to press harder and deeper with your hands. Instead of pushing, consider letting the muscles and other tissues come into your hands. Gentle contact allows the hands to be sensitive, while excessive pushing numbs the fingers and does not welcome the tissues into the hands.

Deeper structures can also be accessed with mild pressure. Paradoxically, the deeper you move into the body, the slower and softer your touch needs to become. Ultimately, palpating at different levels of the body is not a question of pressure, but of intention. A clear intention of what it is you are seeking will make for an easier, smoother journey.

Rolling and Strumming

When outlining the shape or edge of a bone, try rolling your fingers across, rather than along, its surface. This is similar to checking the sharpness of a knife by rolling your fingers across the blade. Do the same with the ropy fibers of muscle tissue. Like strumming the strings of a guitar, this method will help you ascertain the muscle's fiber direction and tensile state. (0.3)

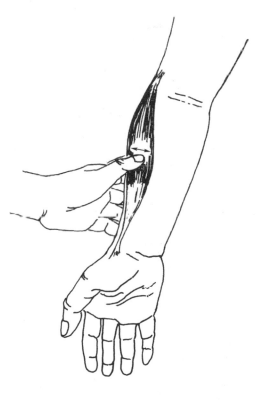

(0.3) Strumming across the fibers of the brachioradialis muscle in the forearm

Movement and Stillness

If you were to compare the texture of smooth paper and rough sandpaper, you would naturally rub your fingers across their surfaces. In contrast, when you lay your hand on an expectant mother's abdomen, hoping to feel the fetus move, you naturally keep your hand still and quiet.

Similarly, when determining the fiber direction of muscles or sculpting the shape of a bone, move your hands along its surface. However, when you want to feel a muscle contract or a bone move, keep your hands still and follow the movement. In other words, if what you are palpating is still, move your hands across it; if it is moving, stay still.

Movement as a Palpation Tool

Throughout the text, you and your partner will be asked to create specific movements of your partner's body. These movements will help clarify the location of structures and the changes occurring in the tissue. The three types of movement applied are active, passive and resisted movements.

Active movement is performed by your partner. He actively moves his body while you palpate or observe the movement. For example, "Ask your partner to slowly flex his elbow while you palpate his biceps brachii muscle." All active movements performed by your partner should be slow and smooth. The changes in tissue are difficult to follow during fast, jerky motions.

Sometimes your partner may be asked to contract and relax a muscle. For example, "To feel the forearm flexors, lay your hand on your partner's forearm and ask him to alternately flex and relax his wrist." The on-and-off aspect of this technique will not only help you locate muscles and tendons, it will also give you the opportunity to feel the difference between contracted and relaxed tissue.

Passive movement is the opposite of active movement: Your partner relaxes while you move his body. For example, when the text says, "Passively flex and extend the elbow," you move the forearm while your partner remains passive and allows the action to occur. (0.4)

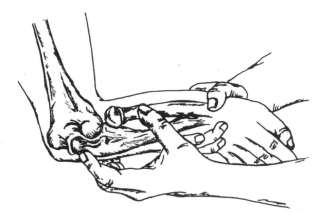

(0.4) Passively moving the forearm to palpate the elbow

An adult has over 600,000 sensory receptors in the skin - more nerve endings than any other part of the body. The fingertips are one of the most sensitive areas, with up to 50,000 nerve endings every square inch. The fingertips are so sensitive that a single touch sensor can respond to a pressure of less than 1/1400 of an ounce - the weight of an average fly.

Resisted movement requires both of you to act: Your partner attempts to perform an action against your gentle resistance. For example, "To feel the forearm flexors in the forearm contract, ask your partner to flex his wrist against your resistance." (0.5) Your partner will not move his forearm, but will simply meet the gentle resistance of your hand. In this text, resisted movements are used to distinguish and compare the lengths, shapes and edges of different muscle bellies and tendons.

When in Doubt, Ask the Body

While palpating, you may be confused or have questions which arise about the body's structures and their whereabouts. When in doubt, ask the body you are palpating. For example you may wonder, "What skinny tendon is this I see running along the top of the foot?" (0.6) Follow it in both directions and see where it leads you. If it runs from the big toe to the ankle and becomes taut when the toe is extended, it is the tendon of extensor hallucis longus (p. 268). You are never alone; the body is waiting to help you.

All of the structures outlined in *Trail Guide to the Body*, with their Latin names, subtle shapes and buried positions, are inside you, your partner and your patients' bodies. These structures have been there for years waiting to be discovered by you. Have faith and you will be able to locate them.

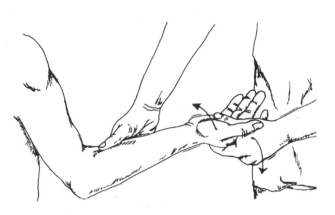

(0.5) Resisted movement to feel the forearm flexors contract

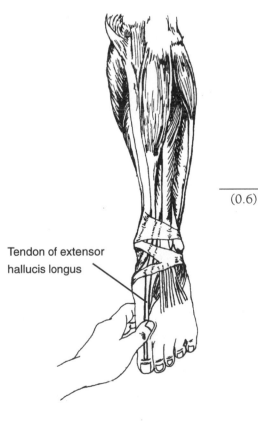

(0.6)

Tendon of extensor hallucis longus

The three principles of palpation:
1) Move slowly. Haste only interferes with sensation.
2) Avoid excessive pressure. Less is truly more.
3) Concentrate and focus your awareness on what it is you are feeling. In other words - be present.

Exploring the Textural Differences of Structures

This section is designed to identify and compare the physical characteristics of the various structures and tissues in the body. To understand the textural differences between structures will only clarify the techniques which you apply in your hands-on practice.

The following explanations describe various structures in a "normal," healthy condition. Of course, the quality or feel of a particular tissue will be as unique as the individual you are palpating. The tissue's basic, structural design, however, will be identical on everyone. For example, a long distance runner may have lean, sinewy bands of muscle tissue; an individual leading a sedentary life-style may have a very different quality to his muscles. The feel of the muscle tissue is different, yet its design and composition are the same.

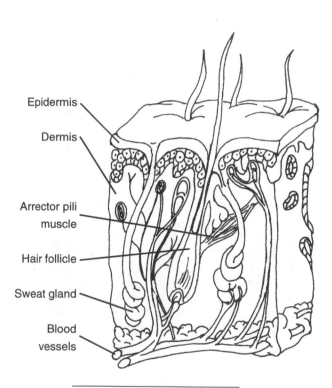

Epidermis

Dermis

Arrector pili
muscle

Hair follicle

Sweat gland

Blood
vessels

(0.7) Cross-section of the skin

Skin

Often disregarded as the simple covering of the body, the skin is the body's single largest organ. (0.7) On an adult male the skin can cover a surface area of nineteen square feet and weigh nearly ten percent of the total body weight. The skin averages about 1/20 of an inch in thickness, with the eyelids having the thinnest skin - less than 1/500 of an inch. The skin is intimately connected with the superficial fascia and deeper tissues and has varying texture, thickness and flexibility throughout the body.

For example, palpate the skin on the back of your hand. Note its thin, delicate and pliable quality. Then explore the palm of your hand and feel the thick, tough layering of skin.

Bone

Bones and bony landmarks (the hills, valleys and bumps on the surface of bones) are easy to distinguish from other tissues because they have a solid feel. Of course, during movement the bones shift and their positions change with the surrounding structures. Sometimes other structures can feel like bone. For example, when a muscle contracts against resistance, its belly and tendons become very hard. Ligaments can have a particularly solid quality. The shape and inflexibility of bones and bony landmarks remain constant, unlike muscles or ligaments, which can transform from a soft to a hard state, and back again.

Muscle

Skeletal muscle, the voluntary contractile tissue that moves the skeleton, is composed of muscle cells (fibers), layers of connective tissue (fascia), and numerous nerves and blood vessels.

If you do not like the skin you have, just wait a month. An average adult sheds about 600,000 particles of skin every hour, amounting to one and a half pounds of skin each year. Altogether, the outer skin changes about every twenty-seven days. Add it up and that is nearly 1000 new skins in a lifetime.

Myofibril

Muscle fiber (endomysium)

Fascicle (perimysium)

Muscle belly (epimysium)

Tendon

Bone

Periosteum

(0.8) Cross-section
of a typical skeletal
muscle

A muscle's design is similar to an orange: a broad
sheet of fascia encases the whole fruit, deeper
layers of fascia separate the orange into "wedges"
(the portions you eat when peeling) and, finally, a
thin coating of tissue surrounds each individual,
tiny "bud" of fruit.

Applying this analogy to a muscle, a layer of
fascia (epimysium) encases the muscle "belly," a
deeper layer (perimysium) wraps the long muscle
fibers into bundles and, finally, each microscopic
muscle fiber is bound in fascia (endomysium). (0.8)
Unlike an orange, a muscle's connective tissues
merge at either end of the muscle to form a strong
tendon. The tendon attaches the muscle to a bone.

Muscle tissue has three specific physical characteristics which can help you distinguish it from
other tissues.

First, *muscle tissue has a striated texture* - similar
to a plank of unsanded wood. This is different than
tendons, which have a smoother feel. The fibrous
quality of a muscle belly is caused by bundles of
muscle fibers running in a particular direction.

Second, *the direction of the muscle fibers* can be
helpful in determining which muscle you are
palpating. Depending on the shape and design of a
muscle (0.9), the direction of its fibers may be
parallel, convergent or diagonal. For example, the
erector spinae muscles (p. 138) have vertical fibers
which run parallel to the spine. Identifying their
fiber direction can help you distinguish the erector
spinae from the oblique and horizontal muscle
fibers of other back muscles.

Last, *muscle tissue is unique because it can be in a
contracted or relaxed state.* When a muscle is relaxed
it often has a soft, malleable feel. When contracted,
it has a firm, solid quality. As the tension in muscle
tissue changes, surrounding tissues like tendons
and fascia also become taut or loose.

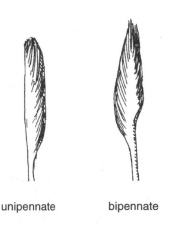

convergent

biceps

fusiform

unipennate

bipennate

multibelly

(0.9) Different shapes of muscle bellies

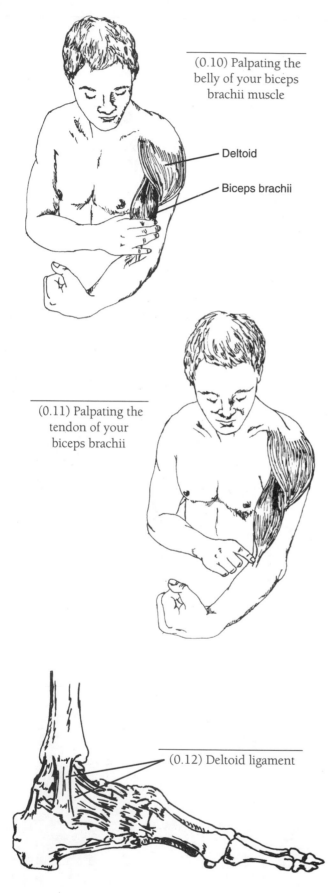

How can you palpate a muscle which is deep to a superficial, overlying muscle? In some areas, the overlying muscle can be shifted to the side. At other times, you can slowly compress your fingerpads beyond the superficial muscle into the deeper tissues, using their different textures and fiber direction as guides. This is similar to palpating through your sweater, shirt and skin to access a muscle in your arm.

Discover the three distinguishing features of muscle tissue by palpating your biceps brachii - the muscle on the front of the arm. (0.10) Keep your arm relaxed and feel for the biceps' ropy fibers. Note how its fiber direction runs distally (down the arm). Contract and relax the biceps and sense how it tightens to a solid mass and relaxes into a soft wad.

(0.10) Palpating the belly of your biceps brachii muscle

Deltoid

Biceps brachii

Tendon

Tendons attach muscle to bone. More accurately, they connect muscle to the periosteum - the connective tissue which surrounds the bone. (0.8) Tendons are composed of dense connective tissue shaped as bundles of parallel collagen fibers. Each end of a muscle has one or more tendons.

Tendons come in a variety of shapes and sizes. Some are short and wide like the gluteus maximus at the buttocks. Others are long and thin such as the cables of your anterior wrist. A broad, flat tendon is called an aponeurosis. An example is the galea aponeurotica (p. 182) which extends across the top of your cranium. Regardless of their shapes, all tendons have a smooth, tough, almost resilient feel to them.

Locate the distal tendon of the biceps brachii by holding your elbow in a flexed position (0.11). First, locate the biceps' muscle belly and follow it distally toward your inner elbow. As you progress, the muscle belly will become more slender and, at the crease of the inner elbow, it will become a smooth, thin tendon. It may feel like a taut strand of cable. Explore around either side of the tendon.

(0.11) Palpating the tendon of your biceps brachii

Ligament

A ligament connects bones together at a joint. Its task is to strengthen and stabilize joints. Like tendons, ligaments are made of dense connective tissue. But unlike a tendon's parallel fiber arrangement, a ligament's fibers have a more uneven design.

The design and length of ligaments vary. Many simply cross a joint and blend in with the deeper

(0.12) Deltoid ligament

---

aponeurosis   **ap**-o-nu-**ro**-sis   Gr. *apo*, from + *neuron* , nerve or tendon
ligament   **lig**-a-ment   L. a band

joint capsule, like the ankle's deltoid ligament. (0.12) Others span a distance between several bones, like the supraspinous ligament of the back (p. 154).

Ligaments often have a dense, taut feel when being palpated. Some have a palpable fiber direction. If you want to distinguish a tendon from a ligament, explore its attachments and variable tension. A tendon connects a muscle belly to a bone, while a ligament attaches a bone to another bone. A tendon will become taut or slack depending on whether it is shortened or lengthened, or if its muscle belly contracts. Ligament will remain taut throughout all movements or states of contraction.

Fascia

Like tendons and ligaments, fascia is a form of dense connective tissue. It is a sheet of fibrous membrane located beneath the skin and around muscles and organs. The fascial system creates a three-dimensional matrix of connective tissue that extends continuously throughout the body from head to toe.

There are two types of fascia: superficial and deep. Superficial fascia is located immediately deep to the skin and covers the entire body. It could be viewed as less of a thin sheet and more of a spacial layering filled with adipose, nerves, blood and lymph vessels, and connective tissue. (0.13) The density of the superficial fascia varies from very thin (on the back of the hand) to quite thick (the sole of the foot).

Deep fascia has a more extensive design. It surrounds muscle bellies, holds them together and separates them into functional groups. It also fills in the spaces between muscles and, like superficial fascia, carries blood and nerve vessels. Portions of the deep fascia permeate into the muscle belly and encase each tiny muscle fiber.

Bone

Periosteum

Interosseous membrane

Muscle tissue

Skin

Superficial fascia

Adipose (fatty) tissue

Deep fascia

(0.13) A cross-section of the forearm showing the arrangement of bone, muscle and fascia

Because of its ubiquitous presence, specific palpation of the fascial system requires an experienced, sensitive touch. Here are three simple exercises which can help you get a basic feel of the fascia and its relationship to other structures.

a) Pull up the skin on the back of your hand. Notice how the skin does not pull up entirely (like when you pull a baggy shirt away from your body). This is because the fascia is holding the skin down. Try this on various parts of your body and notice how it is easier to lift the skin and fascia in some areas, but more difficult in others.

b) This exercise can give you a sense of the continuous sheet fascia forms throughout the body and how pulling at one portion of the sheet can affect another part.

Draw a small "X" on your forearm. Place your fingerpads approximately two inches away from the "X". Using the gentle pressure of your fingerpads, slowly move the skin of your arm in various directions away from the mark. (0.14)

Notice how the "X" stretches and responds more easily when you move in a certain direction, yet may move less when pulled in another direction. As you continue, reposition your fingers further away from the "X", so eventually you are pulling across the skin of the hand.

c) Here is an exercise to comprehend the omnipresent, yet phantomlike nature of fascia. Put a latex glove on your partner's hand followed by a thick winter glove. If you explore your partner's hand, you will immediately detect the texture and thickness of the winter glove and the general shape of the hands and fingers. The latex glove (fascia), however, may be more challenging to distinguish.

(0.14) Exploring the superficial fascia with an "x" drawn on the forearm

Leonardo de Vinci (1452-1519), who dissected bodies secretly at night, was the first to document his anatomical findings. Laid out over 750 drawings, not only are his illustrations detailed and accurate, but they also contain many structural variations seen in the body. The anomalies in the drawings were not a case of Leonardo the artist dominating Leonardo the scientist. As a true renaissance man there can be little question that he drew exactly what he observed.

The structures of the human body do not conform to one standard anatomical model, so it is unjust to assume they are free from variations. Structural differences have been recorded in almost every muscle, bone, major blood vessel and organ in the body. Knowing this can alleviate confusion and perhaps frustration by recognizing that the trail may not follow the guidebook exactly.

Retinaculum

A retinaculum is any structure that holds an organ or tissue in place. Regarding muscular connective tissue, the retinaculum is a transverse thickening of the deep fascia which straps down tendons in a particular location or position. For example, the retinacula of the ankle stabilize tendons which traverse around the sharp curve of the ankle. (0.15)

Most retinacula are superficial and accessible. Distinguishing a retinaculum from its deeper tendons can be determined by comparing their different fiber directions. Retinacula will have transverse fibers which run perpendicular to the deeper tendons.

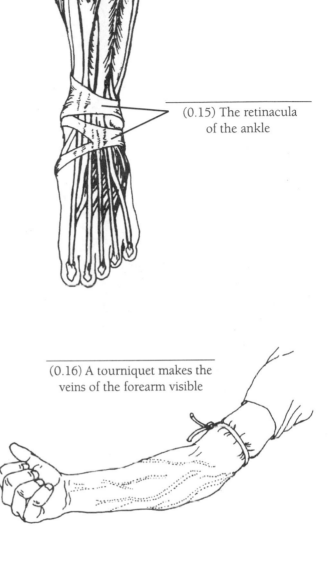

(0.15) The retinacula of the ankle

Artery and Vein

Arteries and veins each have distinct features when you palpate them. The pulse of the heart can be felt when pressing on an artery but not a vein. Arteries are often situated on the protected side of an appendage and buried deep to the musculature. Veins, on the other hand, are superficial and often easily recognized on the dorsum of the hands and feet by their bluish color.

Locating an artery is not only necessary to determine the pulse, but also important when palpating other structures. For example, when palpating the sternocleidomastoid muscle in the neck, it is vital that you be aware of the location of the carotid artery, the chief blood vessel supplying the head and neck, so you avoid pressing on it. When palpating an appendage, if an artery is impeded for a sustained period of time, the distal portion of the appendage will begin to tingle or become numb.

Let your arm hang at your side for a minute, allowing the blood to fill the superficial veins of your hand and forearm. For more dramatic results, gently squeeze your forearm with your opposite hand. The veins will swell with the increased pressure and be clearly visible. (0.16)

(0.16) A tourniquet makes the veins of the forearm visible

William Harvey (1578-1657), regarded as the first experimental scientist, discovered that blood circulated throughout the body. Along with his descriptions of the cardiovascular system, he also explained how veins are equipped with valves which prevent blood from flowing backwards between heart beats.

To prove his theory, Harvey tied a tourniquet around an assistant's arm and allowed the blood to pool in the distal veins. He observed that the valves appeared as small swellings along the path of the veins. Harvey pressed on a valve and pushed the blood out of the vein to the next valve. As he held his finger on the distal valve, the proximal valve prevented blood from flowing backwards and the vein remained empty.

retinaculum	**ret-i-nak**-u-lum	L. halter
retinacula	**ret-i-nak**-u-la	plural for retinaculum

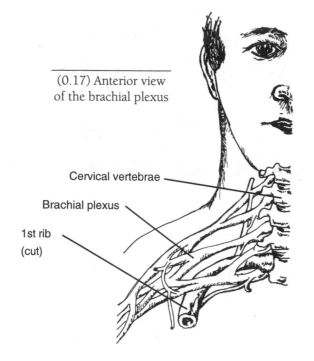

(0.17) Anterior view
of the brachial plexus

Cervical vertebrae

Brachial plexus

1st rib
(cut)

Adipose Tissue

Adipose (fatty) tissue is a form of loose connective tissue. It is deposited at many levels throughout the body including the marrow of long bones, around the kidneys, the padding around joints and behind the eyeballs. Needless to say, some of these areas are outside the reach of this text.

The most palpable location for adipose tissue is in the subcutaneous layer of tissue between the skin and superficial fascia. (0.13) This layer of adipose varies in thickness throughout the body and may have different consistencies. Adipose usually has a gelatinous (jellylike) consistency, making it easy to sink the fingers into and detect deeper structures.

Stand up and squeeze the flesh of your own buttocks to feel adipose tissue. Note the superficial layer of adipose. Then tighten the muscles of your buttocks and feel the textural difference between the adipose and the deeper muscles.

Nerve

Nerves are tube-shaped, mobile and tender when compressed. Although sections of nerves and plexuses (bundles of nerves) can be accessed throughout the body, they are best avoided. Compression or impingement of a nerve may create a sharp, shooting sensation down the appendage or into the local area. (0.17)

Lymph Node

Lymph nodes collect lymph fluid from lymphatic vessels. They are bean-shaped and may range in size from a tiny pea to an almond. Lymph nodes are located throughout the body, with palpable groups of nodes found in the body's creases such as the groin, axilla and neck. (0.18) Healthy lymph nodes are roundish, slightly moveable and nontender. This differs from other glands which are usually larger and have irregular, lumpy surfaces.

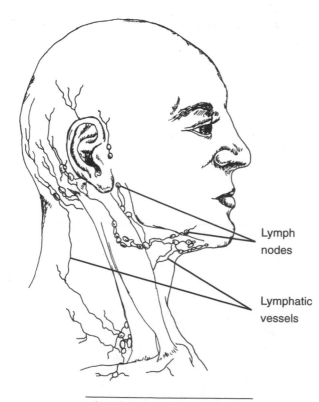

Lymph
nodes

Lymphatic
vessels

(0.18) Cervical lymph nodes

adipose   **a-di-pos**   L. fat, copious
plexus   L. interwoven

Chapter 1

Navigating the Body

The nature of this book demands that our journey explore specific, individual structures and regions. However, before we set out into valleys and hills of the body, some preparation is in order. This chapter will familiarize you with important mapping and navigational terms. It will also show you a "big picture" of the systems of the body which are highlighted in the text. So when the trail guide leads you in a certain direction, you will know which way to go!

(1.1) Anatomical position

Regions of the Body

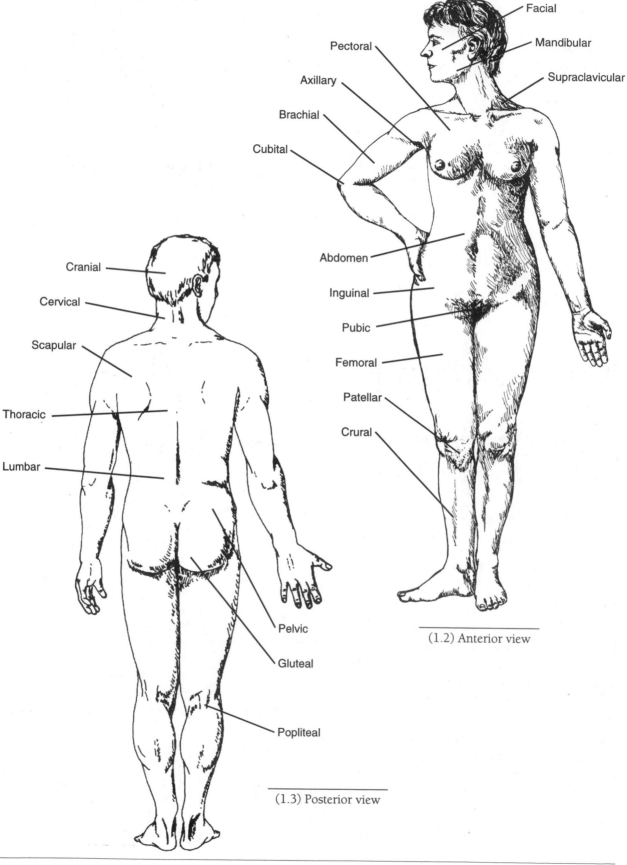

Facial

Pectoral

Mandibular

Axillary

Supraclavicular

Brachial

Cubital

Abdomen

Inguinal

Pubic

Femoral

Patellar

Crural

(1.2) Anterior view

Cranial

Cervical

Scapular

Thoracic

Lumbar

Pelvic

Gluteal

Popliteal

(1.3) Posterior view

Planes of Movement

The body can be divided into three imaginary planes. (1.4) These planes help to clarify and specify movements when the body is in the standard anatomical position (standing erect with the palms facing forward). (1.1)

The *sagittal plane* divides the body into left and right parts. The descriptive terms medial and lateral correlate to the sagittal plane; the actions of flexion and extension occur along this plane. The midline (or midsagittal plane) runs down the center of the body, dividing the sagittal plane in two symmetrical halves.

The *frontal (or coronal) plane* divides the body into front and back portions. The terms anterior and posterior relate to the frontal plane; the actions of adduction and abduction happen along this plane.

Dividing the body into upper and lower parts is the *transverse plane*. Superior and inferior refer to the transverse plane; rotation happens along this plane.

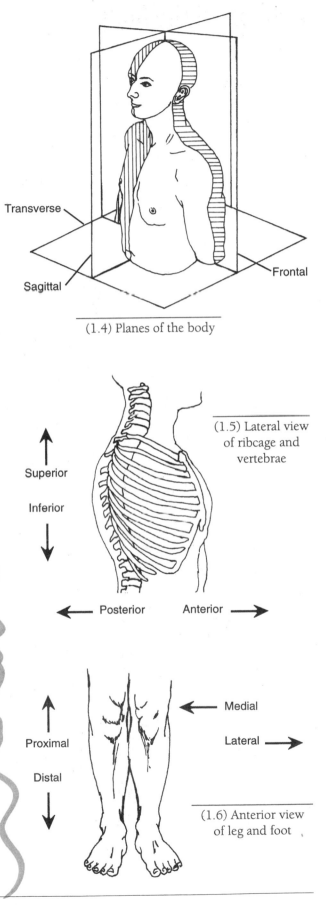

(1.4) Planes of the body

(1.5) Lateral view of ribcage and vertebrae

(1.6) Anterior view of leg and foot

Directions and Positions

Certain specific terms are used to help communicate location, direction and position of body structures. These terms replace more general references like "up there" or "north of here," which are less exact and can be confusing. Each direction is paired up with its complimentary direction.

Superior refers to a structure closer to the head. *Inferior* means closer to the feet. "The nose is superior to the navel." "The navel is inferior to the nose." The terms *cranial* (closer to the head) and *caudal* (closer to the buttocks) are used in reference to structures on the thorax. (1.5)

Posterior concerns a structure further toward the back of the body than another structure. *Anterior* refers to a structure further in front. "The sternum is anterior to the spine." These terms are also referred to as dorsal (posterior) and ventral (anterior). (1.5)

Medial pertains to a structure closer to the midline (or center) of the body. *Lateral* refers to a structure further away from the midline. "The pinkie toe is lateral to the big toe." (1.6)

Distal means a structure further away from a limb's origin or the body's midline. *Proximal* designates a

| sagittal | **saj**-i-tal | L. arrowlike | | transverse | L. around, to turn |
| coronal | ko-**ro**-nal | Gr. crow | | | |

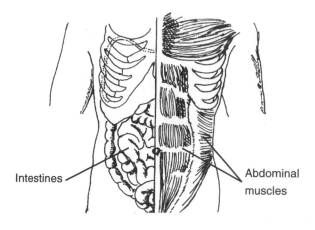

Intestines

Abdominal
muscles

(1.7) Anterior view of abdomen

structure closer to a limb's origin. These directions
are used only when referring to the arms and legs.
"The foot is distal to the thigh." (1.6) "The forearm
is proximal to the hand."

Superficial describes a structure closer to the
body's surface. *Deep* refers to a structure deeper in
the body. "The abdominal muscles are superficial
to the intestines." (1.7)

Movements of the Body

Movement of the body occurs at the joints, where
bones articulate (or connect). Although movement
affects the placement of bones, the terminology of
movement always refers to joints. Bending your
knee is called "flexion of the knee." "Flexion of the
leg" would require an ambulance.

See pages 20 - 23 for a description of movement
at specific joints.

Extension is movement that straightens or opens
a joint. In anatomical position, most joints are
extended. When a joint can extend beyond its
normal range of motion it is called hyperextension.
Flexion is movement that bends a joint or brings
the bones closer together. In a fetal position all the
joints are in a flexed position. (1.8) Both flexion
and extension take place along the sagittal plane.

Prone is the position of laying on the table face
down. Supine ("on your spine") is to lie face up.
Sidelying is just that - lying on your side.

Adduction of a joint brings a limb medially
toward the body's midline ("adding to the body").
Abduction moves a limb laterally away from the
midline ("abduct or carry away"). These actions
happen along the frontal plane and pertain only
to the appendages. To adduct the fingers or toes
is to bring them together; to abduct is to spread
them apart.

Medial rotation and *lateral rotation* occur at the
shoulder and hip joints. When the joint medially
rotates, the limb turns in toward the midline.
Lateral rotation swings the limb away from
the midline.

Circumduction is possible only at the shoulder
and hip joints. It is a combination sequence of
extension, adduction, flexion and abduction;
together these actions form a cone shaped
movement. (1.9) Swimming the backstroke
requires circumduction at the shoulder joint.

Rotation pertains only to the axial skeleton,
specifically the head and vertebral column; for

(1.8) In fetal position
most joints are flexed

| dorsi | **dor**-si | L. back |
| plantar | **plan**-tar | L. pertaining to the sole of the foot |

example, rotating the head and neck while driving a car. These movements happen along the transverse plane.

Lateral flexion occurs only at the axial skeleton. For example, when the head or vertebral column bend laterally to the side.

Supination and *pronation* describe the pivoting action of the forearm. Supination ("carrying a bowl of soup") occurs when the radius and ulna lie parallel to one another. Pronation ("prone to spill it") takes place when the radius crosses over the ulna, turning the palm down. Supination and pronation also occur at the feet.

Opposition happens only at the carpometacarpal joint of the thumb. It occurs when the thumbpad crosses the palm toward the other fingers.

Inversion and *eversion* occur as a combination of movement of several joints of the feet. Inversion ("in") elevates the foot's medial side and brings the sole of the foot medially. Eversion ("evert or away") elevates the foot's lateral side and moves the sole laterally.

Plantar flexion and *dorsiflexion* refer only to the ankle. Plantar flexion is performed by bending the ankle to point your foot into the earth or stepping on a car's gas pedal. Dorsiflexion is the opposite movement, such as bending the ankle to let off the gas pedal.

Protraction and *retraction* pertain to the clavicle, head and jaw. Protraction ("protrude") occurs when one of these structures moves anteriorly. Retraction ("retreat") is movement posteriorly.

Elevation and *depression* refer to the movement of the scapula and jaw. Elevation is movement superiorly. Depression is movement inferiorly.

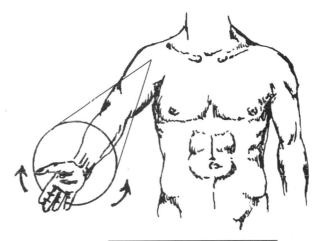

(1.9) Shoulder circumduction

The names of many bones, bony landmarks and muscles may initially look and sound foreign. They are - most anatomical terms are in Latin or Greek. However, the source or story behind the terms can help to explain and clarify their meaning.

Take the phrase "infraspinous fossa of the scapula." The scapula is a flat bone of the shoulder. In Latin, scapula means "shoulder blade" - its common name. Fossa translates as "shallow depression." Infraspinous is a directional term (like north or southwest). It means inferior (infra-) to the spine of the scapula (-spinous). Put this all together and the "infraspinous fossa of the scapula" translates as "the shallow ditch located below the spine of the shoulder blade." Keep an eye peeled for translations and phonetic descriptions at the bottoms of pages.

Spine of the scapula

Infraspinous fossa

Movements of the Body

Spine and Thorax

| Flexion | Extension | Rotation | Lateral flexion |

Neck

| Flexion | Extension | Rotation | Lateral flexion |

Mandible (Jaw)

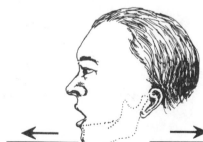

Protraction   Retraction

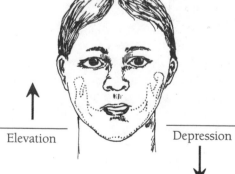

Elevation   Depression

Scapula

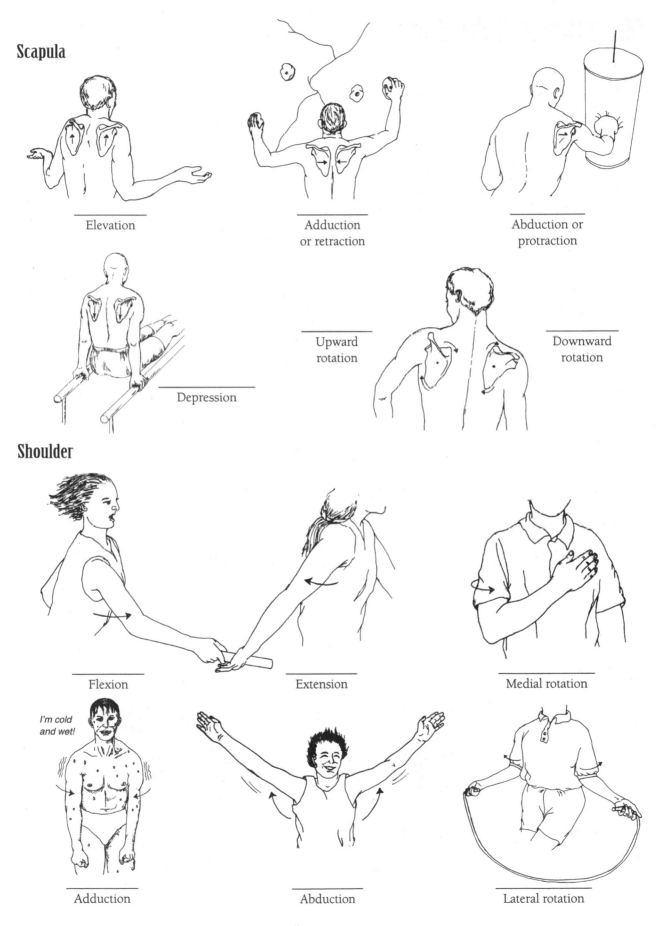

Elevation

Adduction or retraction

Abduction or protraction

Depression

Upward rotation

Downward rotation

Shoulder

Flexion

Extension

Medial rotation

I'm cold and wet!

Adduction

Abduction

Lateral rotation

Elbow and Forearm

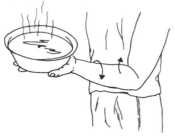

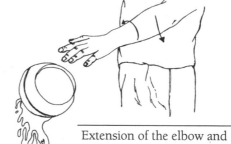

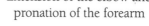

Flexion of the elbow and
supination of the forearm

Extension of the elbow and
pronation of the forearm

Wrist

Extension of the wrist and fingers

Flexion of the wrist and fingers

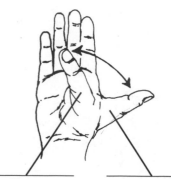

Adduction of the wrist

Abduction of the wrist

Thumb

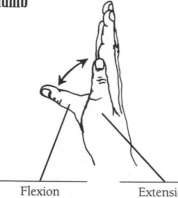

Flexion   Extension

Opposition

Adduction   Abduction

Pelvis

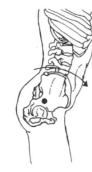

_____
Anterior tilt

_____
Posterior tilt

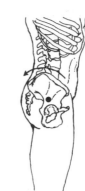

Hip and Knee

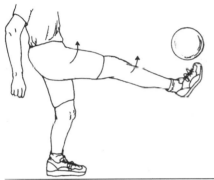

Flexion of hip and extension of knee

Abduction of hip

Internal rotation of hip

Extension of hip and flexion of knee

Adduction of hip

External rotation of hip

Ankle and Foot

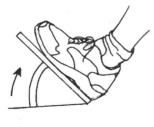

Dorsiflexion of ankle

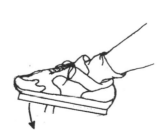

Plantar flexion of ankle

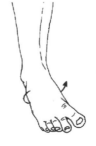

Inversion of foot

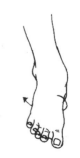

Eversion of foot

Systems of the Body

The Skeletal System

The bones are linked together to form the skeleton. The skeleton is divided into two sections: the axial and the appendicular skeletons. The *axial* skeleton contains the skeleton's center. It includes the cranium, vertebral column, ribs, sternum and hyoid bone.

The *appendicular* ("appendages") skeleton is composed of the arms and legs, including the pectoral girdle (scapula and clavicle) and pelvic girdle (hips).

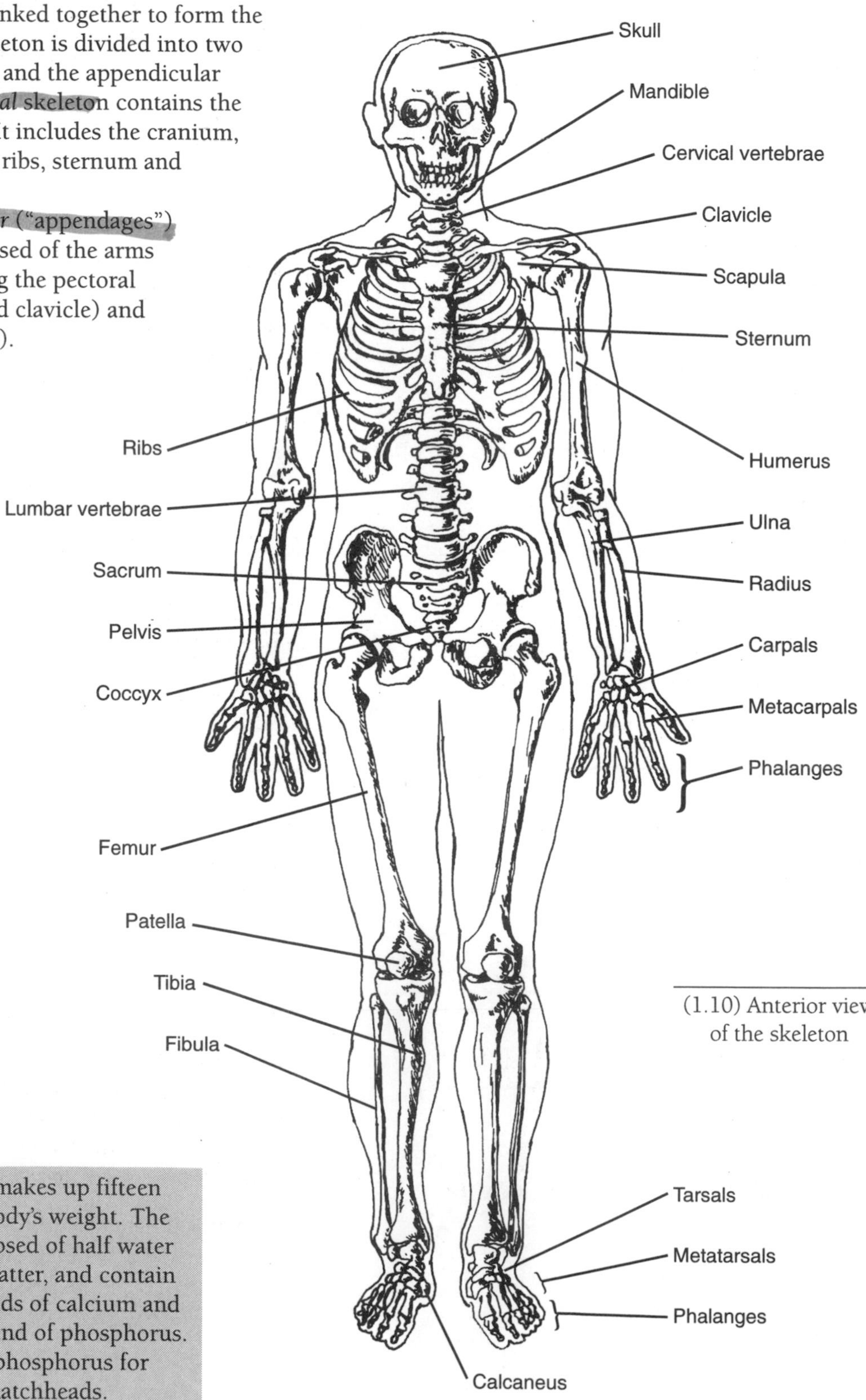

Skull

Mandible

Cervical vertebrae

Clavicle

Scapula

Sternum

Humerus

Ulna

Radius

Carpals

Metacarpals

Phalanges

Ribs

Lumbar vertebrae

Sacrum

Pelvis

Coccyx

Femur

Patella

Tibia

Fibula

(1.10) Anterior view
of the skeleton

Tarsals

Metatarsals

Phalanges

Calcaneus

The skeleton makes up fifteen percent of the body's weight. The bones are composed of half water and half solid matter, and contain nearly two pounds of calcium and more than a pound of phosphorus. That is enough phosphorus for two-thousand matchheads.

| appendicular | **ap**-en-**dik**-u-lar | L. to hang to | skeleton | G. dried up |
| axial | **ak**-se-al | L. axle | | |

Types of Joints

A joint or articulation is the point of contact between bones. How a joint is structured will determine how it functions. All articulations have either a fibrous, cartilaginous or synovial structure. Due to their design, *fibrous* and *cartilaginous* joints are capable of little or no movement. *Synovial* joints, however, contain a joint cavity (absent in fibrous and cartilaginous joints). The space between the bones allows the synovial joints and bones to move.

All synovial joints have the same basic structural components, yet vary in respect to their movement capabilities. There are six types of synovial joints: ball-and-socket, ellipsoid, hinge, saddle, gliding, and pivot.

A *ball-and-socket joint* is self-describing: a spherical surface of one bone fits into the dish-shaped depression of another bone. These joints are capable of movement in every plane such as circumduction at the glenohumeral (shoulder) joint. (1.11)

An *ellipsoid joint* consists of the oval-shaped end of one bone that articulates with the elliptical basin of another bone. It permits flexion/extension and abduction/adduction as seen at the radiocarpal (wrist) joint. (1.12)

Hinge joints allow only flexion and extension, similar to the movements of a door hinge. An example of a hinge joint is the humeroulnar (elbow) joint. (1.13)

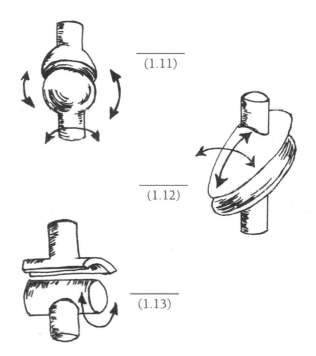

(1.11)

(1.12)

(1.13)

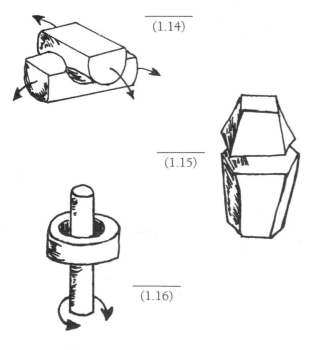

(1.14)

(1.15)

(1.16)

A *saddle joint* is a modified ellipsoid joint composed of convex and concave articulating surfaces - like two saddles. The joint between the trapezium (one of the small carpals in the wrist) and the first metacarpal is an example of a saddle joint. (1.14)

Gliding joints are usually between two flat surfaces and allow the least movement of all types of synovial joints. Only small shifting movements are available at these articulations, such as between the carpals in the wrist or tarsals in the foot. (1.15)

A *pivot joint* is designed to allow one bone to rotate around the surface of another bone. For example, rotation of the head occurs because of the pivot joint between the atlas and the axis. (1.16)

The Muscular System

A muscle's name can give you clues to its specific features. The name reflects either a muscle's shape (rhomboids), location (temporalis), fiber direction (external oblique), action (adductors), or attachment sites (coracobrachialis).

At either end of the muscle is a tendon which attaches the muscle to a bone. Each muscle has an origin and an insertion.

The origin is the attachment to the more stationary bone while the insertion is the connection to the more mobile bone.

A muscle which is the major contributor in generating a movement is called the *prime mover* (or agonist). For example, the deltoid is the

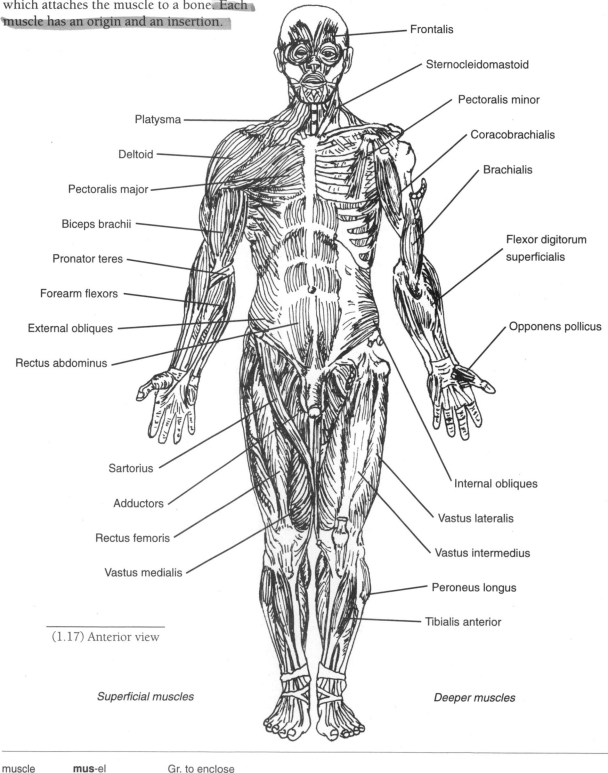

Frontalis

Sternocleidomastoid

Pectoralis minor

Coracobrachialis

Brachialis

Flexor digitorum superficialis

Opponens pollicus

Platysma

Deltoid

Pectoralis major

Biceps brachii

Pronator teres

Forearm flexors

External obliques

Rectus abdominus

Sartorius

Adductors

Rectus femoris

Vastus medialis

Internal obliques

Vastus lateralis

Vastus intermedius

Peroneus longus

Tibialis anterior

(1.17) Anterior view

Superficial muscles

Deeper muscles

| muscle | **mus**-el | Gr. to enclose |
| tendon | **ten**-dun | Gr. to stretch |

prime mover when abducting the shoulder (bringing the arm away from the body). Muscles which work together to create an action and assist the prime mover are called *synergists*. Upon abduction of the shoulder (moving the arm away from the body), the surrounding muscles

of the shoulder stabilize and work as synergists with the deltoid. A muscle which creates the opposite action to the prime mover and synergists is called an *antagonist*. The pectoralis major and latissimus dorsi adduct the arm. Therefore, they are antagonists to the deltoid.

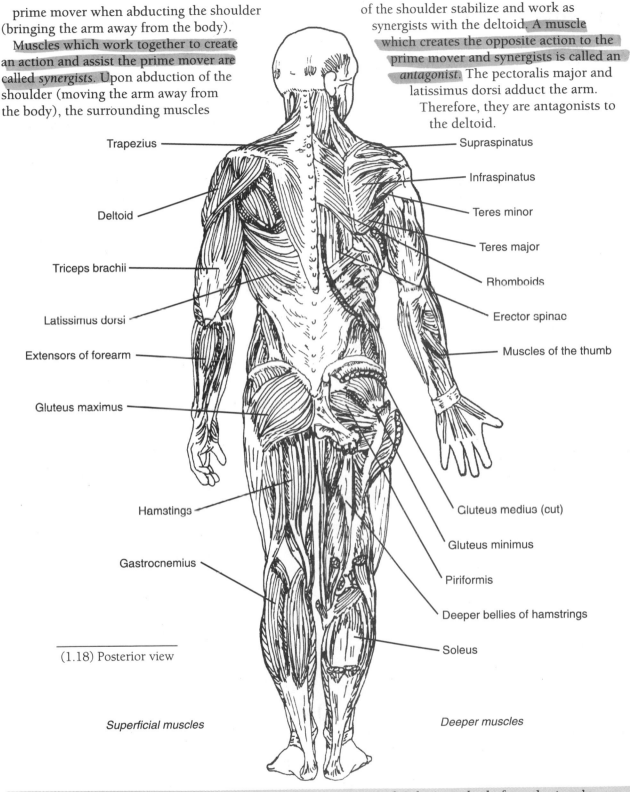

Trapezius

Deltoid

Triceps brachii

Latissimus dorsi

Extensors of forearm

Gluteus maximus

Hamstings

Gastrocnemius

Supraspinatus

Infraspinatus

Teres minor

Teres major

Rhomboids

Erector spinae

Muscles of the thumb

Gluteus medius (cut)

Gluteus minimus

Piriformis

Deeper bellies of hamstrings

Soleus

(1.18) Posterior view

Superficial muscles

Deeper muscles

There are six hundred thirty-nine named muscles in the human body. Yet in the time of Galen (130-200), one of the first great anatomists, few of the muscles had names. Vesalius and other Renaissance contemporaries attempted to introduce nomenclature, yet continued Galen's method of numbering the muscles. It was not until the 18th century, thanks largely to British anatomist William Cowper and Scottish anatomist James Douglas that the specific myological terminology we use today was established.

The Fascial System

The following illustrations
show aspects of the fascia
from both topographical
and cross-section viewpoints.

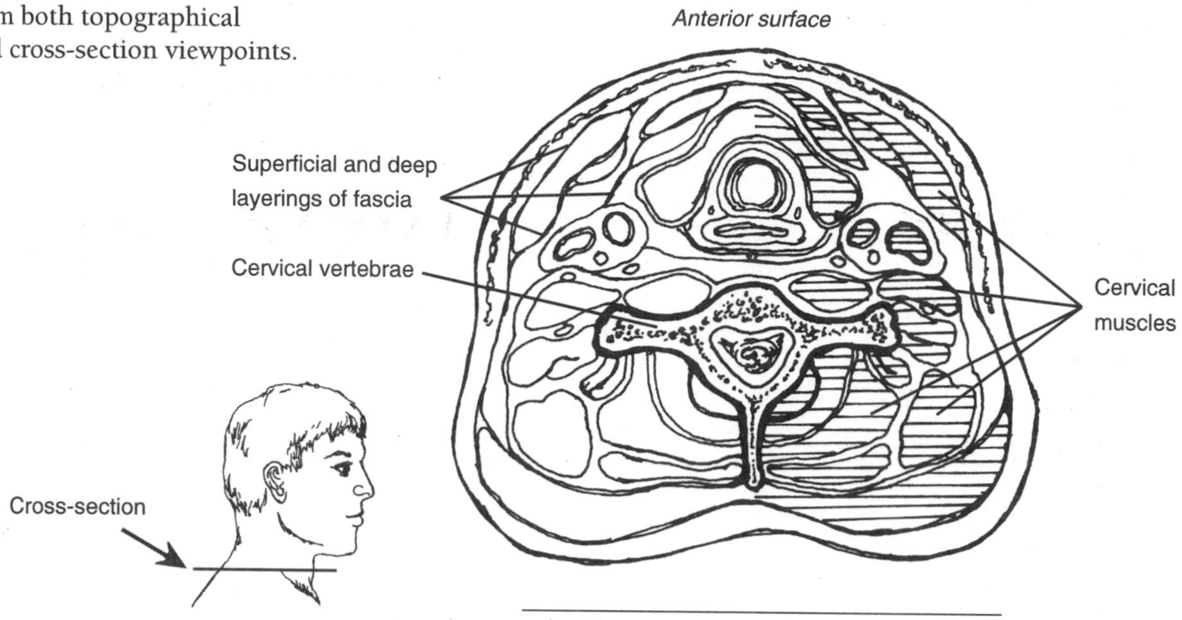

Anterior surface

Superficial and deep
layerings of fascia

Cervical vertebrae

Cervical
muscles

Cross-section

(1.19) Cross-section of the neck showing
layers of fascia encasing groups of muscles

Biceps brachii

Humerus

Axillary
fascia

Forearm fascia

Flexor muscles

Brachial
fascia

Ulna

Antebrachial
fascia

Brachial fascia

Triceps brachii

Palmar
aponeurosis

Radius

Extensor muscles

Interosseous membrane

(1.20) Cross-section of the arm

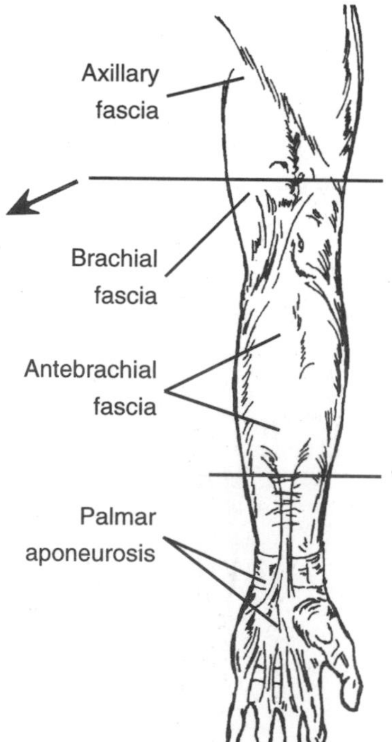

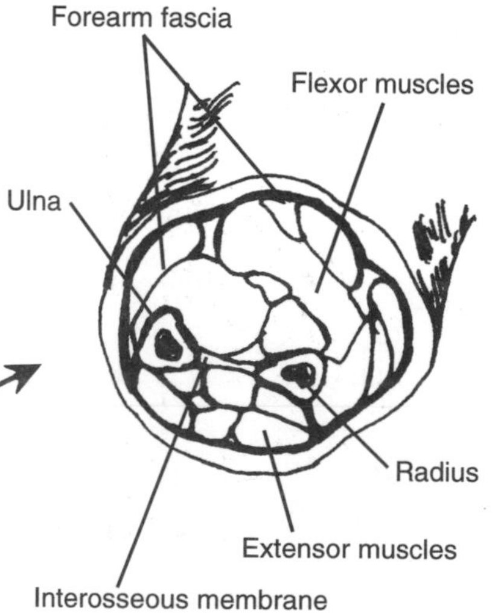

(1.22) Cross-section of the forearm

(1.21) Anterior view of the arm
and forearm, skin removed

| fascia | **fash**-e-a | L. a band |
| retinaculum | **ret**-i-**nak**-u-lum | L. halter |

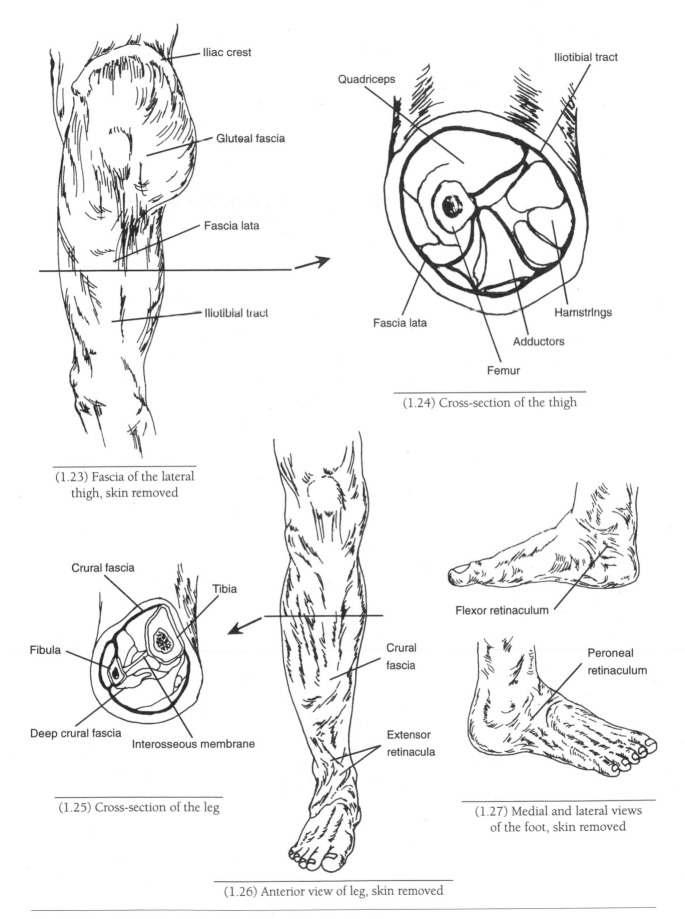

Iliac crest

Gluteal fascia

Fascia lata

Iliotibial tract

(1.23) Fascia of the lateral
thigh, skin removed

Quadriceps

Iliotibial tract

Fascia lata

Hamstrings

Adductors

Femur

(1.24) Cross-section of the thigh

Crural fascia

Tibia

Fibula

Deep crural fascia

Interosseous membrane

(1.25) Cross-section of the leg

Crural
fascia

Extensor
retinacula

(1.26) Anterior view of leg, skin removed

Flexor retinaculum

Peroneal
retinaculum

(1.27) Medial and lateral views
of the foot, skin removed

aponeurosis   **ap**-o-nu-**ro**-sis   Gr. *apo*, from + *neuron*, nerve or tendon

The Cardiovascular System

Arteries and veins are the blood vessels of the cardiovascular system. They create an amazing network that transports blood from the heart, brings it to the body's tissues and then carries it back to the heart.

Arteries carry blood away from the heart. As an artery progresses away from the heart, it divides into smaller branches. Arterioles, the smallest branches, divide into millions of microscopic vessels called capillaries. The walls of the capillaries serve as nutrient and waste exchange sites between the body tissues and the blood. The capillaries then merge back together and create small veins or venules which unite to form larger veins, which carry the blood back to the heart.

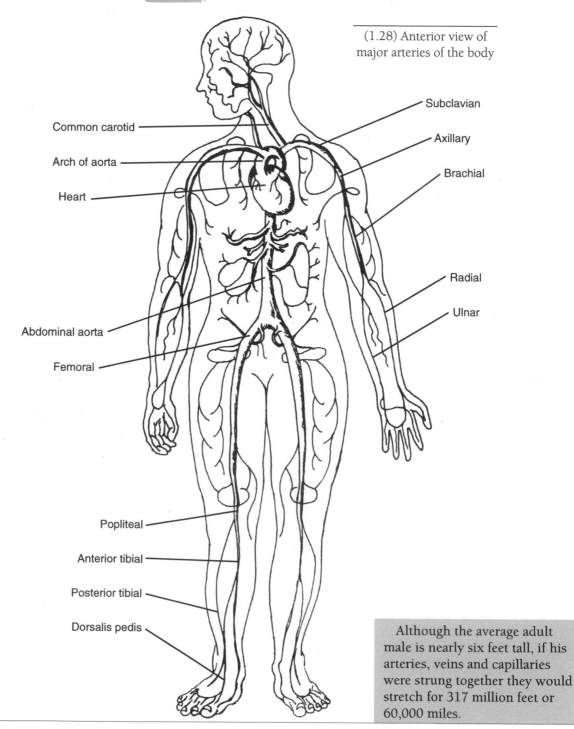

(1.28) Anterior view of major arteries of the body

Common carotid

Arch of aorta

Heart

Abdominal aorta

Femoral

Popliteal

Anterior tibial

Posterior tibial

Dorsalis pedis

Subclavian

Axillary

Brachial

Radial

Ulnar

Although the average adult male is nearly six feet tall, if his arteries, veins and capillaries were strung together they would stretch for 317 million feet or 60,000 miles.

artery   Gr. windpipe
vein   L. vessel

capillaries   L. hairlike

It may seem a little confusing that the names of arteries and veins change as they progress through the body - similar to a road crossing a border into another county. For example, the subclavian artery, axillary artery and brachial artery are the same vessel. Its name changes as it passes through those regions of the body.

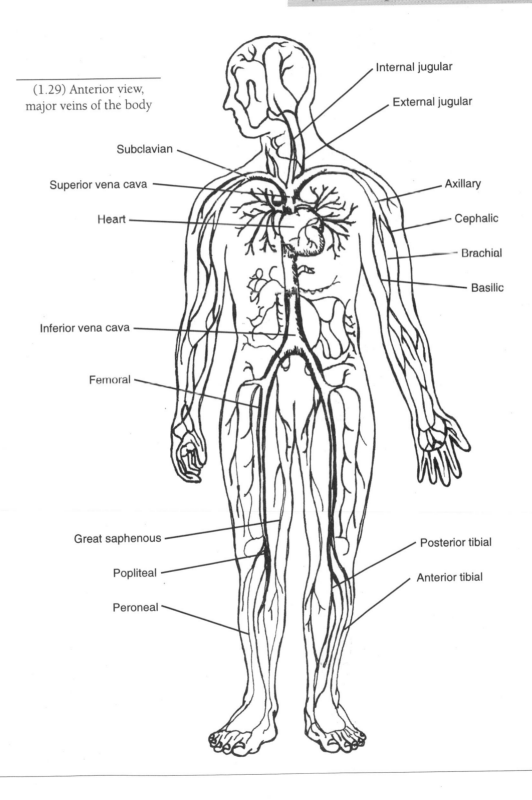

(1.29) Anterior view, major veins of the body

Internal jugular

External jugular

Subclavian

Superior vena cava

Heart

Axillary

Cephalic

Brachial

Basilic

Inferior vena cava

Femoral

Great saphenous

Popliteal

Peroneal

Posterior tibial

Anterior tibial

The Nervous System

The nervous system is the body's functional headquarters. It senses, interprets and responds to the body's needs in order to maintain homeostasis, or equilibrium. The brain and spinal cord compose the central nervous system while the remaining aspects form the peripheral nervous system. Many nerves branch off the spinal cord and exit through the sides of the vertebrae. Some of these nerves regroup to form a plexus. The main plexuses are the cervical, brachial, lumbar and sacral. The individual nerve branches of the plexus split off with names which correspond to the regions they innervate.

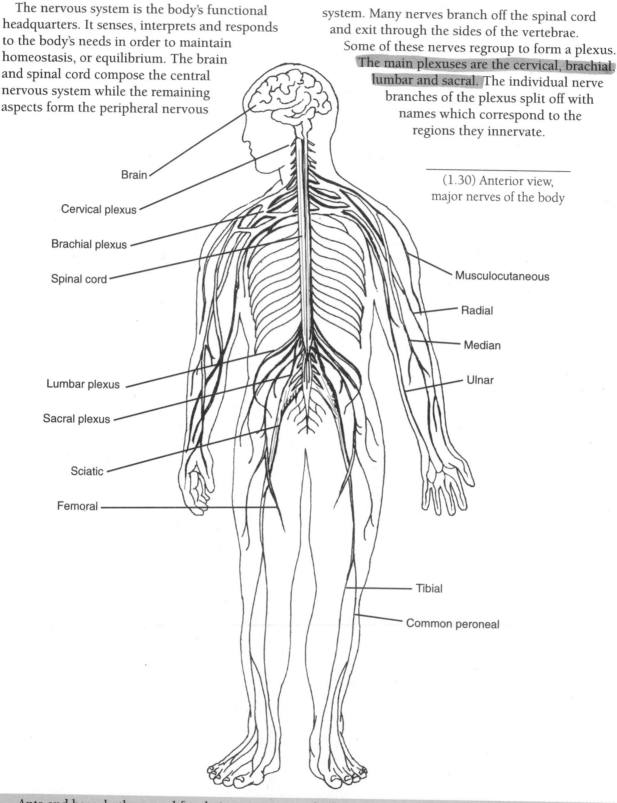

(1.30) Anterior view, major nerves of the body

Brain

Cervical plexus

Brachial plexus

Spinal cord

Lumbar plexus

Sacral plexus

Sciatic

Femoral

Musculocutaneous

Radial

Median

Ulnar

Tibial

Common peroneal

Ants and bees, both revered for their intelligence and diligence, have roughly 250 and 900 nerve cells, respectively, in their entire bodies. Humans, not always demonstrating such qualities, have an estimated 10,000,000,000 nerve cells in the brain alone.

nerve   L. sinew
plexus   L. a braid, interwoven

The Lymphatic System

The lymphatic system is composed of several organs, yellow liquid called lymph, small microscopic vessels called lymphatics and lymph nodes. These structures perform many functions throughout the body such as draining interstitial fluid which escapes from capillaries and transporting it back to the heart. Lymphatic vessels carry fats from the intestines to the blood. Lymphatic tissue also assists the body's immune system to defend against foreign cells, microbes and cancer cells.

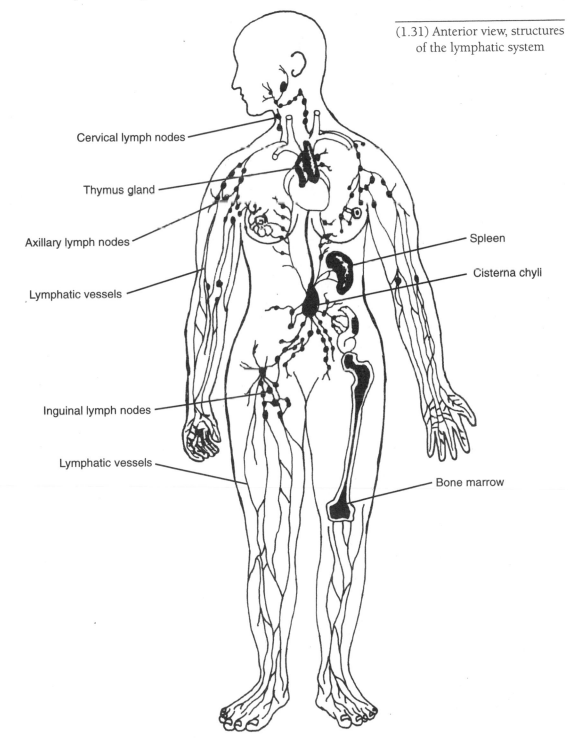

(1.31) Anterior view, structures of the lymphatic system

Cervical lymph nodes

Thymus gland

Axillary lymph nodes

Lymphatic vessels

Inguinal lymph nodes

Lymphatic vessels

Spleen

Cisterna chyli

Bone marrow

| lymph | limf | L. pure spring water |
| interstitial | in-ter-**stish**-al | L. placed, or lying between |

Chapter
Shoulder & Arm 2

Topographical Views

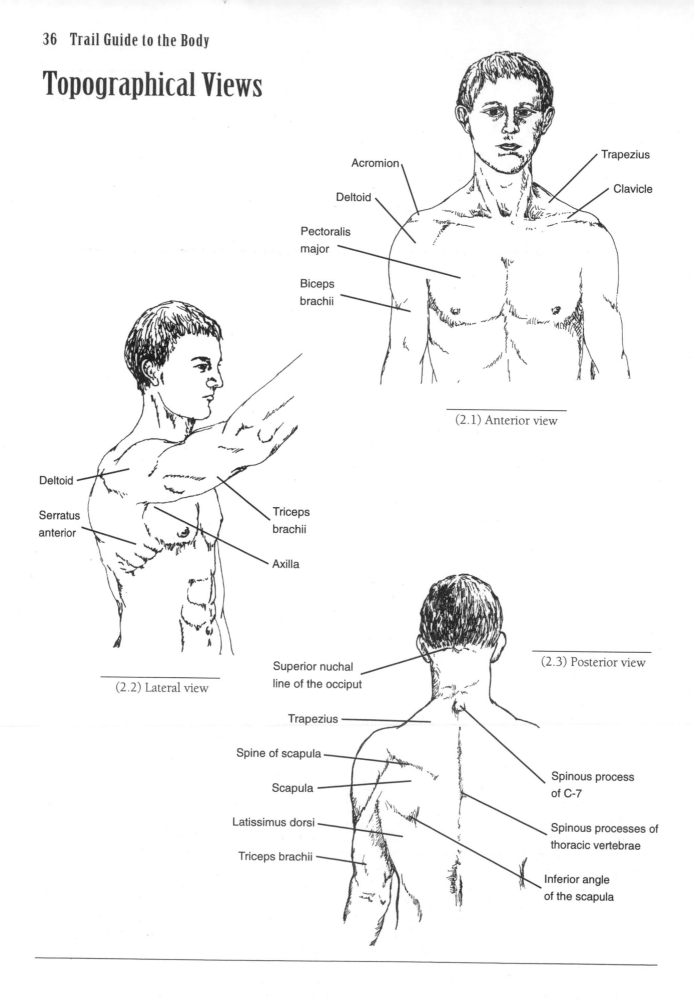

Acromion

Deltoid

Pectoralis
major

Biceps
brachii

Trapezius

Clavicle

(2.1) Anterior view

Deltoid

Serratus
anterior

Triceps
brachii

Axilla

(2.2) Lateral view

Superior nuchal
line of the occiput

Trapezius

Spine of scapula

Scapula

Latissimus dorsi

Triceps brachii

(2.3) Posterior view

Spinous process
of C-7

Spinous processes of
thoracic vertebrae

Inferior angle
of the scapula

Bones of the Shoulder and Arm

The shoulder complex is made up of three bones: the clavicle, scapula and humerus. (2.4)

The *clavicle* or collarbone is superficial and runs horizontally along the top of the chest at the base of the neck. It articulates laterally with the acromion of the scapula (acromioclavicular joint) and medially with the sternum (sternoclavicular joint). Both joints are synovial joints. The sternoclavicular joint is the single attachment site between the upper appendicular and axial skeletons.

The *scapula* is a triangular-shaped bone of the upper back. Along with the clavicle, the scapula plays a vital role in stabilization and movement of the arm. The scapula contains several fossas,

corners and ridges which serve as attachment sites for seventeen muscles.

The scapula glides across the posterior surface of the thorax to form the scapulothoracic joint. Because this articulation does not have any of the usual joint components, it is considered a false joint.

The *humerus* is the bone of the arm. The proximal humerus articulates with the glenoid fossa of the scapula to form the glenohumeral joint. The glenohumeral joint is a synovial, ball and socket joint with a wide range of movement. The proximal humerus and the glenohumeral joint are surrounded by the deltoid muscle and numerous tendons.

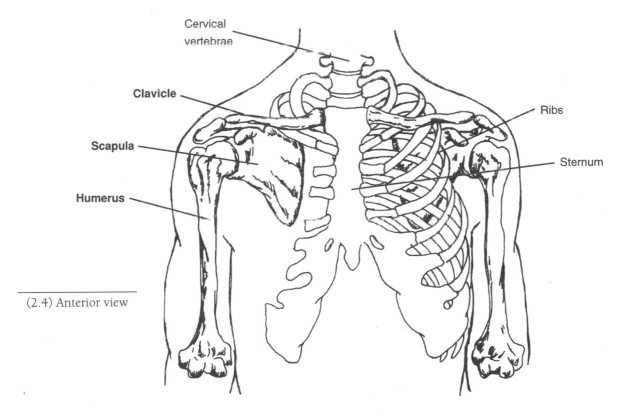

Cervical vertebrae

Clavicle

Scapula

Humerus

Ribs

Sternum

(2.4) Anterior view

The clavicle is the first bone to start ossifying (hardening) in a human fetus, yet is the last to completely develop - often in the late teens or early twenties. This, along with its superficial location, may explain why the clavicle is one of the most commonly broken bones in the body.

A quadruped such as a dog or cat, however, is not concerned about a broken clavicle. Since a quadruped's scapula is positioned on the lateral side of the trunk (as opposed to a human's,

which lies on the posterior side of the trunk), a clavicle is not as necessary for movement of the shoulder complex. Actually, cats have a thin sliver of a clavicle and dogs have a small piece of cartilage.

A bird's clavicles are united to form a furcula. As a single unit, the furcula acts as a strut and offers greater stability to the large pectoral muscles during flight. We separate the furcula when vying for the long end of the "wishbone."

| clavicle | **klav**-i-k'l | L. little key | scapula | **skap**-u-la | L. to dig, like the end of a shovel |
| humerus | **hu**-mer-us | L. upper arm | furcula | **fur**-ku-la | L. a little fork |

Bony Landmarks of the Shoulder

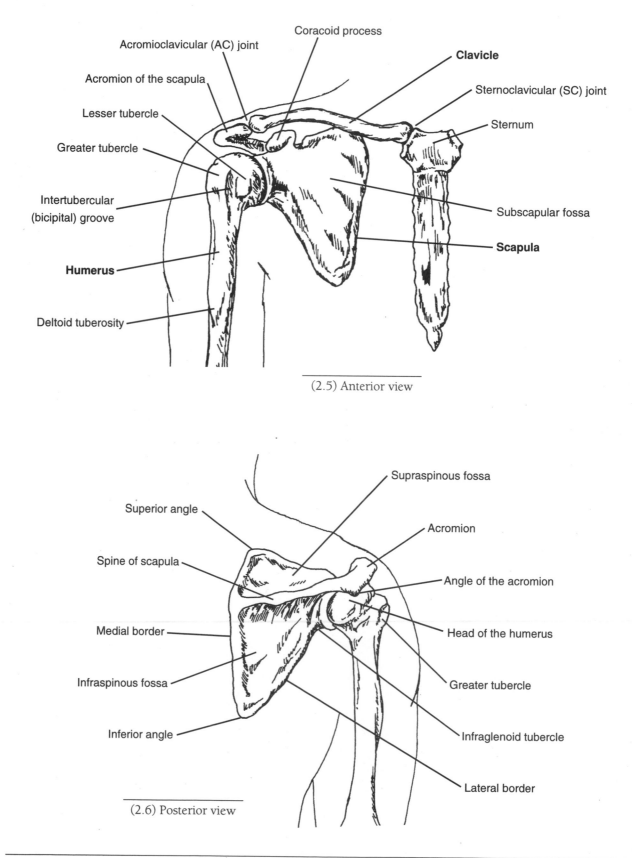

Coracoid process

Acromioclavicular (AC) joint

Acromion of the scapula

Lesser tubercle

Greater tubercle

Intertubercular (bicipital) groove

Humerus

Deltoid tuberosity

Clavicle

Sternoclavicular (SC) joint

Sternum

Subscapular fossa

Scapula

(2.5) Anterior view

Superior angle

Spine of scapula

Medial border

Infraspinous fossa

Inferior angle

Supraspinous fossa

Acromion

Angle of the acromion

Head of the humerus

Greater tubercle

Infraglenoid tubercle

Lateral border

(2.6) Posterior view

process   **pros**-es   L. going before

Bony Landmarks Trails

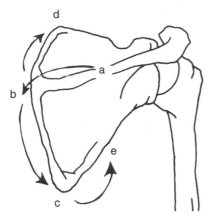

Trail 1 - "Along the Edges" explores the sides and corners of the scapula.

 a) Spine of the scapula
 b) Medial border
 c) Inferior angle
 d) Superior angle
 e) Lateral border

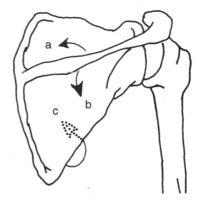

Trail 2 - "In the Trenches" sinks into the three basins of the scapula.

 a) Supraspinous fossa
 b) Infraspinous fossa
 c) Subscapular fossa

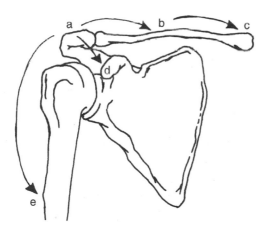

Trail 3 - "Springboard Ledge" leads around to the anterior shoulder, using the scapula's acromion as a jumping off point.

 a) Acromion
 b) Clavicle
 c) Acromioclavicular and
    sternoclavicular joints
 d) Coracoid process
 e) Deltoid tuberosity

Trail 4 - "Two Hills and a Valley" focuses on the three landmarks located along the proximal humerus.

 a) Greater tubercle
 b) Intertubercular groove
 c) Lesser tubercle

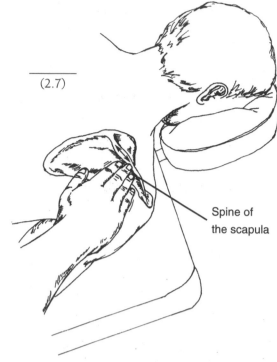

(2.7)

Spine of
the scapula

Because of its central location, the spine of the scapula makes for a great base camp for locating other landmarks. If you become lost or confused while palpating the scapula, return to its spine.

Trail 1 - "Along the Edges"
Spine of the Scapula

The spine of the scapula is a superficial ridge located just off the top of the shoulder. It runs at an oblique angle along the body, stretching from the acromion to the medial border. It is an attachment site for the posterior deltoid and middle and lower fibers of the trapezius (p. 52).

1) Prone. Lay your hand across the upper back and slide your fingertips inferiorly until they roll over the superficial spine. (2.7)
2) Strum your fingers vertically, palpating its width and edges. Also explore its entire length by palpating laterally toward the acromion and medially toward the vertebral column.

✔ *As you strum your fingers over the spine, do you feel a ditch of soft tissue above and below it? If your partner slowly elevates his scapula, does the spine elevate as well?*

Medial Border

The medial border is a long edge of the scapula that runs parallel to the vertebral column. It can measure five to seven inches in length, depending on body type. The medial border is an attachment site for the rhomboids (p. 62) and serratus anterior (p. 65) and is deep to the trapezius.

1) Prone. Place your partner's hand in the small of his back to raise the medial border off the ribs. For more exposure, scoop and raise the shoulder with one hand.
2) Locate the spine of the scapula and glide your fingertips medially until they slide off the spine onto the medial border. (2.8)
3) Follow the medial border inferiorly and superiorly; note that it extends further inferior to the spine of the scapula than superiorly.

✔ *Does the edge you feel run vertically?*

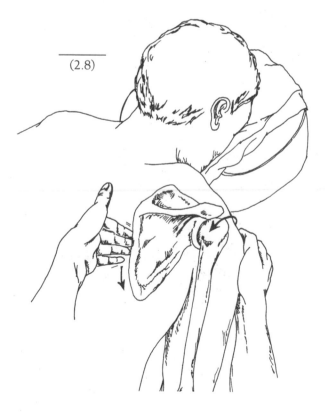

(2.8)

Inferior Angle

There are two angles of the scapula: one on either end of the medial border. The inferior angle is superficial and is located at the medial border's lower end.

1) Prone. Place your partner's hand in the small of his back.
2) Glide your fingers inferiorly along the medial border.
3) At the end of the medial border, the edge of the scapula will turn a corner and start to rise superiorly and laterally. This corner is the inferior angle. (2.9)

✔ *Can you sculpt around the inferior angle and pinch it with your fingertip and thumb?*

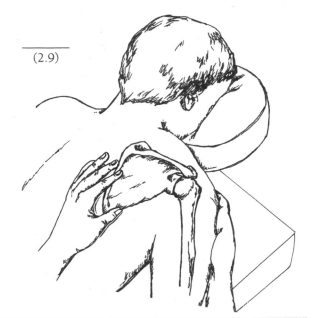
(2.9)

When an individual has a "winged scapula," the inferior angle is the portion of the scapula that visibly protrudes posteriorly.

Superior Angle

The superior angle is located at the superior end of the medial border. It serves as the insertion site of the levator scapula muscle. Because of the angle's location deep to the trapezius muscle (p. 52), it may not be as easy to isolate as the inferior angle.

1) Prone. Place a rolled towel under the shoulder or scoop the shoulder with your hand to raise it off the table. This will shorten the overlying muscles.
2) Locate the medial border. Slide your fingertips superiorly along the medial border. (2.10)
3) You may need to move an inch superior to the spine of the scapula to reach the superior angle.

With your partner sidelying, elevate the scapula toward the ear. As the scapula falls away from the ribcage, the superior angle will be quite palpable.

✔ *Sculpt out the superior angle and note if it is continuous with the medial border. Locate both the inferior angle and the superior angle. Note the distance between them and gently slide the scapula superiorly and inferiorly.*

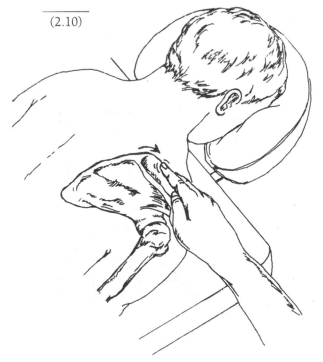
(2.10)

Lateral Border

The lateral border extends superiorly and laterally from the inferior angle up toward the axilla or "arm pit." It is an attachment site for the teres major and teres minor muscles (p. 56) and, due to these thick tissues, may not be as clearly defined as the medial border.

1) Prone. Drape the arm off the side of the table. Slide your thumb from the inferior angle superiorly along the lateral border.
2) Follow the border in the direction of the axilla. If the musculature is too thick to palpate through, try curling your thumb underneath the tissue. (2.11) This is most effective when locating the infraglenoid tubercle (see below).

Try the above method with your partner's hand in the small of his back.

✔ *Is the edge of bone you are palpating continuous with the inferior angle? As you follow it superiorly, does it lead you in the direction of the axilla?*

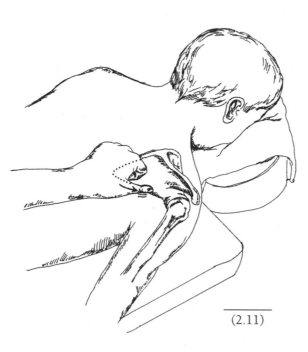

(2.11)

The infraglenoid tubercle is often tender when accessed. Using your broad thumbpad will allow you to palpate specifically without causing pain.

Infraglenoid Tubercle

The infraglenoid tubercle is located at the most superior aspect of the lateral border. The tubercle is not a distinguishable point, but a small spot which serves as an attachment site for the long head of the triceps brachii (p. 71). It lies deep to the teres minor and deltoid muscles.

1) Prone. Locate the lateral border.
2) Slide along the lateral border to its most superior portion. (2.12) You can either compress through the overlaying muscles or curl underneath them to access the landmark directly.

✔ *Are you along the edge of the lateral border? Are you on the posterior side of the armpit?*

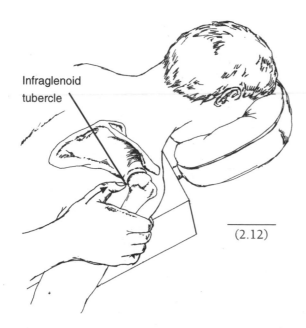

Infraglenoid tubercle

(2.12)

tubercle   **tu**-ber-kl   L. a little swelling

Trail 2 - "In the Trenches"
Infraspinous Fossa

The scapula contains three fossas, or depressions: the infraspinous, supraspinous and subscapular. Each fossa is designed to accommodate a muscle belly and its tendonous attachments. The infraspinous fossa is a triangular area inferior to the spine of the scapula. It is filled with the infraspinatus muscle (p. 56).

1) Prone. Palpate the spine of the scapula, medial border and lateral border to isolate the infraspinous fossa.
2) Cradle the inferior angle with the webbing between your index finger and thumb. Your index finger will rest along the medial border, your thumb along the lateral border.
3) Place a finger of the opposite hand along the length of the spine of the scapula. The triangular-shaped area you isolate is the infraspinous fossa. (2.13)

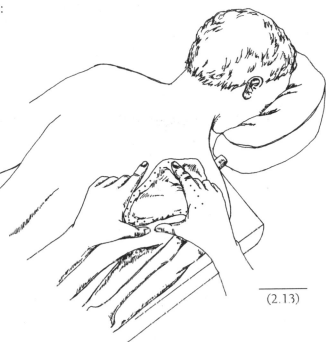

(2.13)

Supraspinous Fossa

The supraspinous fossa is located superior to the spine of the scapula. It is small in size, yet quite deep. The supraspinatus muscle (p. 56) attaches to and lies in this basin, so the supraspinous fossa is difficult to access.

1) Prone. Drop your thumb pad inferiorly and lateral from the superior angle into the fossa or lay your thumb along the spine of the scapula and raise them superiorly into the fossa.
2) Although the fossa is covered by the trapezius and supraspinatus muscles, explore the fossa's size and shape. (2.14)
3) Slide your fingers laterally, noting how the fossa becomes thinner and finally ends at the junction of the acromion and clavicle. Actually, the fossa continues underneath the acromion but is inaccessible.

✔ *Are you superior to the spine of the scapula? If you strum your fingers superiorly and inferiorly can you palpate the supraspinatus fibers running laterally toward the acromion?*

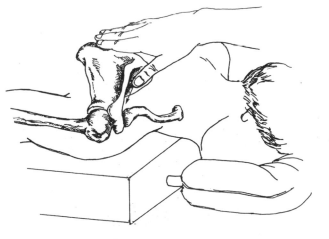

(2.14)

---

fossa   **fos**-a   L. a shallow depression

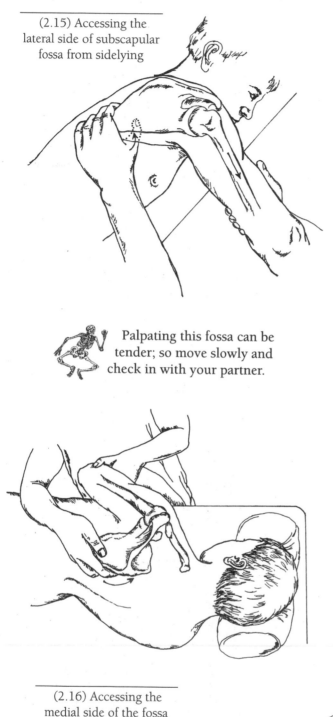

(2.15) Accessing the lateral side of subscapular fossa from sidelying

Palpating this fossa can be tender; so move slowly and check in with your partner.

(2.16) Accessing the medial side of the fossa from sidelying

Subscapular Fossa

The subscapular fossa is located on the scapula's anterior (or underside) surface, next to the ribcage. It is the attachment site for the subscapularis and serratus anterior muscles (p. 65). Due to the numerous muscle bellies surrounding the fossa and the scapula's close proximity to the ribcage, the fossa can be challenging to access.

1) Sidelying. This position allows the scapula to slide away from the ribcage for easier access.
2) Place your thumb at the middle of the lateral border. Be sure to position your thumb anterior to the large mass of muscles along the lateral border.
3) Slowly sink and curl your thumb tip onto the surface of the fossa. (2.15) Use your other hand to maneuver the scapula and arm so your thumb can sink in more easily. You may only be able to sink an inch onto the fossa.

Depending on the tissue's flexibility, the medial portion of the subscapular fossa can be accessed.
1) Sidelying. Place your partner's arm against his side and lay your fingertips along the medial border. With the other hand, move the scapula posteriorly (bringing the medial border off the ribs). (2.16)
2) Slowly curl your fingers through the rhomboid and trapezius muscles (p. 52), under the scapula and onto the fossa.

Can you feel the ribcage and anterior surface of the scapula on either side of your thumb?

(2.17) Palpating the lateral side of the fossa from a prone position

Trail 3 - "Springboard Ledge"
Acromion

The acromion is the lateral extension of the spine of the scapula located at the top of the shoulder. It has a flat surface and articulates with the clavicle's lateral end. The acromion serves as an attachment site for the trapezius and deltoid muscles.

The acromial angle is a small corner that can be felt along the acromion's lateral/posterior aspect.

1) Supine or seated. Locate the spine of the scapula.
2) Follow the spine as it rises superiorly and laterally to the top of the shoulder. Use your fingerpads to explore its flat surface. (2.18)
3) Explore and sculpt around all sides of the acromion and its attachment to the clavicle.

✔ *Is the bone you are palpating superficial and directly on the top of the shoulder? Can you feel the small point of the acromial angle on the posterior edge of the acromion?*

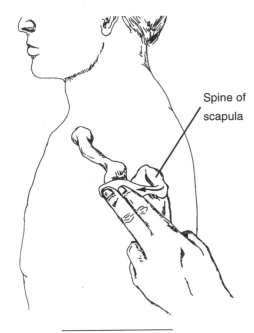

Spine of scapula

(2.18) Lateral view

Clavicle

The clavicle lies horizontally across the upper chest. It is superficial and has a gentle "S" shape. It is an attachment site for numerous muscles. Both ends of the clavicle are superficial and accessible. The lateral end is relatively flat and often rises slightly higher than the acromion. The medial end is round and articulates with the sternum.

1) Seated. Locate the acromion and walk your fingers medially to the shaft of the clavicle.
2) Grasp the clavicle's cylindrical body between your finger and thumb and explore its length from the acromion to the sternum. Observe how its acromial end rises superiorly and its sternal end curves inferiorly. (2.19)

✔ *Have your partner bring his shoulder anteriorly and the shaft of the clavicle will protrude visibly. Can you simultaneously locate both the medial and lateral ends of the clavicle?*

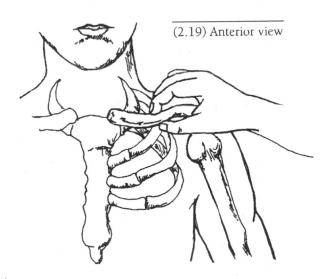

(2.19) Anterior view

With your fingers at either end of the clavicle, ask your partner to elevate, depress, adduct and abduct his scapula. As the scapula moves, notice how the ends of the clavicle shift in position.

acromion   a-**cro**-me-on   Gr. *akron*, extremity + *omos*, shoulder

Acromioclavicular and Sternoclavicular Joints

The *acromioclavicular (AC) joint* is the small articulation between the acromion of the scapula and the acromial end of the clavicle. The anterior and superior surfaces of this thin crevice can be directly palpated.

The *sternoclavicular (SC) joint* is the articulation between the sternal end of the clavicle and the sternum. Unlike the slender, smooth AC joint, the SC joint is wedge-shaped and contains a small, impalpable fibrous disk. At rest, only the inferior portion of the sternal end makes contact with the sternum. When the clavicle is elevated, the sternal end pivots on the sternum.

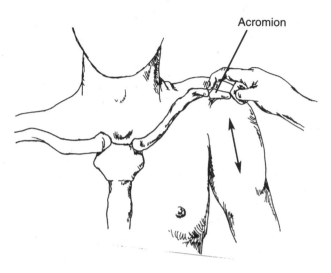

Acromion

(2.20) Accessing the AC joint while partner moves the shoulder

AC joint

1) Supine or seated. Locate the acromion.
2) Glide medially toward the clavicle. Your finger will feel a small "step" as you rise up onto the surface of the clavicle.
3) Backtrack slightly. Just lateral to the step will be the AC joint's slender ditch.

✔ *Does the acromial end of the clavicle lie slightly higher than the acromion? Place a finger where you believe the AC joint to be and ask your partner to slowly elevate and depress his scapula. (2.20) As the scapula rises, do you feel the joint space widen slightly? As it depresses, does the joint space diminish?*

SC joint

1) Slide your fingers medially along the shaft of the clavicle.
2) Just lateral to the body's centerline, the shaft will broaden to become the bulbous sternal end.
3) Locate the SC joint by sliding your finger medially off the sternal end.
4) Passively elevate, depress and abduct the scapula and explore the changes occurring at the SC joint.

✔ *Place a finger where you believe the SC joint to be and ask your partner to slowly elevate and depress his scapula. (2.21) Can you feel the joint space widen and diminish?*

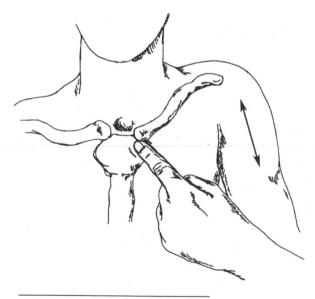

(2.21) Palpating the SC joint while partner moves shoulder

acromioclavicular   a-**kro**-me-o-kla-**vik**-u-lar
sternoclavicular   **ster**-no-kla-**vik**-u-lar

Coracoid Process

The coracoid process of the scapula is the beak-like projection found inferior to the shaft of the clavicle. Depending on the position of the scapula, it is often found in the deltopectoral groove between the deltoid and pectoralis major fibers. The coracoid process can be tender when palpated, so proceed carefully.

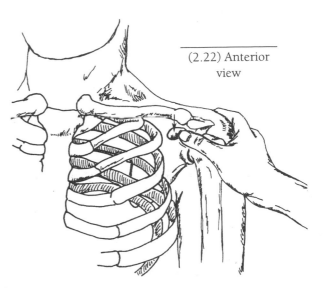

(2.22) Anterior view

1) Supine or seated. Lay your thumb along the lateral shaft of the clavicle.
2) Slide inferiorly off the clavicle no more than an inch and a half. Locate the tip of the coracoid process by compressing your fingerpads into the tissue. (2.22)
3) As the coracoid becomes more apparent, get a better understanding of its shape and size by sculpting a circle around its edges.

✔ *Are you inferior to the shaft of the clavicle? Passively move the scapula with your other hand and feel the coracoid follow.*

Deltoid Tuberosity

The deltoid tuberosity is located on the lateral side of the middle of the humeral shaft. It is a small, low bump that serves as an attachment site for the converging fibers of the deltoid (p. 51).

1) Supine or seated. Locate the acromion.
2) Slide off the acromion and down the lateral aspect of the arm. (2.23)
3) When you reach the halfway point between the shoulder and elbow, there will be a small mound on the lateral side of the arm.

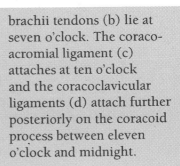

(2.23) Lateral view

✔ *If your partner abducts his shoulder, do the deltoid fibers converge where you are palpating?*

The coracoid process is an attachment site for several tendons and ligaments. The arrangement of these structures can be described in a clockwise fashion. On the right scapula, the pectoralis minor tendon (a) connects at four o'clock, while the coracobrachialis and biceps

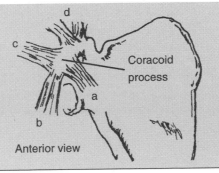

Anterior view

brachii tendons (b) lie at seven o'clock. The coraco-acromial ligament (c) attaches at ten o'clock and the coracoclavicular ligaments (d) attach further posteriorly on the coracoid process between eleven o'clock and midnight.

coracoid	**kor**-a-koyd	Gr. raven's beak
tuberosity	tu-ber-**os**-i-te	L. a swelling

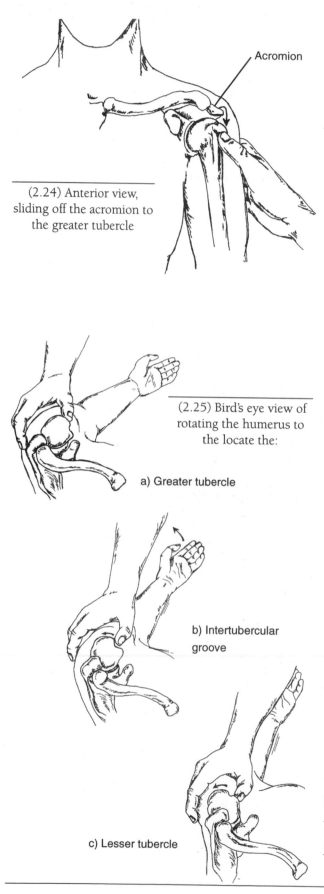

Acromion

(2.24) Anterior view, sliding off the acromion to the greater tubercle

(2.25) Bird's eye view of rotating the humerus to the locate the:

a) Greater tubercle

b) Intertubercular groove

c) Lesser tubercle

Trail 4 - "Two Hills and a Valley"
Greater and Lesser Tubercles
Intertubercular Groove

These three landmarks are located on the proximal humerus deep to the deltoid muscle. The *greater tubercle* is located inferior and lateral to the acromion. It is shaped more like a low mound than a pointy hill. The greater tubercle is an attachment site for three of the four rotator cuff muscles - supraspinatus, infraspinatus and teres minor (p. 56).

The *lesser tubercle* is smaller than the greater tubercle and is an attachment site for the fourth rotator cuff muscle - subscapularis. The *intertubercular groove*, or bicipital groove, is situated between the greater and lesser tubercles, and is roughly a pencil's width in diameter. Within the groove lies the tendon of the long head of the biceps brachii, which can be tender, so palpation in this region should be gentle.

👁 Greater tubercle

1) Seated or supine, shaking hands with your partner. Locate the acromion.
2) Slide off the acromion inferiorly and laterally approximately one inch. (2.24)
3) The solid surface located deep to the deltoid fibers will be the greater tubercle. You may feel a small dip between the tubercle and the acromion.

👁 Intertubercular groove and lesser tubercle

1) Place your thumb on the greater tubercle. (2.25)
2) Begin to laterally rotate the arm. As the humerus rotates, the greater tubercle (a) will move out from under your thumb and be replaced by the slender ditch of the intertubercular groove (b).
3) As you continue to laterally rotate, your thumb will rise out of the groove onto the lesser tubercle (c).

✔ *Place your thumb at the greater tubercle and passively rotate the arm medially and laterally. Do you feel the "bump-ditch-bump" sequence as the three landmarks pass beneath your thumb? Are you horizontal to the level of the coracoid process?*

Muscles of the Shoulder and Arm

The muscles of the shoulder are an amazingly diverse group. Some of them span across the back and ribcage, some attach at the cranium or stretch down to the elbow. All of the muscles create movement of the shoulder complex (formed by the scapula, clavicle and humerus). Some also raise the ribs, extend the head and cervical vertebrae, or bend the elbow. (2.26 - 2.28)

The superficial muscles of the shoulder and back are presented first, followed by the deeper muscles of the back and last, the muscles of the arm. Some muscles are presented together to better understand how they function as groups.

Although the instructions for each muscle specify a position in which to place your partner (seated or supine), exploration of all positions is encouraged to better understand the muscle and its surrounding structures.

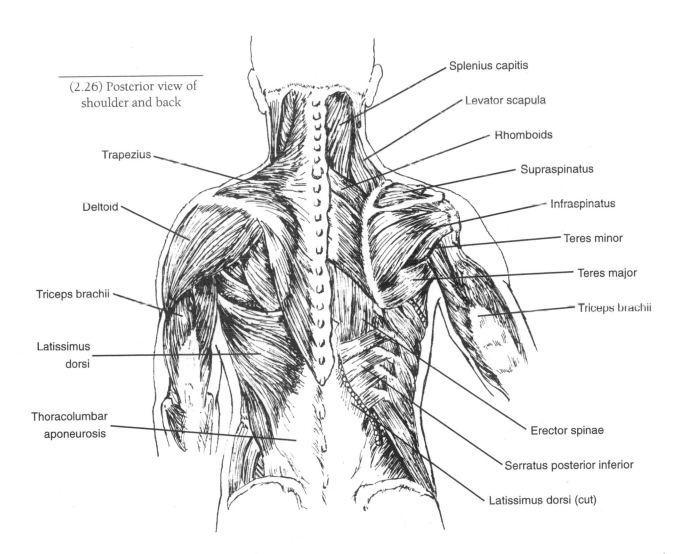

(2.26) Posterior view of shoulder and back

Splenius capitis

Levator scapula

Rhomboids

Supraspinatus

Infraspinatus

Teres minor

Teres major

Triceps brachii

Erector spinae

Serratus posterior inferior

Latissimus dorsi (cut)

Trapezius

Deltoid

Triceps brachii

Latissimus dorsi

Thoracolumbar aponeurosis

Latissimus dorsi, trapezius and deltoid are removed on right side

The trapezius received its present name from the British anatomist William Cowper (c. 1700). Previously, it was called the *musculus cucullaris* (L. muscle hood), since both trapezius muscles together resemble a monk's hood.

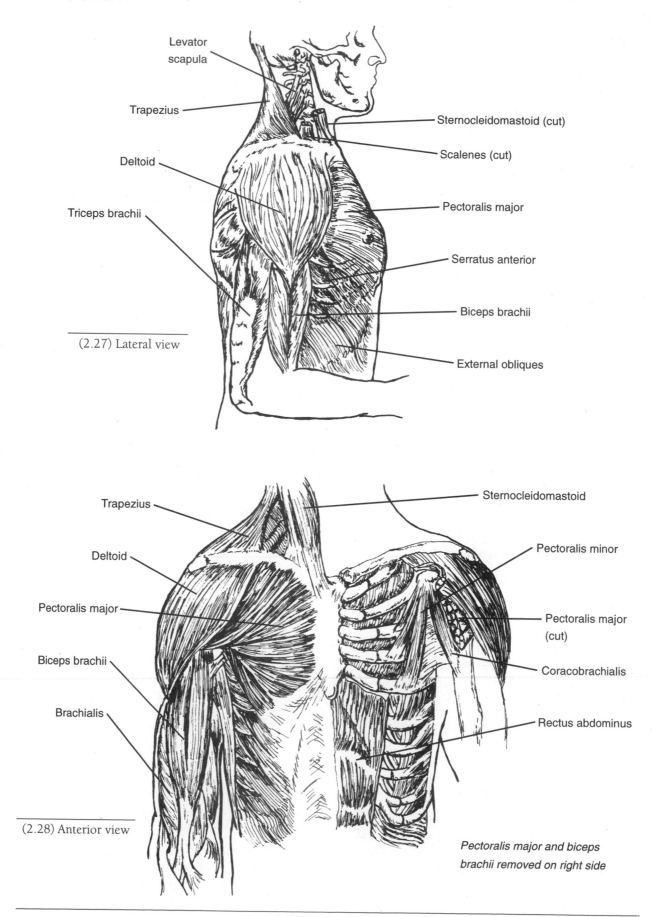

Levator
scapula

Trapezius

Deltoid

Triceps brachii

Sternocleidomastoid (cut)

Scalenes (cut)

Pectoralis major

Serratus anterior

Biceps brachii

External obliques

(2.27) Lateral view

Trapezius

Deltoid

Pectoralis major

Biceps brachii

Brachialis

Sternocleidomastoid

Pectoralis minor

Pectoralis major
(cut)

Coracobrachialis

Rectus abdominus

(2.28) Anterior view

*Pectoralis major and biceps
brachii removed on right side*

Deltoid

The triangle-shaped deltoid is located on the cap of the shoulder. The origin of the deltoid (which is curiously identical to the insertion of the trapezius) curves around the spine of the scapula and clavicle making a "V" shape. From this broad origin, the fibers converge down the arm to attach at the deltoid tuberosity. (2.29)

The deltoid fibers can be divided into three groups: the anterior, middle and posterior fibers. All three groups abduct the humerus, but the anterior and posterior fibers are antagonists in flexion/extension and medial/lateral rotation.

A -

All fibers:
   Abduct the shoulder joint
Posterior fibers:
   Extend and laterally rotate the shoulder joint
Anterior fibers:
   Flex and medially rotate the shoulder joint

O - Lateral one-third of clavicle, acromion and spine of scapula

I - Deltoid tuberosity

N - Axillary from brachial plexus

1) Seated. Locate the spine of the scapula, the acromion and the lateral one-third of the clavicle. Note the "V" shape these landmarks form.
2) Locate the deltoid tuberosity.
3) Palpate between these landmarks to isolate the superficial, convergent fibers of the deltoid.
4) Be sure to explore the deltoid's most anterior and posterior aspects.

✔ *Are the fibers you feel superficial and converging toward the deltoid tuberosity? If your partner alternately abducts and releases, do you feel the fibers contract and relax? (2.30)*

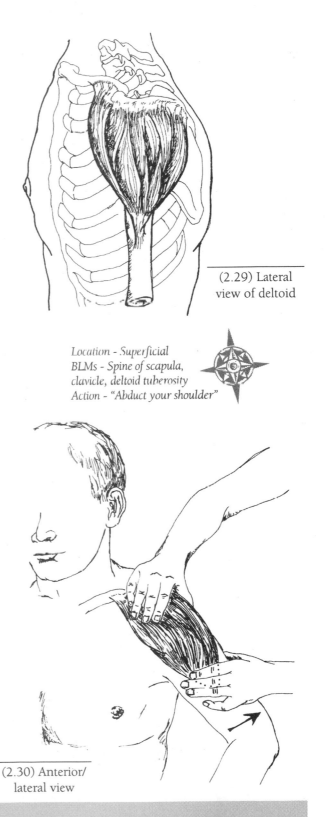

(2.29) Lateral view of deltoid

Location - Superficial
BLMs - Spine of scapula, clavicle, deltoid tuberosity
Action - "Abduct your shoulder"

(2.30) Anterior/ lateral view

Feel the antagonistic abilities of the deltoid's anterior and posterior fibers:
1) Shake hands with your partner. Place your other hand on the deltoid.
2) Keeping his elbow next to his side, ask your partner to internally and externally rotate his arm against your resistance. Can you sense the anterior fibers contract upon internal rotation and relax upon external rotation? (Visa-versa for the posterior fibers?)

deltoid   **del**-toid   Gr. *delta*, letter d (Δ) in Greek alphabet

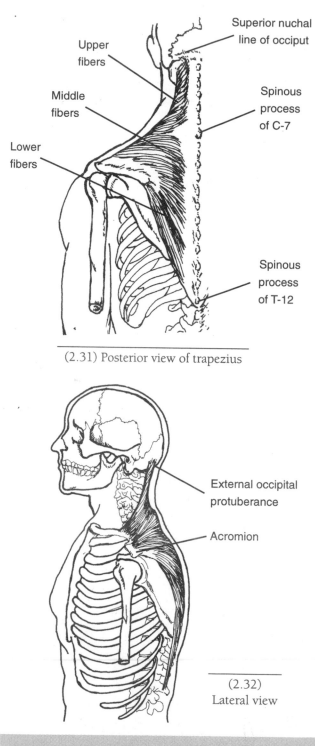

(2.31) Posterior view of trapezius

(2.32)
Lateral view

Trapezius

The trapezius lies superficially along the upper back and neck. Its broad, thin fibers blanket the shoulders, attaching to the occiput (the bone at the base of the head, p. 162), lateral clavicle, scapula and spinous processes of the thoracic vertebrae. (2.31, 2.32) The trapezius fibers can be divided into three sections: upper (descending) fibers, middle fibers, and lower (ascending) fibers. The upper and lower fibers are antagonists in elevation and depression of the scapula, respectively. All fibers of the trapezius are palpable.

A -
Upper Fibers:
Bilateral-
    Extend the head and neck
Unilateral-
    Extend
    Laterally flex
    Rotate the head and neck to the opposite side

    Elevate
    Upwardly rotate the scapula

Middle Fibers:
    Adduct
    Stabilize the scapula

Lower Fibers:
    Depress
    Upwardly rotate the scapula

O - External occipital protuberance, medial portion of superior nuchal line of occiput, ligamentum nuchae and spinous processes of C-7 through T-12

I - Lateral portion of clavicle, acromion and spine of the scapula

N - Spinal accessory and cervical plexus

The upper fibers of the trapezius which run along the posterior neck are surprisingly skinny - perhaps two fingers in diameter.

👁✋ Supine. To feel this portion of the trapezius muscles, stand at the head of the table and ask your partner to extend his head slightly, "Raise your head a quarter inch off the face cradle." Can you isolate the two cigar-shaped bands running down the posterior neck? They should be superficial and extend up to the base of the skull.

Trapezius

👁 Upper fibers of the trapezius

1) Prone. The trapezius' upper fibers are the easily accessible flap of muscle lying across the top of the shoulder.
2) Grasp the superficial tissue on the top of the shoulder and feel the upper trapezius fibers. Take note of their slender quality. (2.33)
3) Follow the fibers superiorly toward the base of the head at the occiput. Then follow them inferiorly to the lateral clavicle.

✔ *Is the muscle you are grasping thin and superficial? Grasp the upper fibers and have your partner elevate his scapula gently toward his ear. Do the muscle fibers become taut?*

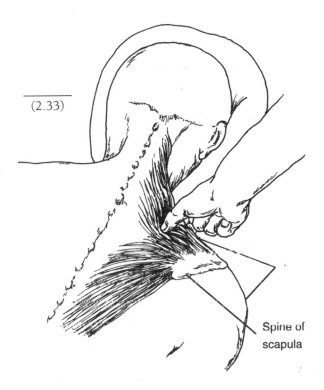

(2.33)

Spine of scapula

👁 Middle fibers of the trapezius

1) Locate the spine of the scapula.
2) Slide medially from the spine of the scapula onto the trapezius and move your fingers across its fibers. The trapezius fibers are superficial and thin, so explore at a superficial level and not into the deeper rhomboids or erector spinae muscles.

✔ *Palpate the middle fibers and ask your partner to adduct his scapula, "Bring your shoulder up off the table." Can you feel any contraction in the fibers?*

👁 Lower fibers of the trapezius

1) Locate the edge of the lower fibers by drawing a line from the spine of the scapula to the spinous process of T-12 (p. 123).
2) Palpate along this line and push your fingers into the edge of the lower fibers. Ask your partner to depress his scapula and feel for the superficial fibers of the trapezius. (2.34)
3) Attempt to lift the lower fibers between your fingers, raising it off the underlying musculature.

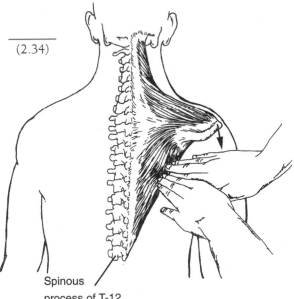

(2.34)

Spinous process of T-12

✔ *Have your partner depress his shoulder and see if you can strum across the lower fibers. Do the lower fibers run at a gentle angle toward the scapula (versus parallel with the vertebral column like the erector spinae muscles)?*

Location - Superficial, upper back
BLMs - Spine of scapula, occiput, acromion, T-12
Action - "Elevate (or) depress your shoulder"

| nuchae | **nu**-kay | L. nape of neck |
| occiput | **ok**-si-put | L. the back of the skull |

Latissimus Dorsi and Teres Major

The *latissimus dorsi* is the broadest muscle of the back. Its thin, superficial fibers originate at the low back, ascend the side of the trunk and merge into a thick bundle at the axilla. (2.35) Both ends of the latissimus dorsi are difficult to isolate; however its middle portion next to the lateral border of the scapula is easy to grasp.

The *teres major* is called "lat's little helper" because it is a complete synergist with the latissimus dorsi. (2.36) It is superficial and located along the scapula's lateral border between the latissimus dorsi and teres minor (p. 56). Although they share names, the teres major and teres minor rotate the arm in opposite directions - the major medially, the minor laterally.

The latissimus dorsi and teres major are sometimes called the "handcuff muscles," since their actions bring the arms into an "arresting" position!

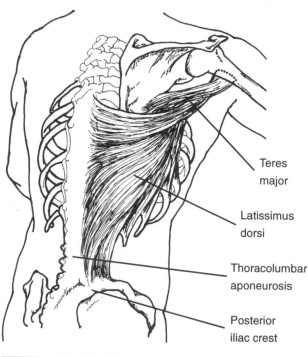

(2.35) Lateral/posterior view of latissimus dorsi

Latissimus Dorsi

A - Extend
 Adduct
 Medially rotate the shoulder joint

O - Spinous processes of last six thoracic vertebrae, last three or four ribs, thoracolumbar aponeurosis and posterior iliac crest

I - Crest of the lesser tubercle of the humerus

N - Brachial plexus

Teres Major

A - Extend
 Adduct
 Medially rotate the humerus

O - Dorsal surfaces of inferior angle and lower half of lateral border of scapula

I - Crest of the lesser tubercle of the humerus

N - Subscapular

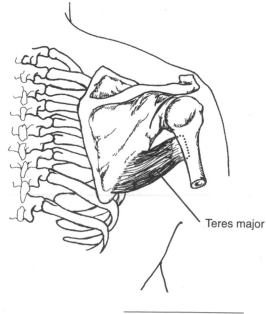

(2.36) Posterior view

Location - Posterior axilla, lateral trunk
BLMs - Lateral border of scapula
Action - "Extend and internally rotate your shoulder"

The latissimus dorsi not only moves the arm, but because of its broad origin can also affect the trunk and spine. When one latissimus contracts, it assists in lateral flexion of the trunk. If the arm is fixed, like when hanging from a bar, the latissimus assists extension of the spine and tilting of the pelvis anteriorly and laterally.

latissimus dorsi	la-**tis**-i-mus **dor**-si	L. widest, back
teres	**te**-reez	L. round

✋👁 Latissimus dorsi

1) Prone with arm off side of table. Locate the scapula's lateral border.

2) Grasp the thick wad of muscle tissue lateral to the lateral border between your fingers and thumb. This is the latissimus dorsi (and perhaps some of teres major). Note how this muscle tissue seems to flair off the side of the trunk.

3) Feel the latissimus fibers contract by asking your partner to medially rotate his arm against your resistance. "Swing your hand up toward your hip." (2.37) As this occurs, follow the latissimus fibers superiorly into the axilla and inferiorly on the ribs.

✔ *To ensure you are not just lifting the skin, grasp the tissue and let it slowly slip out between your fingers. Do you feel the muscle's fibrous texture or just the skin's jellylike quality?*

◆ Latissimus dorsi

1) Supine. Cradling the arm in a flexed position, grasp the tissue of the latissimus located beside the lateral border.

2) Ask your partner to extend his shoulder against your resistance. "Press your elbow toward your hip." This will force the latissimus to contract. (2.38)

✋👁 Teres major

1) Prone with the arm off the side of the table. Locate and grasp the latissimus dorsi fibers between your fingers and thumb.

2) Move your fingers and thumb medially to where you feel the scapula's lateral border. The muscle fibers which lay medial to the latissimus and attach to the lateral border will be the teres major.

3) Follow these fibers toward the axilla to where they blend with the latissimus dorsi.

✔ *Lay your thumb on the inferior aspect of the lateral border and have your partner medially rotate the shoulder joint to distinguish the teres major and latissimus dorsi. (2.39) The fibers of both muscles will contract; those that attach to the lateral border belong to teres major. The fibers further lateral belong to latissimus dorsi.*

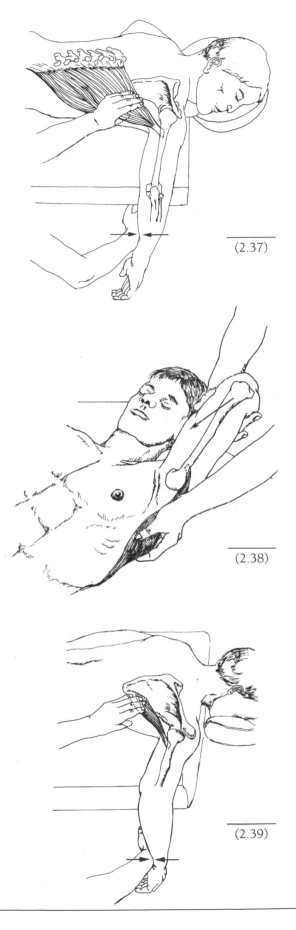

(2.37)

(2.38)

(2.39)

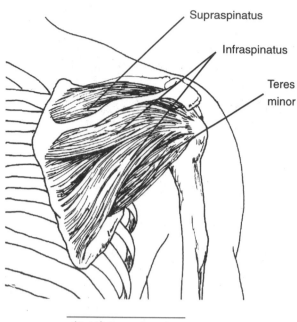

(2.40) Posterior view

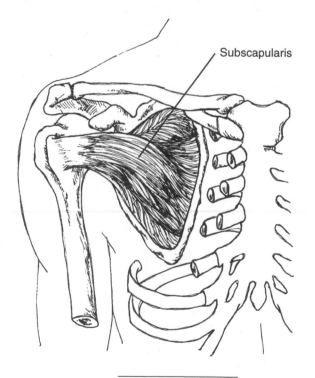

(2.41) Anterior view

Rotator Cuff Muscles

Supraspinatus
Infraspinatus
Teres Minor
Subscapularis

Supraspinatus, infraspinatus, teres minor and subscapularis are known as the rotator cuff muscles. Together they encompass, and therefore stabilize, the glenohumeral joint. All of the rotator cuff muscles are accessible, including their tendons which attach to the humerus.

The chunky *supraspinatus* is located in the supraspinous fossa, deep to the trapezius' upper fibers. Its belly runs underneath the acromion and attaches to the humerus' greater tubercle. (2.40) The supraspinatus assists the deltoid with abduction of the shoulder and is the only muscle of the group not involved in shoulder rotation.

The flat, convergent belly of *infraspinatus* is located in the infraspinous fossa. Most of its belly is superficial with a medial portion deep to trapezius and a lateral portion beneath the deltoid. (2.26) The infraspinatus attaches immediately posterior to the supraspinatus on the greater tubercle. (2.40) It is a synergist with teres minor in lateral rotation of the shoulder.

The *teres minor* is a small muscle squeezed between the infraspinatus and teres major. It is located high into the axilla and can be challenging to grasp. (2.40) Teres minor and teres major are antagonists in rotation of the humerus.

The deep *subscapularis* (2.41), located on the scapula's anterior surface, is sandwiched between the subscapular fossa and serratus anterior muscle (p. 65). With only a small portion of its muscle belly accessible, subscapularis is the only rotator cuff muscle which attaches to the humerus' lesser tubercle. It creates medial rotation of the shoulder.

Supraspinatus
Location - Deep to trapezius
BLMs - Supraspinous fossa
Action - "Abduct your shoulder"

Infraspinatus
Location - Mostly superficial
BLMs - Infraspinous fossa
Action - "Laterally rotate your shoulder"

infraspinatus   **in**-fra-spi-**na**-tus   supraspinatus   **soo**-pra-spi-**na**-tus
subscapularis   sub-skap-u-**la**-ris

Supraspinatus

A -
Abducts the humerus
Stabilize head of humerus
  in glenoid cavity

O - Supraspinous fossa of scapula

I - Greater tubercle of humerus

N - Suprascapular

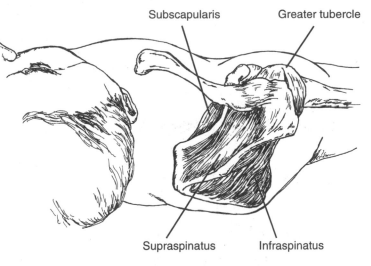

(2.42) Superior view of shoulder

Infraspinatus

A -
Laterally rotate
Adduct
Extend the shoulder joint
Stabilize head of humerus in glenoid cavity

O - Infraspinous fossa of scapula

I - Greater tubercle of humerus

N - Suprascapular from brachial plexus

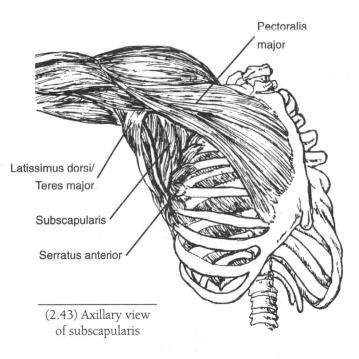

(2.43) Axillary view
of subscapularis

Teres Minor

A -
Laterally rotate
Adduct
Extend the shoulder joint
Stabilize head of humerus in glenoid cavity

O - Superior half of lateral border of scapula

I - Greater tubercle of humerus

N - Axillary

Subscapularis

A -
Medially rotate the shoulder joint
Stabilize head of humerus in glenoid cavity

O - Subscapular fossa of the scapula

I - Lesser tubercle of the humerus

N - Subscapular

Subscapularis
Location - Between scapula and ribcage
BLMs - Subscapular fossa
Action - "Medially rotate your shoulder"

Teres Minor
Location - Posterior axilla
BLMs - Upper lateral border of scapula
Action - "Laterally rotate your shoulder"

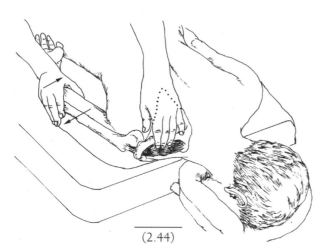

(2.44)

👁 Supraspinatus

1) Prone. Locate the spine of the scapula. Slide your fingers up into the supraspinous fossa.
2) Palpate through the trapezius and onto the supraspinatus fibers. As you palpate, note how the fibers run parallel to the spine.
3) Follow the belly laterally until it tucks under the acromion.

✔ *Can you differentiate the fibers of the trapezius and the deeper supraspinatus? With the arm alongside the body, have your partner alternate between abducting slightly and relaxing the shoulder. (2.44) Can you feel the supraspinatus tighten and soften underneath the inactive trapezius?*

👁 Infraspinatus

1) Prone, with the forearm off the side of the table. Locate the spine, medial border and lateral border of the scapula.
2) Form a triangle around the infraspinatus by laying a finger along each of these landmarks.
3) Palpate in the triangle and strum across the infraspinatus fibers. Follow them laterally as they converge underneath the deltoid to attach to the humerus.

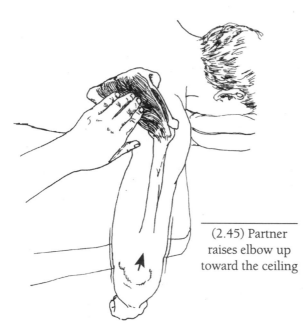

(2.45) Partner raises elbow up toward the ceiling

✔ *With the forearm off the side of the table, ask your partner to alternately raise his elbow one inch toward the ceiling and relax. (2.45) Do you feel the infraspinatus contract and tighten during this action?*

👁 Teres minor

1) Prone, with the arm off the side of the table. Locate the lateral border of the scapula; specifically, its superior half. Slide laterally off the lateral border onto the surface of the teres minor.
2) Compress into and across its tube-shaped belly. Move inferiorly and compare it in size to the teres major. Also, reach your thumb up into the axilla and grasp the belly of the teres minor like you would a hamburger. (2.46)
3) Ask your partner to laterally rotate his shoulder. "Swing your hand up toward your head." Bringing the elbow toward the ceiling also forces the teres minor to contract.

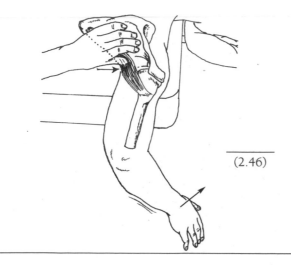

(2.46)

✔ *Does the muscle you are palpating attach along the superior half of the scapula's lateral border?*

👁✋ Feel the opposite rotational abilities
of the teres major and teres minor (2.47):
1) Prone. Lay your hand on the surfaces
of the teres major and minor.
2) Ask your partner to alternately medially
and laterally rotate his arm. (Be sure he does
not raise his elbow, because then they both
contract.) Can you feel the teres major contract
while the teres minor softens upon internal
rotation? Vice versa for external rotation?

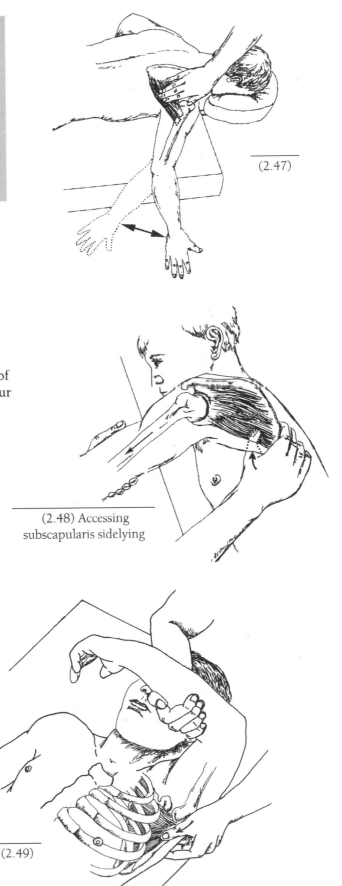

(2.47)

👁✋ Subscapularis

1) Sidelying. Flex the shoulder and pull the arm
anteriorly as much as possible. This will allow
easier access to the scapula's anterior surface.
2) Hold the arm with one hand while the thumb of
the other locates the lateral border. Hint: Slide your
thumb underneath the latissimus dorsi and teres
major fibers versus going through them. (2.48)
3) Slowly and gently curl your thumb onto the
subscapular fossa. You may not immediately feel
the subscapularis fibers, but if your thumb is on
the anterior surface of the scapula, you are
accessing a portion of the fibers.

(2.48) Accessing
subscapularis sidelying

◈ 1) Supine. Cradle the arm in a flexed
position and locate the lateral border.
2) Slowly sink your thumbpad onto the
subscapular fossa, adjusting the arm
and scapula as you progress. (2.49)

✔ *Ask your partner to medially rotate his
shoulder. Do you feel the subscapularis fibers
contract beneath your thumb? Explore the
subscapularis by moving your thumb further
superiorly or inferiorly.*

(2.49)

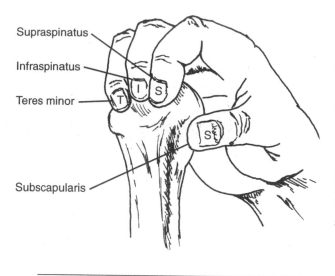

Supraspinatus

Infraspinatus

Teres minor

Subscapularis

(2.50) The rotator cuff tendon attachment sites

Rotator Cuff Tendons

In anatomical position, the tendons of the rotator cuff muscles are difficult to access. (2.50) The supraspinatus and infraspinatus tendons are situated deep to the acromion, while the tendons of the subscapularis and teres minor lie deep to the thick belly of the deltoid. This dilemma, however, can be overcome and each tendon isolated by placing the humerus in a specific position (outlined below). Since the rotator cuff tendons lie against the surface of the greater or lesser tubercles of the humerus, they cannot be separately distinguished from the underlying bone.

👁 Supraspinatus tendon

1) The attachment of the tendon will be located just distal to the acromion on the greater tubercle.
2) Supine, with the arm at the side of the body. Locate the acromion and slide inferiorly to the surface of the greater tubercle. (2.51)
3) Sink your thumb tip through the deltoid fibers. Using firm pressure, roll your thumb across the small mound of the supraspinatus tendon.

✔ *Are you palpating on the surface of the greater tubercle?*

1) Seated. Place your partner's arm behind his back. This position will medially rotate and extend the humerus.
2) Passively extend the arm as far as is comfortable for your partner. This position brings the supra-spinatus tendon out from under the acromion, just anterior and inferior to the acromioclavicular (AC) joint.

✔ *Is the arm medially rotated and hyperextended as far as comfortably possible? Are you palpating inferior to the AC joint?*

(2.51) Anterior/ lateral view

👁 Infraspinatus and teres minor tendons

1) Prone with arm off the side of the table. Locate the bellies of these muscles.

2) Strumming across their fibers, follow their bellies laterally as they pass under the acromion. Palpating through the deltoid, roll across their slender tendonous attachments at the greater tubercle. (2.52)

3) Turn your partner supine. Locate the tendonous attachment of supraspinatus. Move posteriorly along the greater tubercle and feel for the small tendonous attachments of the infraspinatus and teres minor.

✔ *Are you palpating deep to the deltoid fibers? Do you feel the solid surface of the greater tubercle beneath your fingers?*

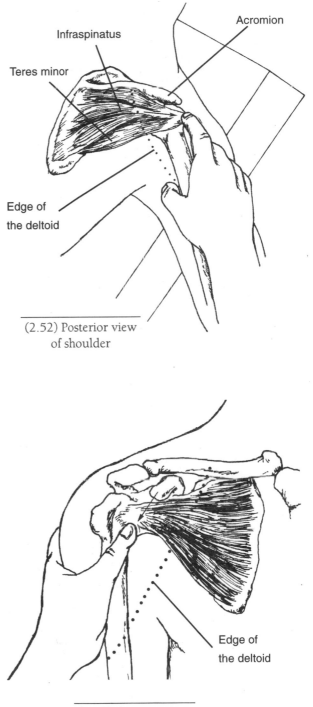

(2.52) Posterior view
of shoulder

👁 Subscapularis tendon

1) Seated or supine. Place the arm next to the trunk in anatomical position.

2) Locate the coracoid process of the scapula. Slide one inch inferior and lateral from the coracoid. You will be between the two tendons of the biceps brachii.

3) Palpate through the deltoid fibers, exploring the deeper tissue which lies along the lesser tubercle of the humerus. (2.53) This is the location of the subscapularis tendon. Explore for more of the tendon by moving medially off the lesser tubercle.

✔ *Is the arm positioned next to the body? Are you palpating deep to the deltoid fibers? Can you feel the solid surface of the lesser tubercle?*

(2.53) Anterior view
of shoulder

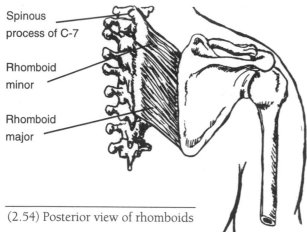

Spinous
process of C-7

Rhomboid
minor

Rhomboid
major

(2.54) Posterior view of rhomboids

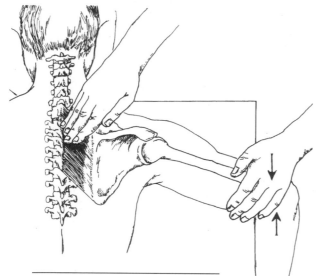

(2.55) Partner raises elbow off the
table while you create resistance

Location - Deep to trapezius
BLMs - Medial border of scapula,
spinous processes C-7 to T-5
Action - "Adduct your scapula"

Rhomboid Major and Minor

The rhomboid muscles are located between the
scapula and vertebral column. Named after their
geometric shape, the major is larger than the minor.
(2.54) Both muscles are difficult to distinguish
individually. They have thin fibers which lie deep
to the trapezius and superficial to the erector
spinae muscles (p. 138).

A -
Adduct
Elevate
Downwardly rotate the scapula

O - Rhomboid Major: Spinous processes
of T-2 to T-5
Rhomboid Minor: Spinous processes
of C-7 and T-1

I - Medial border of the scapula between spine
and inferior angle

N - Dorsal scapular from brachial plexus

👁 1) Prone. Locate the medial border of the
scapula and the spinous processes of C-7
through T-5 (p. 122).
2) Palpating through the thin trapezius, explore the
area you have identified and strum vertically across
the fibers of the rhomboids. Palpate all sides of the
rhomboids. On some individuals you can press
your fingers into the lower border of the
rhomboid major and locate its edge.

✔ *Are you deep to the trapezius fibers? Do the fibers*
you are palpating run at an oblique angle? Ask your
partner to raise his elbow toward the ceiling as you
create gentle resistance. (2.55) Although this action
will engage the superficial trapezius, can you feel the
deeper rhomboids contract?

Palpating between the medial
border of the scapula and spinous
processes of the thoracic vertebrae
gives you an opportunity to explore
layers of muscle tissue. The superficial trapezius, intermediate rhomboids and deep erector spinae

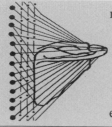

muscles all have different fiber directions.
Explore in this area and see if you can
differentiate between the perpendicular
fibers of the trapezius and rhomboids.
Try also differentiating between the
rhomboids and vertical fibers of the
erector spinae.

Levator Scapula

The levator scapula is located along the lateral and posterior sides of the neck. Its inferior portion is deep to the upper trapezius. However, as the levator ascends the lateral side of the neck, its fibers come out from under the trapezius to become superficial. (2.57) Its belly is approximately two fingers in diameter with fibers which naturally twist around themselves. (2.56)

The levator scapula attaches to the transverse processes of the cervical vertebrae (p. 126). Located on the lateral side of the neck, these small protuberances extend down from the earlobe. All the transverse processes have the same width except for the processes of C-1 which are much wider.

The brachial plexus, a large group of nerves which innervate the arm, exits from the transverse processes of the cervical vertebrae. When accessing the processes to locate the origin of the levator scapula, begin by using your fingerpads to avoid compressing a nerve.

The levator is completely accessible by either palpating through the upper fibers of the trapezius or directly from the side of the neck.

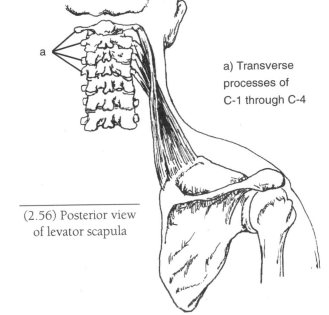

a) Transverse processes of C-1 through C-4

(2.56) Posterior view of levator scapula

A -
Unilaterally:
Elevate
Downwardly rotate the scapula
Laterally flex the head and neck
Bilaterally:
Extend the head and neck

O - Transverse processes of first through fourth cervical vertebrae

I - Medial border and superior angle of scapula

N - Dorsal scapula nerve and cervical nerves

The levator scapula is situated between the splenius capitis and posterior scalene (p. 173) on the lateral neck. (2.57) During palpation, the levator can be distinguished from these neighboring muscles because it moves the scapula. No other muscles deep to the upper trapezius or muscles attaching to the lateral cervical vertebrae are capable of this action.

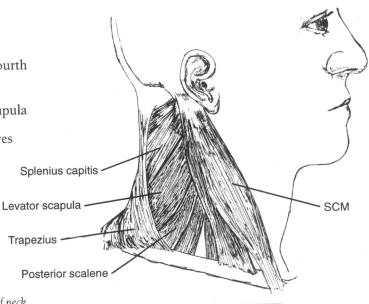

Splenius capitis

Levator scapula

Trapezius

Posterior scalene

SCM

(2.57) Lateral view

Location - Deep to trapezius, superficial on side of neck
BLMs - Superior angle, cervical transverse processes
Action - "Elevate your scapula"

---

levator   le-**va**-tor   L. lifter

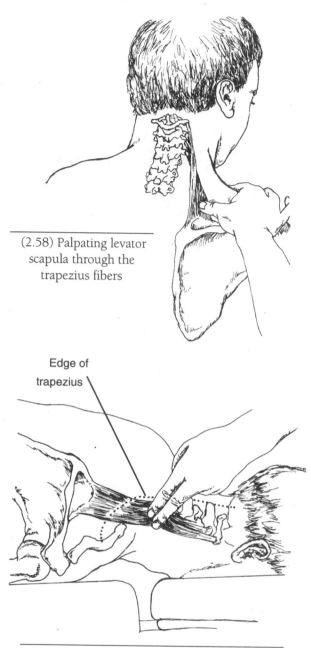

(2.58) Palpating levator scapula through the trapezius fibers

Edge of trapezius

(2.59) Lateral view of neck, strumming across the superficial fibers of levator scapula

1) Prone, supine or sidelying. Palpating through the trapezius, locate the superior angle of the scapula and the upper region of the medial border.
2) Place your fingers just off the superior angle and firmly strum across the belly of the levator. The fibers will likely have a ropy texture. (2.58)
3) Follow these fibers superiorly as they extend to the lateral side of the neck to the transverse processes of the cervical vertebrae (p. 126).

Palpating the levator scapula on the lateral side of the neck:
1) Prone, supine or sidelying. Locate the upper fibers of the trapezius.
2) Roll two fingers anteriorly off the trapezius and press into the tissue of the neck.
3) Gently strum your fingers anteriorly or posteriorly across the levator fibers. (2.59) Often, you will feel a distinct band of tissue that leads superiorly toward the lateral neck and inferiorly under the trapezius.
4) Place your fingertips on the levator and ask your partner to alternately elevate and relax his scapula. Do you feel the levator scapula contract and relax beneath your fingertips?

Can you differentiate the levator fibers from the trapezius fibers? Do the fibers you are palpating lead toward the lateral side of the neck?

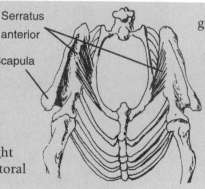

Serratus anterior (p. 65) has a different function on a dog and other quadrupeds. Unlike humans, a dog (right, anterior view) carries part of his body weight on his front legs. The serratus muscles form a sling from either scapula to the thorax that cradles and supports the weight of the trunk and stabilizes the pectoral

Serratus anterior

Scapula

girdle against the thorax.

On humans, the serratus anterior is responsible for primarily abducting the scapula or resisting a push against the shoulder. If you get down on your hands and knees to do a push-up, you will see (and feel!) how this position forces your serratus muscles to function like a dog's.

Serratus Anterior

Always well-developed on superheroes, the serratus anterior lies along the posterior and lateral ribcage. Its oblique fibers extend from the ribs underneath the scapula and attach to the medial border of the scapula. (2.60) The majority of the serratus is deep to the scapula, latissimus dorsi or pectoralis major. However, below the armpit a portion of serratus is superficial and easily accessible. This muscle is unique in its ability to abduct the scapula, making it an antagonist to the rhomboids.

Palpating along the sides of the ribs can tickle, so use slow, firm pressure. Also, if you are accessing the left serratus, it may be easier to stand on the right side of the table.

A -

With the origin fixed:
    Abduct the scapula
    Hold the medial border of scapula
       against the rib cage
If scapula is stabilized:
    May act in forced inspiration

O - Surfaces of upper eight or nine ribs

I - Anterior surface of medial border of scapula

N - Long thoracic from brachial plexus

1) Supine. Abduct the arm slightly and locate the lower edge of the pectoralis major (p. 66) and the anterior border of the latissimus dorsi.
2) Place your fingerpads along the side of the ribs between the pectoralis major and latissimus dorsi.
3) Strum your fingers across the ribs and palpate for the serratus anterior fibers. To differentiate between the ribs and the serratus' fibers (both have a similar "speed bump" shape), remember that the ribs are deep and have a solid texture while the serratus fibers are superficial and malleable.

✔ *Feeling the serratus anterior contract: (2.61) Ask your partner to flex his shoulder so his hand is raised toward the ceiling. Place one hand upon the serratus fibers and ask him to alternately abduct his scapula and relax. "Reach toward the ceiling, and then relax." Do you feel the serratus fibers contract and soften? Can you follow the fibers along the ribs to where they tuck underneath the latissimus dorsi?*

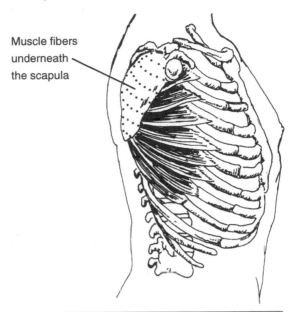

Muscle fibers underneath the scapula

(2.60) Lateral view of serratus anterior

Location - Partly superficial, inferior to axilla
BLMs - Lateral side of ribcage
Action - "Abduct your scapula"

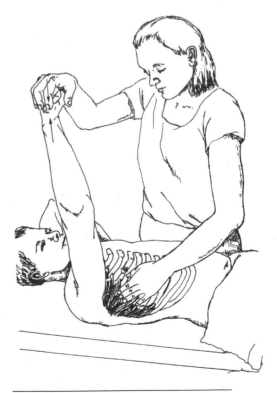

(2.61) Isolating the serratus while partner presses hand toward the ceiling

serratus   ser-**a**-tus   L. a notching

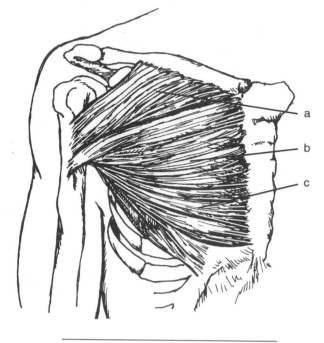

(2.62) Anterior view of pectoralis major

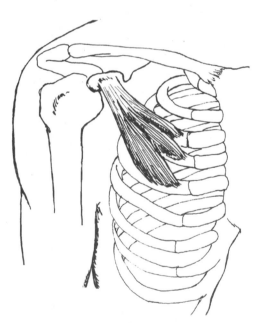

(2.63) Pectoralis minor,
deep to pectoralis major

Pectoralis Major and Minor

The *pectoralis major* is a broad, powerful muscle located on the chest. It has convergent, superficial fibers which are accessible except for the part beneath breast tissue. Pectoralis major is divided into three segments - the clavicular (a), sternal (b), and costal (c) fibers. (2.62) The upper and lower fibers perform opposite actions - flexing and extending at the shoulder joint, respectively - making the pectoralis major an antagonist to itself.

The *pectoralis minor* lies next to the ribcage deep to the pectoralis major. (2.63) Its fibers run perpendicular to the pectoralis major fibers from the scapula's coracoid process to the upper ribs. During aerobic activity the pectoralis minor helps to elevate the ribcage for inhalation. The major vessels serving the arm, the brachial plexus, axillary artery and vein, cross underneath the pectoralis minor, creating the potential for neurovascular compression by this muscle. (2.64)

Pectoralis Major

A -
Muscle as a whole:
    Adduct
    Medially rotate the shoulder joint
    May assist in elevating the thorax
       in forced inspiration
Upper Fibers:
    Flex
    Medially rotate
    Horizontally adduct the shoulder joint
Lower Fibers:
    Extend
    Adduct the shoulder joint

O - Medial half of clavicle, sternum, cartilage of ribs one through six

I - Crest of greater tubercle of humerus

N - Anterior thoracic from brachial plexus

Palpate around mammary (breast) tissue, not directly into it when exploring the pectoralis major and minor. Accessing the minor can be achieved either by pressing through or sliding underneath the thick

pectoralis major. The second method is more specific and will be outlined here. The pectoralis minor can be sensitive, so palpate slowly, allowing your thumb to sink into the tissue.

pectoralis   **pek**-to-**ra**-lis   L. breast or chest

Pectoralis Minor

A -
Depress
Abduct
Tilt the scapula anteriorly
Assist in forced inspiration

O - Third, fourth and fifth ribs

I - Coracoid process of scapula

N - Anterior thoracic from brachial plexus

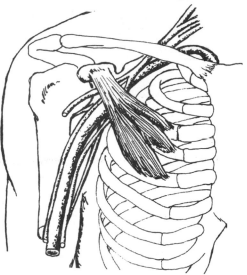

(2.64) Brachial plexus and axillary
artery passing beneath pectoralis minor

👁 Pectoralis major

1) Supine with arm abducted. Sit facing
your partner.
2) Locate the medial shaft of the clavicle and
move inferiorly onto the clavicular fibers.
3) Explore the surface of the pectoralis major.
Follow the fibers laterally as they blend with the
deltoid and attach at the greater tubercle.
4) Grasp the belly of the pectoralis by sinking
your thumb into the axilla. Ask your partner
to medially rotate his shoulder. "Press your
hand toward your belly." (2.65) Note the
contraction of the pectoralis.

✔ *Do the clavicular fibers run parallel with the
anterior deltoid? As you grasp the belly, do you sense
its thickness and how it lies across the ribcage?*

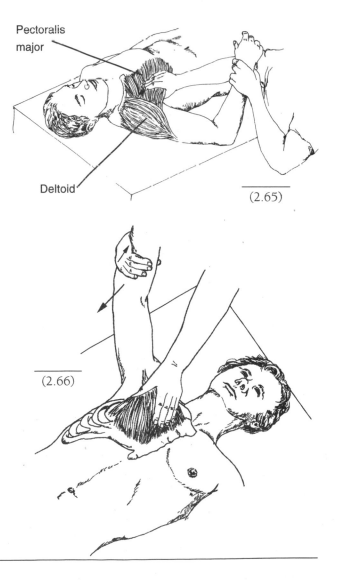

Pectoralis
major

Deltoid

(2.65)

(2.66)

◈ To feel the antagonistic movements of the
pectoralis major's upper fibers (which flex the
humerus) and the middle and lower fibers
(which extend the humerus):
1) Supine with arm flexed, hand up toward the
ceiling. Lay one hand on the pec major fibers.
2) Ask your partner to flex his shoulder as your
other hand creates resistance. (2.66) Do you feel
the upper fibers contract while the lower fibers
remain soft?
3) Ask him to extend his shoulder against your
resistance and note the lower fibers contracting.

When choosing between the "white or dark meat" of a cooked bird, be sure to thank its intramuscular connective tissues. Dark and white meat are present in all mammals, yet more distinct in birds. The reason is that light-colored musculature is rich in muscle fibers and poor in sacroplasm - the tissue which surrounds muscle fibers, while dark meat has the opposite composition. And if you are fond of the "breast," chew on this fact: a bird's pectoralis majors make up 20-35% of its body weight.

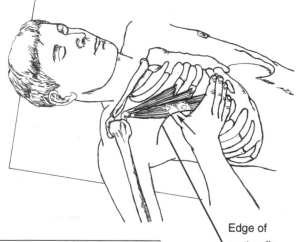

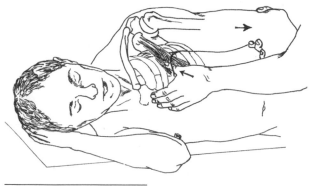

(2.67) Palpating pectoralis minor from a supine position

Edge of pectoralis major

Pectoralis minor

1) Supine. Abduct the arm and place your thumb at the lateral edge of the pectoralis major.
2) Slowly and gently slide your thumb under the major, following along the surface of the ribs.
3) Eventually your thumb will come in contact with the small wall of muscle lying next to the ribs. (2.67) This is the side of the pectoralis minor. If you do not feel its tissue, visualize its location next to the ribs.

✔ *Ask your partner to depress his scapula. "Ever-so-slightly press your shoulder down toward your hip." When he depresses, do you feel the pectoralis minor contract? Do the fibers run toward the coracoid process?*

Sidelying for pectoralis major and minor

1) Support the arm in a flexed position and pull it anteriorly. This brings the pectoralis major off the chest wall and allows the breast tissue to fall away from the area you are palpating.
2) Grasp the pectoralis major, exploring its mass from the ribs to the humerus.
3) To locate pectoralis minor, slowly slide your thumb under the pectoralis major, following along the surface of the ribs. (2.68) Your thumb will press into the side the pectoralis minor. Ask your partner to gently depress his scapula.

(2.68) Locating pectoralis minor from sidelying

Pectoralis Major
Location - Superficial, chest
BLMs - Clavicle and sternum
Action - "Adduct your shoulder"

Pectoralis Minor
Location - Deep to pectoralis major
BLMs - Coracoid process, ribs 3 - 5
Action - "Depress your scapula"

Subclavius

As its name suggests, subclavius is located underneath the clavicle. Its fibers run parallel to the clavicle, deep to the pectoralis major, and can be challenging to specifically isolate. (2.69)

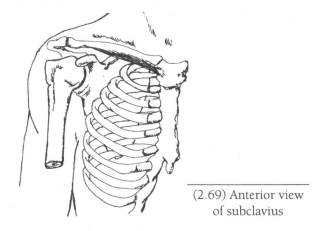

(2.69) Anterior view of subclavius

A -
Draws clavicle down and forward
Elevates first rib
Stabilizes the sternoclavicular joint

O - First rib and cartilage

I - Inferior, lateral aspect of the clavicle

N Fifth and sixth cervicals

1) Sidelying. Support the arm in a flexed position and pull it anteriorly. This position brings the clavicle and pectoralis major off the ribcage and allows your finger or thumb to curl even further around the clavicle.
2) Place your thumb and fingers at the center of the clavicle.
3) Slowly curl your thumb and finger around the clavicle and try to pinch the subclavius between them. You may not access a muscle belly, but instead feel a slight density of tissue tucked under the clavicle. (2.70)
4) Try this method from a supine position.

✔ *Can you detect a slender strip of tissue deep to the clavicle? Can you distinguish between the superficial pectoralis major fibers (heading toward the axilla) and the subclavius fibers (parallel to the clavicle)?*

On quadrupeds (four-legged animals), the subclavius is quite large and plays a important role in stabilizing the clavicle and shoulder girdle during locomotion. A human's subclavius, on the other hand, is a small, tertiary muscle.

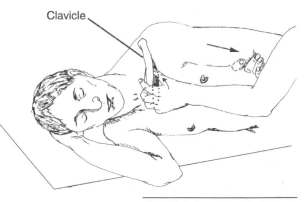

Clavicle

(2.70) Palpating subclavius from sidelying position

Location - Deep to pectoralis major, under clavicle
BLMs - Shaft of clavicle
Action - "Depress your clavicle"

Present in roughly 5% of the population, the sternalis is a thin, superficial muscle lying on the sternum. Its vertical fibers run from the manubrium down to the level of the seventh costal cartilage. The function of the sternalis is unknown. Palpate the surface of your partner's sternum and explore for a sternalis.

Pectoralis major

Sternalis

Rectus abdominus

subclavius   sub-**kla**-ve-us

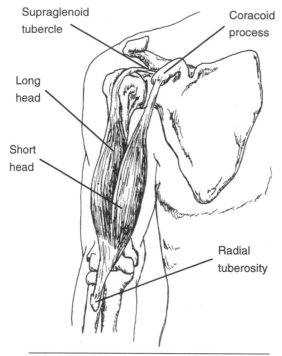

Supraglenoid tubercle

Coracoid process

Long head

Short head

Radial tuberosity

(2.71) Anterior view of the biceps brachii

In addition to a long and a short head, the biceps may have a head which attaches to the humerus. Reported in less than 10% of cases, this extra head originates along the medial humerus next to coracobrachialis before joining the short head.

(2.72) Feeling biceps contract as partner tries to bend his elbow

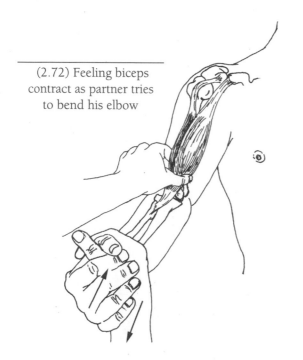

Biceps Brachii

The biceps brachii lies superficial on the anterior arm. It has a long and a short head which merge together to form a long, oval belly. The tendon of the long head passes through the intertubercular groove of the humerus (p. 48). The groove helps to stabilize the tendon as it rises over the top of the shoulder. (2.71)

The distal tendon of the biceps dives into the antecubital space (inner elbow) to attach at the radius, allowing the biceps to be the primary muscle of forearm supination. The majority of the biceps brachii is easily palpable.

A -
Flex the elbow
Supinate the forearm
Flex the shoulder joint

O -
Short head: Coracoid process of scapula
Long head: Supraglenoid tubercle of scapula

I - Tuberosity of the radius and aponeurosis of the biceps brachii

N - Musculocutaneous

1) Supine. Bend the elbow and shake hands with your partner.
2) Ask your partner to flex his elbow against your resistance. Palpate the anterior surface of the arm and locate the hard, round belly of the biceps.
3) Follow the belly distally to the inner elbow. Note how the muscle belly thins, becoming a solid, distinct tendon.
4) Follow the biceps proximally to where it tucks beneath the anterior deltoid.

✔ *Ask your partner to flex his elbow and see if you can sculpt out the biceps' distal tendon and distinguish it from the deeper brachialis muscle. (2.72) Shake hands with your partner and ask him to alternately pronate and supinate his forearm against your resistance. Do you feel the muscle belly and tendon contract upon supination?*

Location - Superficial, anterior arm
BLMs - Coracoid process
Action - "Flex your elbow"

Triceps Brachii

The triceps brachii is the only muscle located on the posterior arm. It creates extension at the elbow and shoulder making it an antagonist at both joints to the biceps brachii. The triceps has three heads: long, lateral and medial. (2.73) The long head extends off the infraglenoid tubercle of the scapula (p. 42), weaving between the teres major and minor. The lateral head lies superficially beside the deltoid while the medial head lies mostly underneath the long head. All three heads converge to a thick, distal tendon proximal to the elbow.

Aside from its proximal portion, which is deep to the deltoid, the triceps is completely superficial and easily accessible.

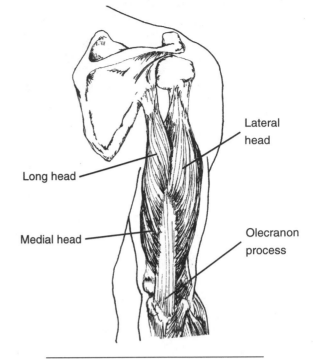

(2.73) Posterior view of triceps brachii

A -
All heads:
   Extend the elbow
Long head:
   Extend
   Adduct the shoulder joint

O - Long head: Infraglenoid tubercle of scapula
Lateral head: Posterior surface of proximal
   half of humerus
Medial head: Posterior surface of distal
   half of humerus

I - Olecranon process of ulna

N - Radial

1) Prone. Bring the arm off the side of the table and palpate the posterior aspect of the arm. Outline the edge of the posterior deltoid and then explore the size and shape of the triceps.
2) Locate the olecranon process to outline the distal tendon of the triceps. Then ask your partner to extend his elbow as you apply resistance at his forearm. (2.74) Slide your other hand off the olecranon process proximally and onto the broad triceps tendon.
3) With your partner still contracting, widen your fingers and palpate the medial and lateral heads on either side of the tendon.

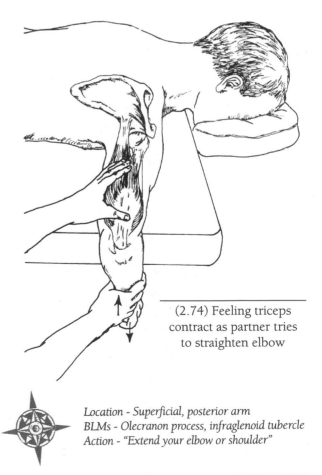

(2.74) Feeling triceps contract as partner tries to straighten elbow

✔ *Can you feel the medial and lateral triceps heads bulge on either side of the distal tendon? Does the muscle tighten when your partner extends his elbow?*

Location - Superficial, posterior arm
BLMs - Olecranon process, infraglenoid tubercle
Action - "Extend your elbow or shoulder"

---

triceps brachii   **tri-seps bra-key-i**   L. "three-headed muscle of the arm"

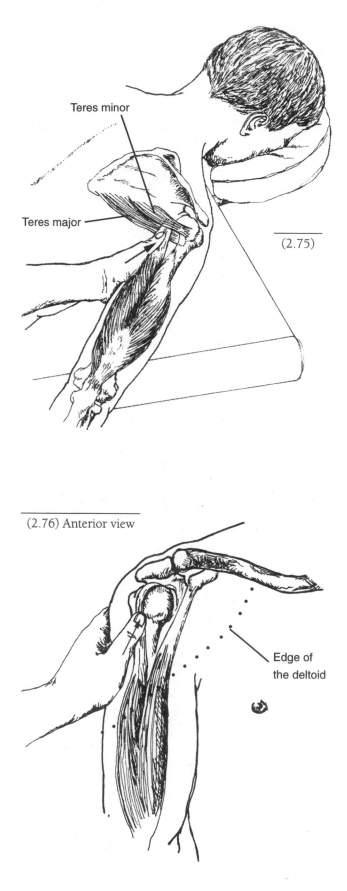

Teres minor

Teres major

(2.75)

(2.76) Anterior view

Edge of
the deltoid

The tendon of the long head of the triceps brachii

A helpful hint for locating the long head of the triceps is the fact that it is the only band of muscle on the posterior arm that runs superiorly along the proximal and medial aspect of the arm. The deltoid fibers run at a more diagonal direction than the long head of the triceps.

👁 1) Prone. Place one hand on the proximal elbow and ask your partner to bring his elbow toward the ceiling against your resistance. This action will contract the long head of the triceps.
2) Locate its belly along the proximal and medial aspect of the arm. Follow the muscle proximally by strumming across the belly. Note how it disappears underneath the posterior deltoid toward the infraglenoid tubercle.
3) With the arm relaxed, press through the posterior deltoid and search for its skinny tendon as it attaches to the infraglenoid tubercle.

✔ *The long head of the triceps crosses over the teres major, and under the teres minor. (2.75) Can you follow the long head up to the division of the teres muscles? Have your partner medially and laterally rotate his shoulder to differentiate the teres muscles (p. 54, 56).*

The tendon of the long head of the biceps brachii

Because the biceps tendon is situated in the intertubercular groove of the humerus and runs parallel to the superficial deltoid fibers, it can be difficult to specifically isolate.

👁 1) Locate the intertubercular groove (p. 48). Laterally rotating the arm may make it easier to locate the tendon. (2.76)
2) Ask your partner to gently flex his shoulder against your resistance to feel the biceps tendon tighten in the intertubercular groove.

Coracobrachialis

The coracobrachialis is a small, tubular muscle located in the axilla. (2.77) Sometimes known as the "arm pit" muscle, it is a secondary flexor and adductor of the shoulder. In anatomical position the coracobrachialis is deep to the pectoralis major and anterior deltoid and lies anterior to the axillary artery and brachial plexus. Abducting the shoulder (opening up the axilla) causes the belly of coracobrachialis to become superficial and palpable.

A -
Flex
Adduct the shoulder joint

O - Coracoid process of scapula

I - Medial surface of middle shaft of humerus

N - Musculocutaneous

1) Supine. Laterally rotate and abduct the shoulder to 45°. Locate the fibers of the pectoralis major. This tissue forms the axilla's anterior wall and will be a handy reference point to locate coracobrachialis.
2) Lay one hand along the medial side of the arm and move your fingerpads into the arm pit.
3) Have your partner adduct gently against your resistance. (2.78) Isolate the solid edge of the pectoralis major then slide off the pectoralis fibers posteriorly (into the axilla) and explore for the slender, contracting belly of coracobrachialis. Its belly may be visible upon adduction.

✔ *Is the muscle you are palpating on the medial side of the upper arm? Does its belly lie posterior to the overlying flap of the pectoralis major? Can you strum along its cylindrical belly?*

Location - Medial side of humerus, axilla
BLMs - Coracoid process
Action - "Adduct your shoulder"

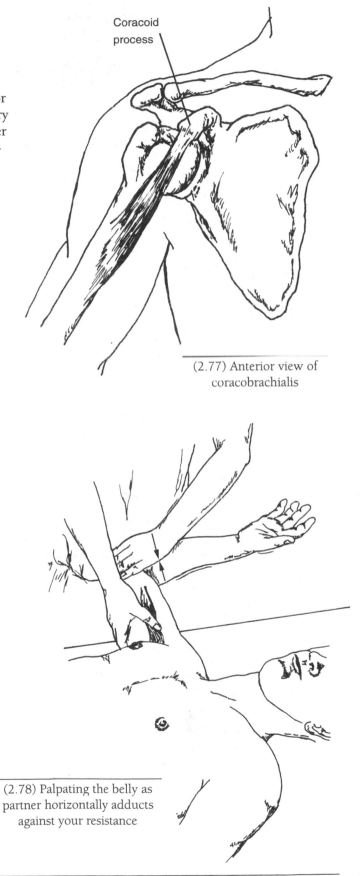

Coracoid process

(2.77) Anterior view of coracobrachialis

(2.78) Palpating the belly as partner horizontally adducts against your resistance

coracobrachialis   kor-a-ko-**bra**-ke-**al**-is

Ligaments, Nodes and Vessels of the Shoulder

Axilla

The axilla is the cone-shaped area commonly called the arm pit. (2.79) It is formed by four walls: a lateral wall (biceps brachii and coracobrachialis), a posterior wall (subscapularis and latissimus dorsi), an anterior wall (pectoralis major), and a medial wall (rib cage and serratus anterior). There are several important vessels which pass through the axillary region including the brachial artery and the brachial plexus (nerves). (2.80)

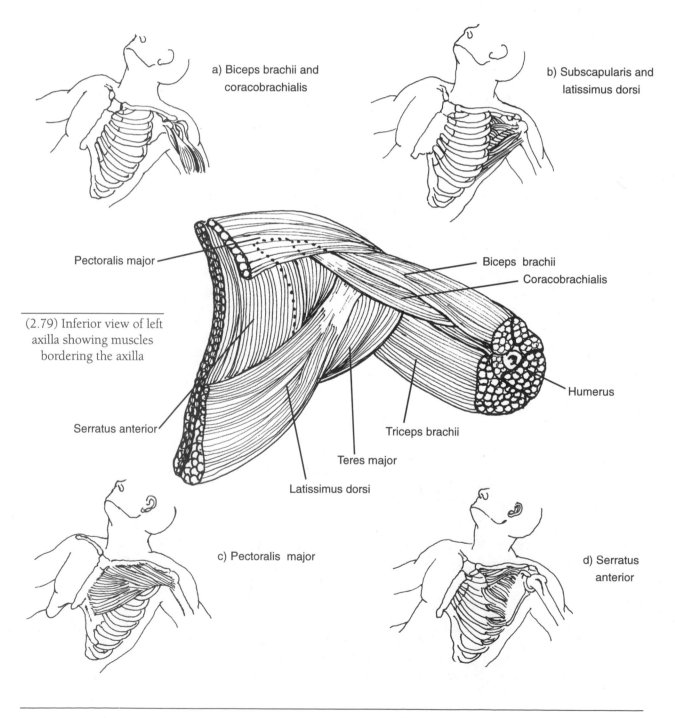

a) Biceps brachii and coracobrachialis

b) Subscapularis and latissimus dorsi

Pectoralis major

Biceps brachii

Coracobrachialis

(2.79) Inferior view of left axilla showing muscles bordering the axilla

Humerus

Serratus anterior

Triceps brachii

Teres major

Latissimus dorsi

c) Pectoralis major

d) Serratus anterior

axilla   **ak**-sil-a
axillary   **ak**-si-**lar**-ee

Compression or impingement of the brachial plexus or one of its nerves can create a sharp, shooting sensation down the arm. If this occurs, immediately release and adjust your position posteriorly. Also, ask your partner for feedback.

Median nerve

Brachial artery

Medial antebrachial cutaneous nerve

Basilic vein

Ulnar nerve

Brachial veins

(2.80) Inferior view of left axilla showing vessels which run through the axillary region

Axillary Lymph Nodes

The axillary lymph nodes are also located in the axilla. When palpating the axilla, use a deliberate yet gentle touch to avoid tickling your partner. Also, move slowly and use gentle pressure to avoid impinging the artery and nerves.

1) Supine or seated. Abduct the arm and slowly sink two fingers up into the axilla. Then bring the arm back to the side of the body to soften the axillary tissue further.
2) Slide your fingers up to the top of the axilla and then medially toward the rib cage. Often there will be a few lymph nodes located against the ribs. (2.81)
3) Move to the lateral side of the axilla and use light pressure against the humerus to locate the strong pulse of the brachial artery. The vessel will be positioned between the stringy coracobrachialis and long head of the triceps brachii muscle.

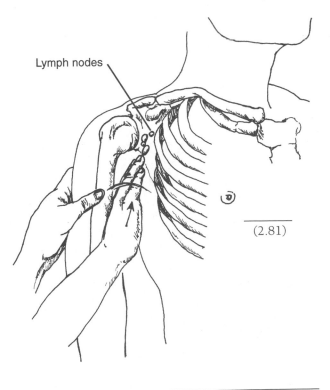

Lymph nodes

(2.81)

brachial   **bray**-ke-al   L. relating to the arm
gland   L. acorn

Ligaments

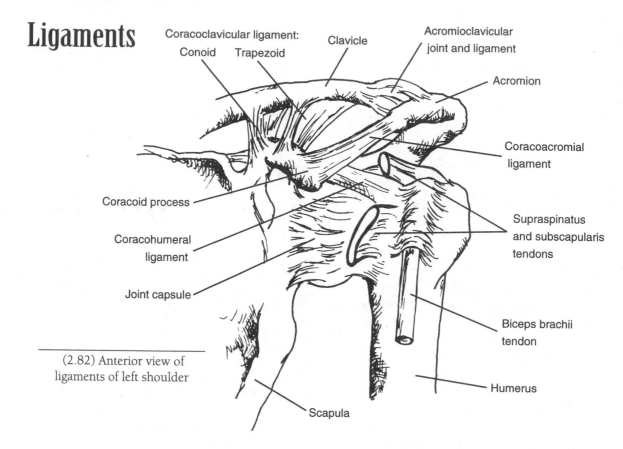

Coracoclavicular ligament:
Conoid   Trapezoid

Clavicle

Acromioclavicular
joint and ligament

Acromion

Coracoacromial
ligament

Coracoid process

Coracohumeral
ligament

Joint capsule

Supraspinatus
and subscapularis
tendons

Biceps brachii
tendon

Humerus

Scapula

(2.82) Anterior view of
ligaments of left shoulder

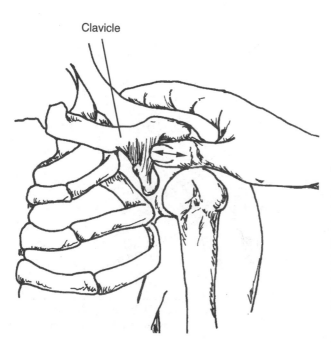

Clavicle

(2.83) Anterior view

Coracoclavicular Ligament

The coracoclavicular ligament is comprised of
two portions: the conoid and trapezoid ligaments.
Both ligaments stretch from the coracoid process
of the scapula to the inferior surface of the clavicle.
(2.82) Together they provide stability for the
acromioclavicular joint and form a strong tie
between the scapula and clavicle. The coracoclavicular ligament can be accessed by palpating
between the clavicle and coracoid process or
curling around the posterior aspect of the clavicle.

1) Supine. Abduct and internally rotate
the shoulder. This position brings the ligaments
more to the surface.
2) Locate the coracoid process of the scapula
and the shaft of the clavicle.
3) Palpate in this space between these landmarks.
Roll your thumbpad across its fibers. (2.83) Unlike
the superficial pectoralis major fibers, the ligaments
will feel like solid, taut bands.

✔ *Passively move the shoulder girdle in several
directions and see if a particular position allows
you greater access to the ligaments.*

coracoclavicular	**cor**-a-**co**-cla-**vic**-u-lar	ligament	**lig**-a-ment L. a band
coracoacromial	**cor**-a-**co**-a-**cro**-mi-ul		

Coracoacromial Ligament

Unlike most ligaments which hold two bones together, the coracoacromial ligament attaches the scapula's coracoid process to its acromion. (2.82) Along with the acromion, the ligament forms the coraco-acromial arch across the top of the shoulder. The arch helps to protect the rotator cuff tendons and subacromial bursae from direct trauma from the acromion. The wide band of the coracoacromial ligament lies deep to the deltoid and is accessible.

1) Supine or seated. Locate the coracoid process.
2) Locate the anterior edge of the acromion.
3) Palpating deep to the deltoid fibers, explore between these landmarks for the wide band of the coracoacromial ligament. Strum your finger across its fibers. (2.84)
4) To bring the ligament more to the surface, try extending the arm. This position will roll the humeral head anteriorly and press the ligament forward.

Are you between the acromion and the coracoid process? Place one finger on the ligament and passively move the shoulder girdle in various positions. Can you feel how the ligament's relationship to the surrounding tissues changes with different positions of the shoulder?

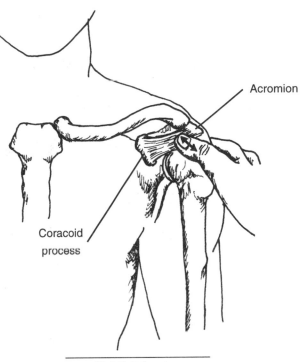

(2.84) Palpating
coracoacromial ligament

Brachial Artery

The brachial artery is a continuation of the axillary artery and runs between the biceps and triceps brachii. Its pulse can be felt between these muscles on the medial side of the arm. (2.80) Before the brachial artery branches off to the radial and ulnar arteries, its pulse can be felt at the elbow, just medial to the biceps brachii tendon.

1) Seated or supine. Abduct the arm and place two finger pads on the medial side of the arm. A helpful guide is the shallow dip which forms between the biceps and triceps. (2.85)
2) Gently press to feel the brachial pulse.
3) The brachial pulse can also be located just medial to the distal tendon of the biceps brachii.

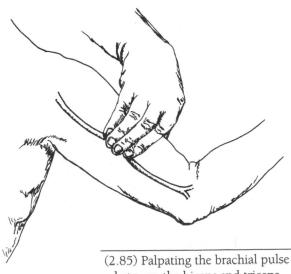

(2.85) Palpating the brachial pulse
between the biceps and triceps

Chapter
Forearm & Hand 3

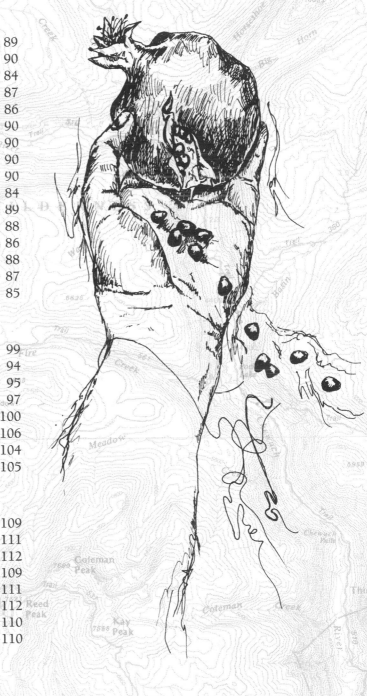

Topographical Views

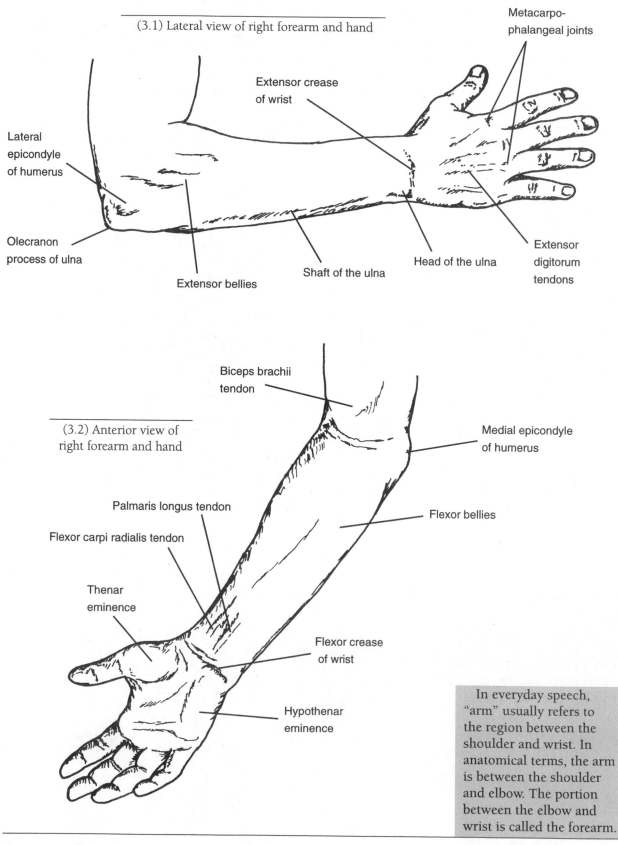

Metacarpo-
phalangeal joints

(3.1) Lateral view of right forearm and hand

Extensor crease
of wrist

Lateral
epicondyle
of humerus

Olecranon
process of ulna

Extensor bellies

Shaft of the ulna

Head of the ulna

Extensor
digitorum
tendons

Biceps brachii
tendon

(3.2) Anterior view of
right forearm and hand

Medial epicondyle
of humerus

Palmaris longus tendon

Flexor carpi radialis tendon

Flexor bellies

Thenar
eminence

Flexor crease
of wrist

Hypothenar
eminence

In everyday speech,
"arm" usually refers to
the region between the
shoulder and wrist. In
anatomical terms, the arm
is between the shoulder
and elbow. The portion
between the elbow and
wrist is called the forearm.

thenar	**the**-nar	Gr. palm
hypothenar	hi-**poth**-e-nar	Gr. *hypo*, under or below

Bones of the Forearm and Hand

The *humerus* is the bone of the arm. Its proximal end articulates with the scapula to form the glenohumeral joint. Its distal end joins with the ulna and radius at the elbow. The elbow has two joints: the humeroulnar and humeroradial.

The *radius* and *ulna* compose the bones of the forearm. The ulna is superficial and has a palpable edge that extends from the elbow to the wrist. The radius ("thumb side") is lateral to the ulna and is partially buried in muscle. Pronation and supination of the forearm are created by the radius pivoting around the ulna at the proximal and distal radioulnar joints (3.3).

The three groups of bones in the wrist and hand are the carpals, metacarpals and phalanges. The *carpals* are eight, pebble-sized bones that form two rows (proximal and distal), each row containing four carpal bones. (3.6) Located distal to the "flexor crease" of the wrist, the carpals are accessible from the dorsal and palmar surfaces of the hand.

The *metacarpals* are five long bones spanning the palm of the hand. The metacarpal's proximal end is the base, the long midsection is the shaft and the distal end is the head. (3.4, 3.5) The metacarpals are easily palpable along the hand's dorsal surface. They are deep to muscles on the palmar side.

The *phalanges* are the bones of the fingers. The thumb has two phalange bones and the fingers have three. All sides of the phalanges are accessible.

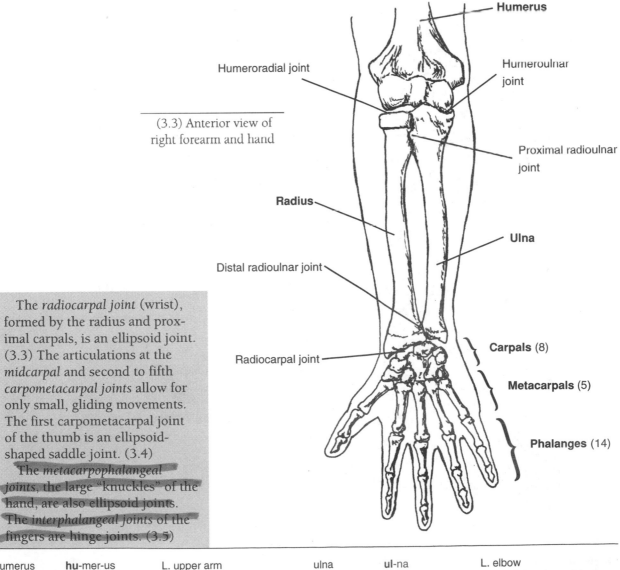

(3.3) Anterior view of right forearm and hand

Humerus

Humeroradial joint

Humeroulnar joint

Proximal radioulnar joint

Radius

Ulna

Distal radioulnar joint

Carpals (8)

Radiocarpal joint

Metacarpals (5)

Phalanges (14)

The *radiocarpal joint* (wrist), formed by the radius and proximal carpals, is an ellipsoid joint. (3.3) The articulations at the *midcarpal* and second to fifth *carpometacarpal joints* allow for only small, gliding movements. The first carpometacarpal joint of the thumb is an ellipsoid-shaped saddle joint. (3.4) The *metacarpophalangeal joints*, the large "knuckles" of the hand, are also ellipsoid joints. The *interphalangeal joints* of the fingers are hinge joints. (3.5)

humerus	**hu**-mer-us	L. upper arm	ulna	**ul**-na	L. elbow
radius	**ray**-dee-us	L. ray			

Bony Landmarks of the Forearm and Hand

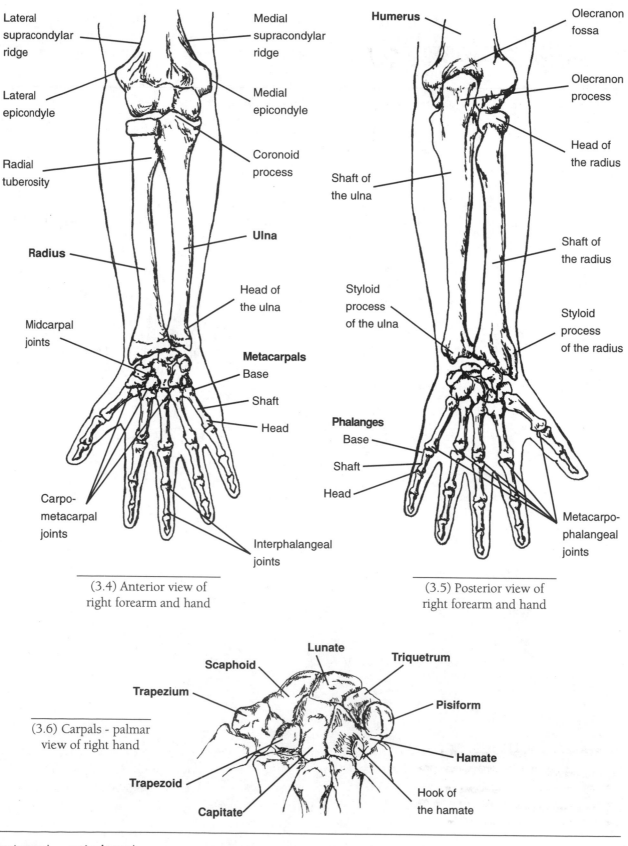

Lateral supracondylar ridge

Medial supracondylar ridge

Lateral epicondyle

Medial epicondyle

Radial tuberosity

Coronoid process

Radius

Ulna

Head of the ulna

Midcarpal joints

Metacarpals

Base

Shaft

Head

Carpo-metacarpal joints

Interphalangeal joints

(3.4) Anterior view of right forearm and hand

Humerus

Olecranon fossa

Olecranon process

Head of the radius

Shaft of the ulna

Shaft of the radius

Styloid process of the ulna

Styloid process of the radius

Phalanges

Base

Shaft

Head

Metacarpo-phalangeal joints

(3.5) Posterior view of right forearm and hand

Lunate

Scaphoid

Triquetrum

Trapezium

Pisiform

(3.6) Carpals - palmar view of right hand

Hamate

Trapezoid

Hook of the hamate

Capitate

metacarpal   **met**-a-**kar**-pul

phalange   fa-**lan**-jee   Gr. closely knit row

Bony Landmark Trails

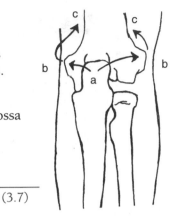

Trail 1 - "Knob Hill" explores the elbow and distal humerus. (3.7)

  a) Olecranon process and fossa
  b) Epicondyles of humerus
  c) Supracondylar ridges
     of humerus

(3.7)

Trail 2 - "The Razor's Edge" follows the length of the superficial ulna. (3.8)

  a) Olecranon process
  b) Shaft of the ulna
  c) Head of the ulna
  d) Styloid process of the ulna

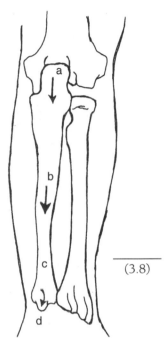

(3.8)

Trail 3 - "Pivot Pass" travels the length of the radius, the bone which creates the pivoting action of the forearm. (3.9)

  a) Medial epicondyle
  b) Head of the radius
  c) Shaft of the radius
  d) Styloid process of the radius

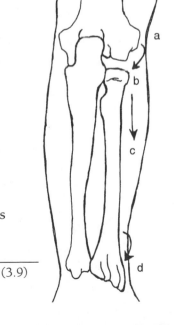

(3.9)

Trail 4 - "Walking On Your Hands" explores the small carpal bones in the wrist as well as the bones and joints of the hand.

Some translations may cause you to scratch your head and wonder what early anatomists were thinking. The carpals, luckily, cause no such distress.

capitate - L. head-shaped
hamate - L. hooked
lunate - L. crescent-shaped
scaphoid - L. boat-shaped

pisiform - L. pea-shaped
trapezium - Gr. little table
trapezoid - Gr. table-shaped
triquetrum - L. three-cornered

carpal   **kar**-pul   Gr. pertaining to the wrist

Trail 1 - "Knob Hill"
Olecranon Process and Fossa

The olecranon process (or elbow) is located on the proximal end of the ulna and articulates with the distal humerus. Its large surface is the attachment site for the triceps brachii. It forms the "point" of the elbow and can be easily located.

The olecranon fossa is a large cavity on the posterior, distal end of the humerus designed to accommodate the olecranon process when the elbow is extended. Located deep to the triceps brachii tendon, the fossa is only partially accessible.

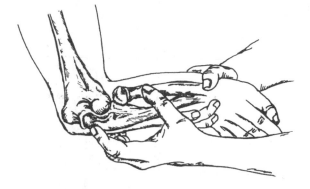

(3.10) Palpating the olecranon process

👁 Olecranon process

1) Partner seated. Shake hands with your partner and explore the large, superficial knob at the elbow.
2) Palpate and explore its angular surface and sides.
3) Passively flex and extend the elbow, noticing how the olecranon process feels in various positions. (3.10)

👁 Olecranon fossa

1) With the elbow slightly flexed, locate the olecranon process.
2) Roll your finger proximally around the top of the process, pressing through the triceps tendon and into the fossa.
3) Due to the presence of the tendon and the close proximity of the process, only a small crescent-shaped ditch will be accessible. (3.11)

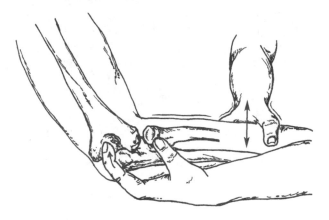

(3.11) Locating the olecranon fossa

✔ *When locating the fossa, are you proximal to the tip of the olecranon process? If you flex and extend the elbow slightly, do you feel a change in the fossa's shape and size?*

Medial and Lateral Epicondyles

As the humerus extends down the arm, its distal end broadens medially and laterally. Directly medial from the olecranon process is the medial epicondyle. It is superficial and has a protruding, spherical shape designed to accommodate the tendons of the wrist and hand flexors.

The lateral epicondyle is smaller than its medial counterpart and is located lateral to the olecranon process. It is an attachment site for the tendons of the wrist and hand extensors.

The ulnar nerve (p. 110), which creates the "funny bone" sensation when hit, courses between the medial epicondyle and the olecranon process.

fossa	**fos**-a	L. a shallow depression	process	**pros**-es	L. going before
olecranon	o-**lek**-ran-on	Gr. elbow			

1) Shake hands with your partner and locate the olecranon process.

2) Slide your finger medially off the olecranon. You will encounter a small ditch before rising up onto the large, superficial medial epicondyle. Explore its bulbous shape. (3.12)

3) Return to the olecranon. Slide laterally to the lateral epicondyle. Note that it is smaller than the medial epicondyle.

✔ *Set a finger on each epicondyle and slowly flex and extend the elbow. The surrounding muscle tissue may move, but the epicondyles should remain stationary. Do they?*

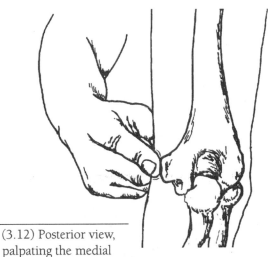

(3.12) Posterior view, palpating the medial epicondyle

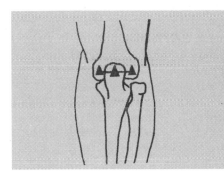

When the elbow is extended, the olecranon process and the epicondyles will form a straight line. If the elbow is flexed, the three landmarks form a triangle. Can you palpate these changes?

Medial and Lateral Supracondylar Ridges of Humerus

These two ridges extend proximally from the respective epicondyles. Both serve as attachment sites for the forearm muscles. The lateral supracondylar ridge is located superficially, while the medial ridge sinks into the arm and is situated close to the ulnar nerve.

1) Shake hands with your partner and locate the medial epicondyle.

2) Move proximally from the epicondyle. The bony ridge which extends from the epicondyle is the medial supracondylar ridge. (3.13) Roll your fingers back and forth across the ridge to sense its distinct edge.

3) Explore the lateral supracondylar ridge.

✔ *Can you follow the ridges a few inches proximally before they disappear under the muscles of the arm?*

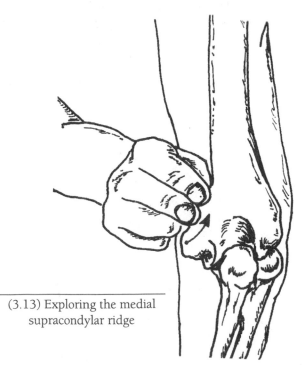

(3.13) Exploring the medial supracondylar ridge

| epi- | Gr. above, upon | condyle | Gr. *kondylos*, knuckle |
| lateral | L. to the side | | |

Trail 2 - "The Razor's Edge"
Shaft of the Ulna

The long, straight shaft of the ulna extends from the olecranon process to the head of the ulna. Although numerous muscles lie beside the shaft, it has a superficial, palpable edge that runs along the arm's medial, posterior aspect.

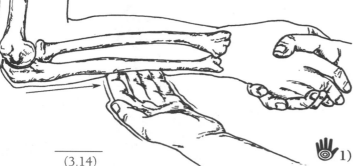

(3.14)

1) Shaking hands with your partner, locate the olecranon process.
2) Slide your fingers distally along the shaft. To define its shape and location, roll your fingers across its edge. (3.14)
3) Follow it down the length of the forearm.

✔ *Is the bone you are palpating superficial? Does it stretch the length of the forearm?*

Head of the Ulna

The shaft of the ulna swells to form the head of the ulna. The head is a superficial knob visible along the posterior, medial side of the wrist and can disrupt the placement of a watchband.

1) Slide your fingers distally along the shaft of the ulna.
2) Just proximal to the wrist, the shaft will bulge to become the head of the ulna. Palpate all sides of the head's bulbous shape. (3.15)

✔ *Is the knob you are palpating connected to the shaft of the ulna? Is it on the posterior and medial side of the arm?*

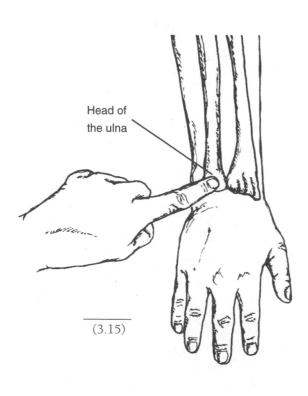

Head of
the ulna

(3.15)

Styloid Process of the Ulna

Both the ulna and the radius have a styloid process at their distal ends. The radius' styloid (p. 88) is larger and extends further distal. The ulna's styloid is sharper and more pronounced. Both are superficial, and the tendons of the forearm muscles pass beside them.

The styloid process of the ulna is a toothlike projection pointing distally off the head of the ulna. It is located on the posterior, medial side of the wrist.

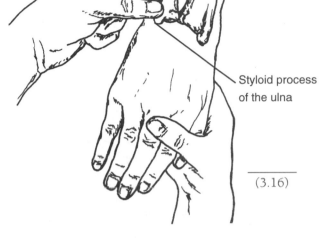

1) Shake hands with your partner. Passively adduct the wrist to soften the surrounding tendons.
2) Use your thumb and fingertip to grasp the posterior aspect of the ulnar head. (3.16)
3) Slide distally off the head to palpate the small tip of the styloid process.

Styloid process of the ulna

(3.16)

✔ *Is the bone you are palpating connected to the head of the ulna (as opposed to a separate carpal bone)? If you slowly flex and extend the wrist, it should remain stationary. Does it?*

Trail 3 - "Pivot Pass"
Head of the Radius

The head of the radius is located distal to the humerus' lateral epicondyle. It forms the radius' proximal end and has a circular, bell shape. The head is stabilized by the annular ligament (p. 109) and is a pivoting point for supination and pronation of the forearm. Although it is deep to the supinator and extensor muscles, the head's posterior, lateral aspect can be accessed.

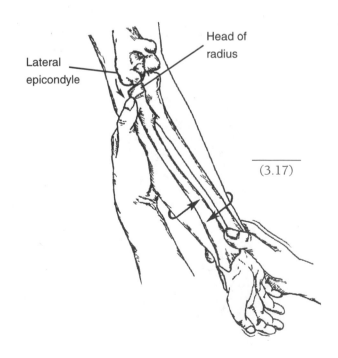

1) Shake hands and locate the lateral epicondyle.
2) Slide distally off the epicondyle, across the small ditch between the humerus and radius, and onto the head of the radius.
3) The head of the radius is the only bony structure in this vicinity. Explore its ring-shaped, superficial surface.

Lateral epicondyle

Head of radius

(3.17)

✔ *Are you distal to the lateral epicondyle? Place your thumb on the head and, with your other hand, slowly supinate and pronate the forearm. (3.17) Do you feel the head's rotating movement under your thumb?*

Shaft of the Radius

The shaft of the radius is located on the lateral side (thumb side) of the forearm. Unlike the superficial edge of the ulnar shaft, most of the shaft of the radius is buried in muscle tissue. Its distal portion, however, is superficial and can be directly accessed.

👁 1) Hold your partner's hand in a supinated position (palm up) with the elbow flexed at 90˚.
2) Locate the head of the radius. Slide distally off the head, noting how the radius sinks beneath the forearm muscles. Continue down the forearm and feel the radius become superficial at the wrist. (3.18)
3) Along the distal forearm, explore all sides of the superficial shaft of the radius.

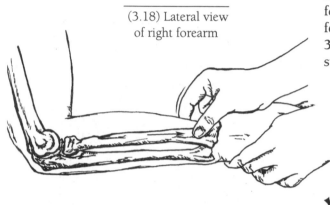

(3.18) Lateral view of right forearm

✔ *Is the bone you are palpating along the lateral side of the arm? Place one hand upon the radial shaft, while the other hand slowly supinates and pronates the forearm. Do you feel the shaft of the radius pivot around the shaft of the ulna?*

Styloid Process of the Radius

As opposed to the toothlike styloid process of the ulna, the styloid process of the radius is a wider, more substantial mound of bone. Located on the lateral side of the radius, the styloid process is surrounded by the extensor tendons and the attachment site for the brachioradialis.

👁 1) With the arm supinated (palm up), grasp the distal radial shaft between your thumb and finger.
2) Slide distally, noting how the radius broadens in all directions.
3) Palpate along the lateral side of the radius (thumb side) to the tip of the styloid process. (3.19)

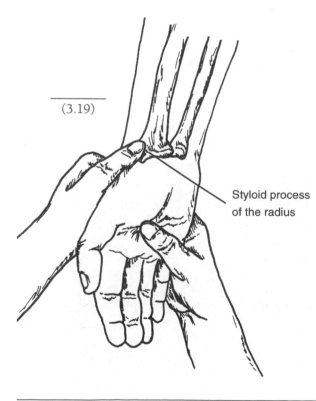

(3.19)

Styloid process of the radius

✔ *Are you proximal to the "flexor crease of the wrist?" Is the portion of bone you are palpating surrounded by several, thin tendons? If you passively flex and extend the wrist, the styloid should remain stationary. Does it?*

| carpal | **kar**-pul | hamate | **ham**-ate |
| metacarpal | **met**-a-**kar**-pul | lunate | **lu**-nate |

Trail 4 - "Walking On Your Hands"
Carpals

Pisiform
Hook of the Hamate

There are eight carpal bones located at the wrist. Small and each uniquely shaped, the carpals are closely wedged together between the distal radius and ulna and the metacarpals. They are located distal to the flexor crease at the wrist and lay deep to numerous flexor and extensor tendons. Aside from the pisiform (see below), palpation of individual carpal bones can be quite challenging. Here they are palpated as a group.

👁1) Supporting the wrist in both hands, locate the styloid processes of the ulna and radius.
2) Explore distal to the styloid processes (along the creases of the wrist) for the region of the carpals.
3) Passively move the wrist in all directions and note how the carpals shift and undulate slightly like small stones in a pouch. (3.20)

✔ *Are you distal to the flexor crease of the wrist? Passively flex and extend the wrist. Can you sense how the carpals press into the dorsum and palmar side of the hand, respectively?*

Pisiform

The knobby pisiform is an attachment site for the flexor carpi ulnaris (p. 100). It protrudes along the medial, palmar surface of the wrist, just distal to the flexor crease. It is the only superficial bump in this vicinity of the wrist.

👁1) Locate the flexor crease of the wrist.
2) Move medially (pinky side) and slightly distal to the crease.
3) Using your fingerpad, explore under the thick tissue of the palm for the nugget-shaped pisiform. (3.21)

✔ *Passively flex the wrist and notice how the pisiform can be wiggled from side to side. Extend the wrist and observe how it becomes immobile. Ask your partner to actively adduct her wrist. Can you feel the tendon of flexor carpi ulnaris come down the medial wrist and attach to the pisiform?*

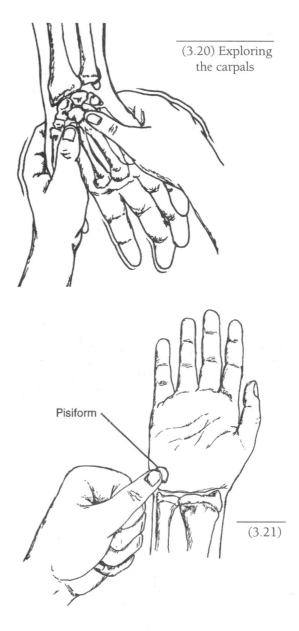

(3.20) Exploring the carpals

Pisiform

(3.21)

The pisiform is much larger on quadrupeds like dogs (right), protruding posteriorly above the heel of the front paw. This arrangement allows the flexor carpi ulnaris muscle that attaches to the pisiform greater leverage and power to flex the wrist when running on all fours. A human's pisiform is only a pea-sized knob. It is, however, still useful for karate "chops" and kneading bread dough.

navicular	na-**vik**-u-lar
phalanx	**fal**-anks
pisiform	**pi**-si-form
scaphoid	**skaf**-oyd

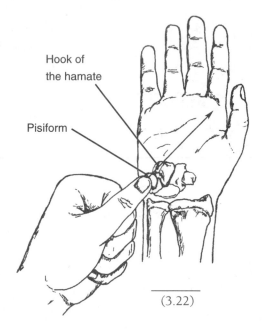

Hook of
the hamate

Pisiform

(3.22)

Hook of the Hamate

The hamate is located just distal and lateral to the pisiform and has a small protuberance or "hook" that is palpable on the palmar surface of the hand. Along with the pisiform, the hook of the hamate serves as an attachment site for the flexor retinaculum. It is often tender when being palpated.

1) Locate the pisiform. Draw an imaginary line from the pisiform to the base of the first finger. **2)** Using your thumbpad, slide off the pisiform and along this line. (3.22) Approximately three-quarters of an inch from the pisiform explore for a subtle mound beneath the padding of the hand.

To locate the carpometacarpal joint (a), ask your partner to flex her fingers and wrist. Roughly an inch or two distal to the extensor crease at the wrist will be a

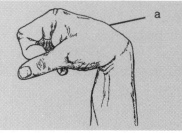

series of bumps across the dorsum of the hand. These bumps are the bases of the metacarpals which articulate with the carpals to form the carpometacarpal joints.

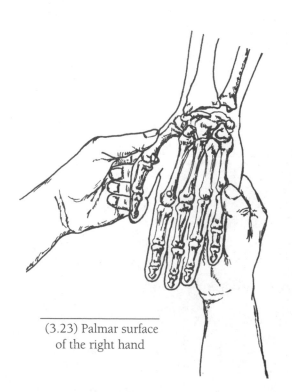

(3.23) Palmar surface
of the right hand

Metacarpals and Phalanges

1) Explore the dorsal side of the hand for the superficial shafts of the metacarpals.
2) Palpate the metacarpals from the palmar surface and note how deep they lie on that side of the hand. (3.23)
3) Move distally and explore where the heads of the metacarpals join with the phalanges to make the large metacarpophalangeal (knuckle) joints.
4) Explore the phalanges and tissues of the fingers. Explore the slender tendons, ligaments and connective tissues in the fingers. Note also the absence of muscle tissue.

The fingers contain no muscles, only the tendons of the digitorum muscles and strong ligaments which hold the phalanges of each finger together.

trapezium	tra-**pee**-ze-um	
triquetrum	tri-**kwe**-trum	
styloid	**sti**-loyd	Gr. a pillar

Muscles of the Forearm and Hand

The muscles of the forearm primarily affect the wrist and fingers. Many have small, fusiform bellies in the forearm which connect to space-efficient tendons. These tendons extend distally into the wrist and hand.

The forearm is crowded with muscle bellies and tendons and can be challenging to palpate. To simplify matters, these muscles will be arranged into four primary categories:

1) Muscles which act primarily at the elbow:
- brachialis
- brachioradialis

2) Muscles which move the wrist and/or fingers ("carpi," "digitorum" or "palmaris" muscles). This group can be further subdivided into four smaller groups:
- extensors of the wrist and fingers
- flexors of the wrist and fingers
- adductors of the wrist
- abductors of the wrist

(Some muscles which act upon the wrist are capable of moving the wrist in two directions. Flexor carpi ulnaris, for example, flexes and adducts the wrist.)

3) Muscles which create the pivoting action between the radius and ulna:
- supinator
- pronator teres

4) Muscles which maneuver the thumb or "pollicis" muscles.

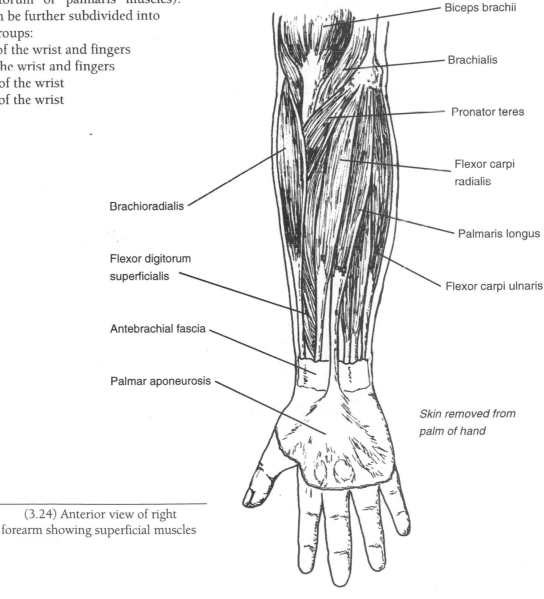

Biceps brachii

Brachialis

Pronator teres

Flexor carpi radialis

Palmaris longus

Flexor carpi ulnaris

Brachioradialis

Flexor digitorum superficialis

Antebrachial fascia

Palmar aponeurosis

Skin removed from palm of hand

(3.24) Anterior view of right forearm showing superficial muscles

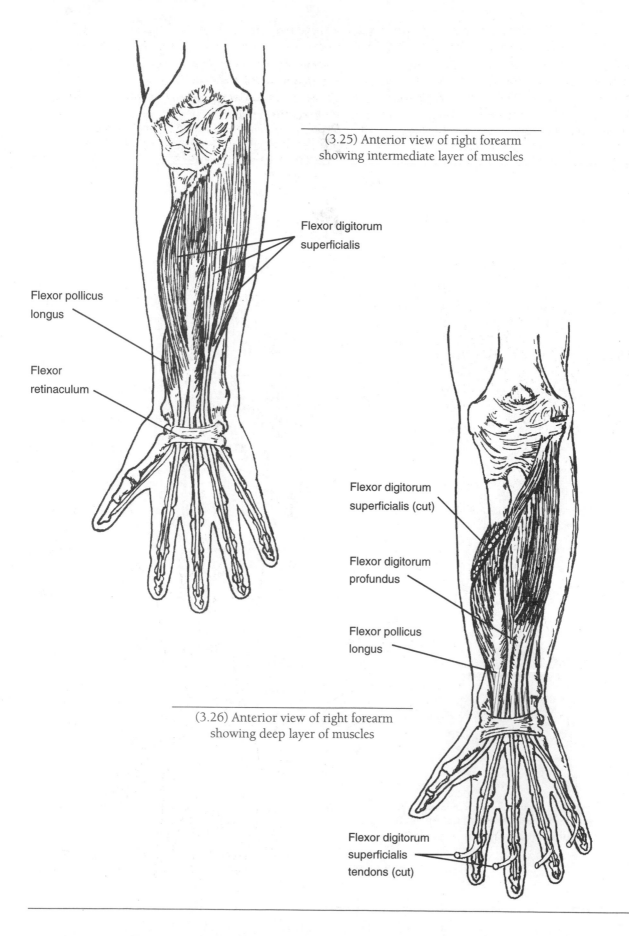

(3.25) Anterior view of right forearm showing intermediate layer of muscles

Flexor digitorum superficialis

Flexor pollicus longus

Flexor retinaculum

Flexor digitorum superficialis (cut)

Flexor digitorum profundus

Flexor pollicus longus

(3.26) Anterior view of right forearm showing deep layer of muscles

Flexor digitorum superficialis tendons (cut)

The names of the forearm muscles can be a mouthful. Yet their names can be very helpful in understanding a muscle's function, location and more. Take for example, the muscle "extensor carpi radialis longus." What does its name reveal?

1) It is specified as an **extensor**. This indicates there is also a **flexor** carpi radialis.

2) "Carpi" means it affects the **carpals** (wrist joint). This means there is a different muscle that moves the **digits** (extensor digitorum).

3) It runs along the **radial** side of the arm (the extensor carpi **ulnaris** is on the ulnar side of the arm.)

4) If there is a **longus**, there must be an extensor carpi radialis **brevis** as well.

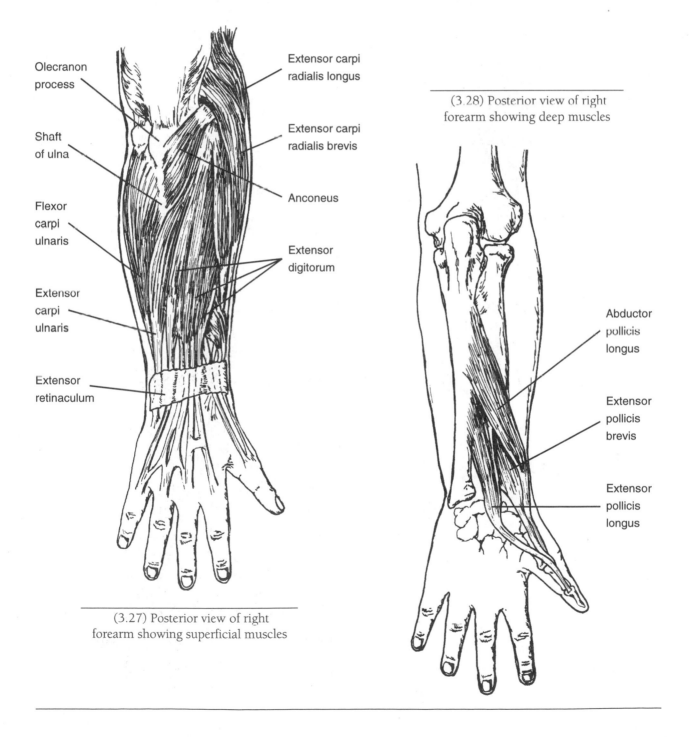

Olecranon process

Shaft of ulna

Flexor carpi ulnaris

Extensor carpi ulnaris

Extensor retinaculum

Extensor carpi radialis longus

Extensor carpi radialis brevis

Anconeus

Extensor digitorum

(3.28) Posterior view of right forearm showing deep muscles

Abductor pollicis longus

Extensor pollicis brevis

Extensor pollicis longus

(3.27) Posterior view of right forearm showing superficial muscles

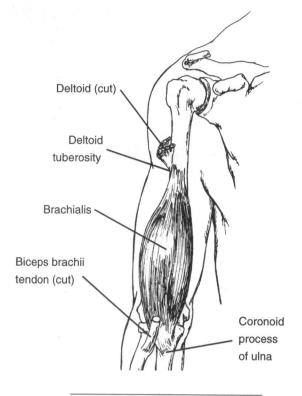

Deltoid (cut)

Deltoid
tuberosity

Brachialis

Biceps brachii
tendon (cut)

Coronoid
process
of ulna

(3.29) Anterior view of right arm

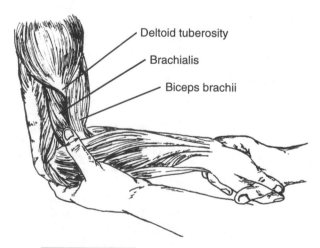

Deltoid tuberosity

Brachialis

Biceps brachii

(3.30) Lateral
view of right arm

Location - Deep to biceps
BLMs - Distal biceps tendon or lateral arm
Action - "Flex your elbow"

Brachialis

The brachialis is a strong elbow flexor that lies deep to the biceps brachii (p. 70) on the anterior arm. It has a flat, yet thick belly. (3.29) Ironically, the brachialis' girth only helps to bulge the biceps further from the arm, making brachialis the biceps' best friend.

Although it lies underneath the biceps, portions of brachialis are accessible. Its lateral edge, sandwiched between the biceps and triceps brachii, is superficial and palpable. The distal aspect of the brachialis is also accessible as it passes along either side of the biceps tendon.

A -
Flex the elbow

O - Distal half of anterior surface of humerus

I - Tuberosity and coronoid process of ulna

N - Musculocutaneous

1) Shake hands with your partner and flex the elbow to 90°. It is important to recognize which muscle tissue is the biceps brachii and which is brachialis. Ask your partner to flex her elbow against your resistance and isolate the edges of the round biceps brachii belly.
2) With the arm relaxed, slide laterally half an inch off the distal biceps. The edge of the brachialis can be detected by rolling your fingers across its surface. As you strum across its solid edge, you will feel a pronounced "thump." (3.30)
3) Continuing to strum across its edge, follow it distally to where it disappears into the elbow.
4) Locate the distal biceps tendon. Palpate along either side of the biceps tendon for portions of the deeper brachialis.

Locate the deltoid tuberosity. Slide distally straight down the lateral side of the arm and explore for the edge of the brachialis.

✔ *Can you roll across a distinct wad of muscle on the lateral side of the arm? Can you follow it distally toward the inner elbow? Locate the triceps and biceps brachii. Are the brachialis fibers between them on the lateral arm?*

brachial	**bre**-ke-al	L. relating to the arm
brachialis	**bra-key-al-is**	

Brachioradialis

The brachioradialis is superficial on the lateral side of the forearm. It has a long, oval belly which forms a dividing line between the flexors and extensors of the wrist and hand. Its muscle belly becomes tendonous halfway down the forearm. It is the only muscle that runs the length of the forearm but does not cross the wrist joint. (3.31) For palpation, resisted flexion of the elbow causes brachioradialis to visibly protrude on the forearm and become clearly palpable.

A -
Flex the elbow
Assists in pronation and supination of the
   forearm when these movements are resisted

O - Lateral supracondylar ridge of humerus

I - Styloid process of radius

N - Musculocutaneous and radial

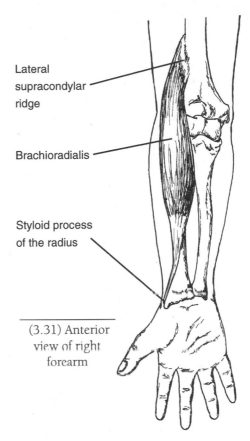

Lateral
supracondylar
ridge

Brachioradialis

Styloid process
of the radius

(3.31) Anterior
view of right
forearm

1) Shake hands with partner and flex the elbow to 90°. With the forearm in a neutral position (thumb toward the ceiling), ask your partner to flex her elbow against your resistance.
2) Look for the brachioradialis bulging out on the lateral side of the elbow. If it is not visible, locate the lateral supracondylar ridge of the humerus and slide distally.
3) With your partner still contracting, use your other hand to palpate the superficial, tubular belly of brachioradialis. (3.32) Try to pinch its belly between your fingers and follow it as far distally as possible.
4) As it becomes more tendonous, strum across its distal tendon toward the styloid process of the radius.

✔ *Upon resisted flexion of the elbow does the belly you are palpating contract and bulge out? Is it superficial? Does it extend off the lateral epicondyle of the humerus?*

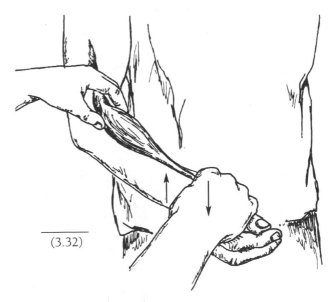

(3.32)

Location - Superficial
BLMs - Lateral epicondyle, styloid process of radius
Action - "Flex your elbow against my resistance"

brachioradialis   **bra**-key-o-**ra**-de-a-lis

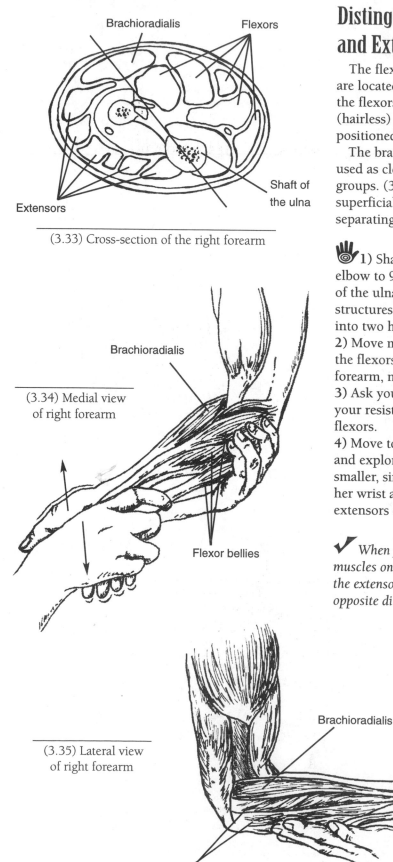

Brachioradialis

Flexors

Extensors

Shaft of the ulna

(3.33) Cross-section of the right forearm

Brachioradialis

(3.34) Medial view of right forearm

Flexor bellies

(3.35) Lateral view of right forearm

Brachioradialis

Extensor bellies

Distinguishing Between the Flexor and Extensor Groups of the Forearm

The flexors and extensors of the wrist and hand are located in the forearm. In anatomical position, the flexors are located on the anterior/medial (hairless) side of the forearm. The extensors are positioned on the posterior/lateral (hairy) side.

The brachioradialis and shaft of the ulna can be used as clear dividing lines between these muscle groups. (3.33) Both of these structures run superficially down opposite sides of the forearm, separating the forearm flexors and extensors.

1) Shake hands with your partner and flex the elbow to 90°. Locate the brachioradialis and shaft of the ulna (p. 86). Palpate the length of these structures, observing how they divide the arm into two halves.

2) Move medially from the shaft of the ulna onto the flexors of the forearm. Explore this half of the forearm, noting the girth of the flexors.

3) Ask your partner to slightly flex her wrist against your resistance. (3.34) Note the contraction of the flexors.

4) Move to the lateral side of the shaft of the ulna and explore the extensor bellies. (3.35) Notice their smaller, sinewy bellies. Ask your partner to extend her wrist against your resistance, feeling the extensors contract.

✔ *When your partner curls (flexes) her wrist, do the muscles on the hairless side of the arm contract? Do the extensors contract when the hand moves in the opposite direction (extension)?*

Extensors

Extensor Carpi Radialis Longus and Brevis
Extensor Digitorum
Extensor Carpi Ulnaris

The four extensors primarily create extension of the wrist and fingers. They are situated between the brachioradialis and the shaft of the ulna along the arm's lateral, posterior surface. All of these muscles are superficial and accessible, yet challenging to specifically isolate. Originating on the lateral side of the humerus, the bellies of the extensors become tendonous approximately two inches proximal to the wrist joint. (3.36) As a group they are smaller and more sinewy than the forearm flexors.

Extensor carpi radialis longus and brevis (discussed here as one muscle) are lateral/posterior to the brachioradialis. Extensor carpi ulnaris (as its name suggests) lies beside the ulnar shaft. Extensor digitorum is located between these muscles and has four long, superficial tendons stretching along the dorsal surface of the hand and fingers.

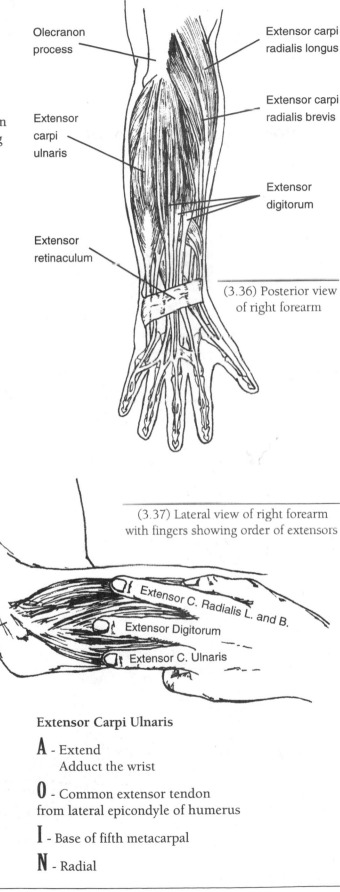

(3.36) Posterior view of right forearm

(3.37) Lateral view of right forearm with fingers showing order of extensors

Extensor Carpi Radialis Longus and Brevis

A - Extend
Abduct the wrist
Assist in flexion of elbow

O - Lateral supracondylar ridge of humerus

I - Base of second metacarpal (longus), Base of third metacarpal (brevis)

N - Radial

Extensor Digitorum

A - Extend four fingers
Assist in extension of the wrist

O - Common extensor tendon from lateral epicondyle of humerus

I - Middle and distal phalanges of each four fingers

N - Radial

Extensor Carpi Ulnaris

A - Extend
Adduct the wrist

O - Common extensor tendon from lateral epicondyle of humerus

I - Base of fifth metacarpal

N - Radial

brevis	L. short
digit	L. finger

carpi   **kar**-pi

✋👁 Extensor group

1) Shake hands with partner and flex the elbow to 90°. Locate the brachioradialis and shaft of the ulna.
2) Lay the flat of your hand between these landmarks and ask your partner to alternately extend and relax her wrist against your resistance. (3.35)
3) Explore the slender, sinewy fibers of these muscles and note how they contract upon extension.

✔ *Are you between the brachioradialis and ulnar shaft? Do the muscles contract on extension of the wrist?*

Brachioradialis

Extensor carpi radialis
longus and brevis

(3.38)

👁 Extensor carpi radialis longus and brevis

1) Shake hands and flex the elbow to 90°. Locate the brachioradialis. Slide laterally off its belly onto the extensor carpi radialis fibers.
2) Ask your partner to alternately abduct and relax her wrist against your resistance. Sense how the fibers tighten with this movement. (3.38)
3) Follow their muscle fibers distally as far as possible to where they become tendonous.

✔ *Differentiate between the extensor carpi radialis muscles and brachioradialis by asking your partner to alternately abduct and relax her wrist against your resistance. During this action the brachioradialis (which does not cross the wrist) will remain passive, while the extensor carpi radialis muscles will contract.*

Brachioradialis and extensor carpi radialis longus and brevis are sometimes known as "the wad of three." Together they form a long mass of muscle which extends distally from the supracondylar ridge of the humerus.

✋👁 To locate the wad, shake hands with your partner and palpate just lateral to the inner elbow. The wad will be the thick, mobile tissue which can be easily grasped between your fingers and thumb. Follow it distally as far as possible.

"Wad of three"

☝Extensor digitorum

1) Shake hands with partner and flex the elbow to 90°. Slide laterally off the extensor carpi radialis fibers.
2) As you move around the arm, palpate the digitorum's flat surface and roll across its fibers.
3) Isolate its belly by asking your partner to extend her wrist and fingers. (3.39) Follow the belly distally to its tendons along the back of the hand.

✔ *Ask your partner to wiggle her fingers as if she were typing. Do you feel an undulating contraction of the extensor digitorum?*

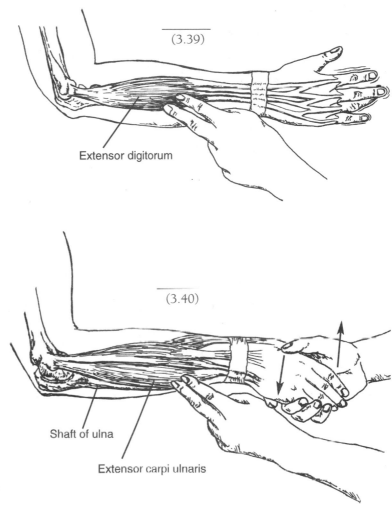

(3.39)

Extensor digitorum

☝Extensor carpi ulnaris

1) Shake hands with your partner and flex the elbow to 90°. Locate the shaft of the ulna.
2) Slide laterally off the shaft onto the slender belly of the extensor carpi ulnaris.
3) Ask your partner to adduct her wrist against your resistance. (3.40) Note how the tissue directly lateral to the ulna tightens with this movement.
4) Follow the tendon distally past the head of the ulna.

(3.40)

Shaft of ulna

Extensor carpi ulnaris

Extensors
Location - Superficial on lateral, posterior forearm
BLMs - Lateral epicondyle, shaft of the ulna
Action - "Extend your wrist and fingers"

The anconeus (p. 93) is a weak elbow extensor located lateral to the olecranon process. Triangular-shaped, it originates at the lateral epicondyle of the humerus and fans out to attach on the shaft of the ulna. The anconeus is superficial, yet can be difficult to differentiate from the surrounding extensors.

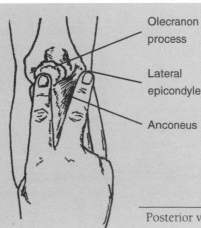

Olecranon process

Lateral epicondyle

Anconeus

☝1) Locate the olecranon process, proximal shaft of the ulna and lateral epicondyle of the humerus.
2) Lay your index finger along the proximal ulna and the tip of your middle finger upon the epicondyle. The "V" formed by your fingers is the outline of the anconeus.

Posterior view of right elbow

anconeus   an-**ko**-nee-us   Gr. elbow

Flexors

Flexor Carpi Radialis
Palmaris Longus
Flexor Carpi Ulnaris
Flexor Digitorum Superficialis
Flexor Digitorum Profundus

The five flexors included in this section primarily create flexion at the wrist or fingers. They are located on the forearm's anterior/medial surface between the brachioradialis and the ulnar shaft. Most of the flexors originate as one mass from the common flexor tendon at the medial epicondyle of the humerus. (3.42) As the flexor bellies extend down the forearm, they become thin tendons starting roughly two inches proximal to the wrist.

As a group, the flexors are thicker and more pliable than the extensors. Although the flexors are easily accessed together, isolating specific muscle bellies can be challenging.

The flexors are arranged in three layers. The superficial layer is formed by the long bellies of flexor carpi radialis, palmaris longus and flexor carpi ulnaris. (3.41) Flexor carpi radialis is medial to the pronator teres and brachioradialis. Flexor carpi ulnaris lies close to the ulnar shaft and has a distinct tendon attaching to the pisiform. The palmaris longus, which is sometimes absent, runs between flexor carpi radialis and flexor carpi ulnaris and attaches to the palmar aponeurosis (p. 111). Portions of all three muscles can be isolated for palpation.

The middle and deep layers contain the wide bellies of flexor digitorum superficialis and flexor digitorum profundus, respectively. (3.43 and 3.44) Each digitorum muscle has four thin tendons which pass through the carpal tunnel (p. 111) and attach at the phalanges. The digitorum bellies are difficult to directly access, but their density can be felt beneath the superficial flexors.

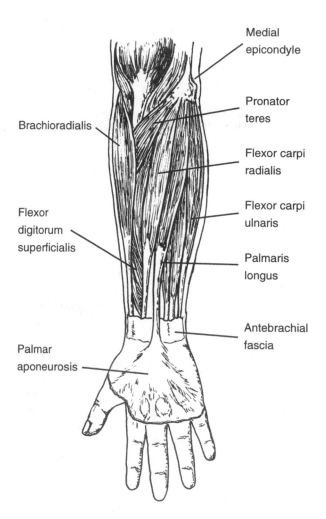

Medial epicondyle

Pronator teres

Flexor carpi radialis

Flexor carpi ulnaris

Palmaris longus

Antebrachial fascia

Brachioradialis

Flexor digitorum superficialis

Palmar aponeurosis

(3.41) Anterior view of right forearm showing superficial layer of flexors

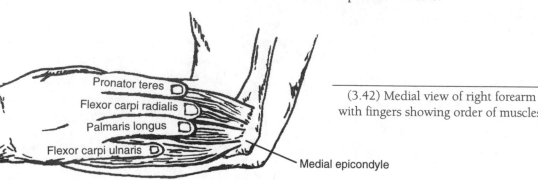

Pronator teres

Flexor carpi radialis

Palmaris longus

Flexor carpi ulnaris

Medial epicondyle

(3.42) Medial view of right forearm with fingers showing order of muscles

profundus   pro-**fun**-dus   L. located deeper than its reference point
superficialis   **soo**-per-fish-ee-a-lis

Flexor Carpi Radialis

A - Flex
Abduct wrist
Flex elbow

O - Common flexor tendon from medial epicondyle

I - Base of second and third metacarpal

N - Median

Palmaris Longus

A - Tenses the palmar fascia
Flex wrist
Flex elbow

O - Common flexor tendon from medial epicondyle

I - Flexor retinaculum and palmar aponeurosis

N - Median

Flexor Carpi Ulnaris

A - Flex
Adduct wrist
May flex elbow

O - Common flexor tendon from medial epicondyle

I - Pisiform

N - Ulnar

Flexor digitorum
superficialis

Flexor pollicus
longus

Flexor
retinaculum

(3.43) Anterior view of
intermediate layer

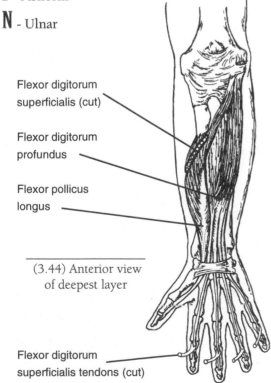

Flexor digitorum
superficialis (cut)

Flexor digitorum
profundus

Flexor pollicus
longus

(3.44) Anterior view
of deepest layer

Flexor digitorum
superficialis tendons (cut)

Flexor Digitorum Superficialis

A - Flex proximal interphalangeal joints
of second through fifth digits
Flex metacarpophalangeal joints
Flex wrist

O - Common flexor tendon from medial epicondyle
of humerus, ulnar collateral ligament, coronoid
process of ulna

I - By four tendons into sides of middle phalanges
of second through fifth digits

N - Ulnar

Flexor Digitorum Profundus

A - Flex distal interphalangeal joints of four fingers
Flex proximal interphalangeal and
metacarpophalangeal joints
May flex wrist

O - Anterior and medial surfaces
of proximal three-quarters of ulna

I - By four tendons into bases of
distal phalanges, anterior surface

N - Ulnar

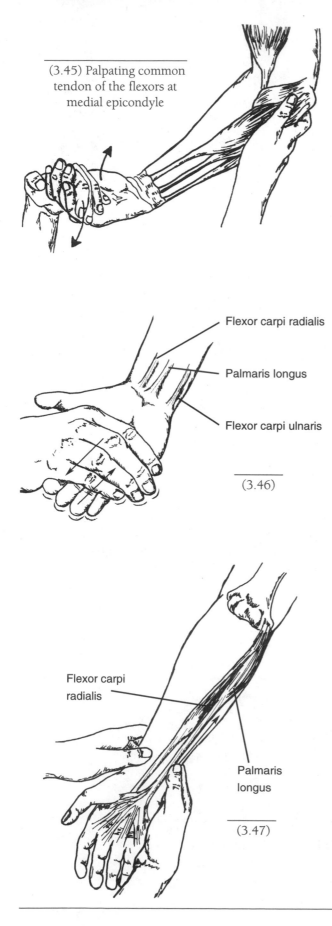

(3.45) Palpating common tendon of the flexors at medial epicondyle

Flexor carpi radialis

Palmaris longus

Flexor carpi ulnaris

(3.46)

Flexor carpi radialis

Palmaris longus

(3.47)

👁Flexor group

1) Shake hands with partner and flex the elbow to 90°. Locate the brachioradialis and shaft of the ulna.
2) Lay the flat of your hand between these landmarks on the arm's anterior surface and ask your partner to alternately flex and relax her wrist against your resistance. (3.45)
3) Explore their chubby bellies from the medial epicondyle to their distal tendons at the wrist.

✔ *Are you between the brachioradialis and ulnar shaft? Do the muscles contract on flexion of the wrist?*

👁Flexor carpi radialis and palmaris longus

1) Flex the elbow to 90˙ and supinate the forearm. Begin at the distal tendons. Ask your partner to flex her wrist against your resistance.
2) At the center of the wrist will be two superficial tendons - flexor carpi radialis and palmaris longus. (3.46) The palmaris may be missing, but if both are present, the palmaris will be furthest medial.
3) With your partner contracting, roll across the tendons and follow them proximally as they expand into muscle bellies. (3.47)
4) Ask your partner to alternately abduct and relax her wrist to create a distinct contraction of flexor carpi radialis.

✔ *Are the tendons/muscle bellies superficial? If you palpate the belly of flexor carpi radialis, is it superficial and medial to the pronator teres (p. 104)? Is the palmaris longus medial to the flexor carpi radialis? Follow the bellies toward the elbow. Do they merge together at the medial epicondyle of the humerus?*

The palmaris longus is absent in 11% of the population. The palmar aponeurosis, however, is always present. The palmaris longus may vary from a mere tendonous band to having a distal muscle belly with a long, proximal tendon. On occasion, there may be two palmaris longus muscles. Its insertion site is also irregular. It may attach to the fascia of the forearm, the tendon of the flexor carpi ulnaris, the flexor retinaculum, the pisiform or the scaphoid.

👁✋ Flexor carpi ulnaris

1) Shaking hands, flex the elbow to 90˚ and supinate the forearm. Begin at the distal tendon by locating the pisiform (p. 89).
2) Slide proximally off the pisiform to the slender, superficial tendon of flexor carpi ulnaris. (3.48)
3) As your partner alternately adducts and relaxes her wrist against your resistance, follow the tendon proximally, strumming across its surface. Feel how it widens into muscle belly and heads toward the medial epicondyle. (Note: Unlike the extensor carpi ulnaris, the flexor carpi ulnaris lies roughly a finger width away from the ulnar shaft.)

✔ *Do you feel the muscle contract upon adduction? Is the tendon/muscle belly superficial and along the arm's anterior/medial surface? Is it medial to the palmaris longus?*

Pisiform

(3.48)

👁✋ Flexor digitorum superficialis and profundus

1) Shake hands with your partner and flex the elbow to 90°. To access portions of the digitorum bellies, palpate through or between the carpi ulnaris, carpi radialis and palmaris longus bellies. Passive flexion of the wrist softens the tissue and may allow for easier access.
2) Along the anterior surface of the wrist, explore on either side of the superficial tendons of the flexor carpi radialis and palmaris longus and locate a few of the deeper digitorum tendons. (3.49)
3) Along the ulnar shaft, the medial aspect of the digitorums is superficial. Locate this tissue by palpating alongside the ulnar shaft as you ask your partner to alternately press together her pinkie and thumb tips. Explore just beside the shaft of the ulna for a portion of the digitorums which will contract and relax.

✔ *Is the tissue you are accessing deep to the first layer of flexors?*

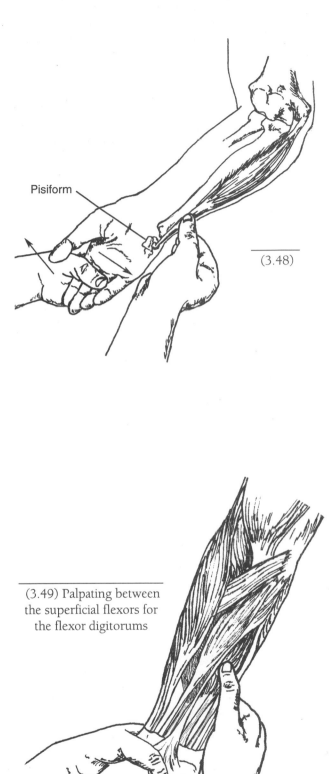

(3.49) Palpating between the superficial flexors for the flexor digitorums

Flexors
Location - Superficial and deep on anterior, medial forearm
BLMs - Medial epicondyle, shaft of the ulna
Action - "Make a fist"

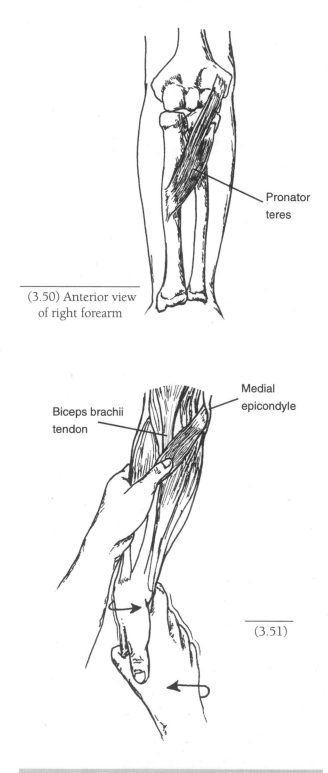

(3.50) Anterior view
of right forearm

Pronator
teres

Medial
epicondyle

Biceps brachii
tendon

(3.51)

Pronator Teres

   Located on the anterior surface of the forearm,
the round pronator teres is tucked between the
brachioradialis and forearm flexors. (3.24) It is
partially superficial and the only muscle in its
vicinity with oblique fibers. (3.50) The pronator
teres is an antagonist to biceps brachii and
supinator ("carrying a bowl of soup") and
creates pronation of the forearm ("prone to spill
it"). The distal tendon of biceps brachii is situated
just lateral to the pronator teres and provides a
good landmark for locating the pronator's fibers.

A -
Pronate the forearm
Assists to flex the elbow

O - Medial epicondyle of humerus, common
flexor tendon and coronoid process of ulna

I - Middle of lateral surface of radius

N - Median

1) Shake hands with your partner and flex the
elbow to 90°. Locate the distal tendon of the biceps
brachii. For assistance, ask your partner to flex her
elbow a little.
2) Slide distally off the tendon into the valley
between the brachioradialis and forearm flexors.
Sink your thumb into this space.
3) Explore for the finger-wide pronator belly
running obliquely from the medial elbow across to
the radius. Strum across its oblique fibers. (3.51)
4) Follow it toward the medial epicondyle (noting
how it blends into the other flexors) and the middle
radius (feel how it tucks under the brachioradialis).

✔ *Shaking hands with your partner, ask her to
pronate against your resistance. Does the belly of the
muscle you are palpating form a solid contraction?
Do the fibers you are palpating run diagonally toward
the middle of the radius?*

   There are two primary supinators - biceps
brachii and supinator - and two pronators -
pronator teres and pronator quadratus, a small
muscle located deep in the wrist. You might
assume that this symmetry would balance the
strength between the pronators and supinators.
But in reality, the size and power of the biceps
brachii tips the scales in favor of supination.

*Location - Distal to the biceps brachii tendon
BLMs - Medial epicondyle, shaft of radius
Action - Resisted pronation of forearm*

pronate   **pro**-nate   L. to bend forward

Supinator

Located on the lateral side of the elbow, the short supinator is deep to the forearm extensors and superficial to the head of the radius. (3.52, 3.53) As its name suggests, it supinates the forearm and is an antagonist to the pronator teres. It has a slender muscle belly and can be difficult to isolate specifically.

A - Supinate the forearm

O - Lateral epicondyle of humerus, radial collateral ligament and annular ligament

I - Lateral surface of proximal shaft of radius

N - Radial

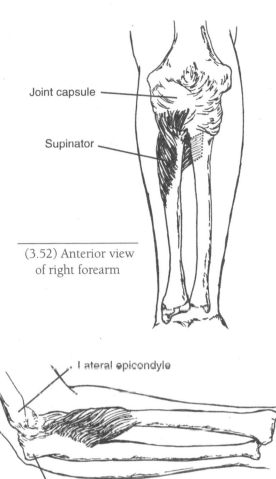

Joint capsule

Supinator

(3.52) Anterior view of right forearm

Lateral epicondyle

Olecranon process

(3.53) Lateral view of right forearm

1) Shake hands with your partner and flex the elbow to 90°. Locate the lateral epicondyle of the humerus and the proximal shaft of the radius.
2) Place your fingerpads between these landmarks and palpate through the extensor fibers for the deep supinator belly. (3.54)
3) Ask your partner to alternately supinate and relax her forearm against your resistance. The brachioradialis may contract with this movement, but the brachioradialis will be felt superficially, while the supinator is deep to the extensors.

(3.54)

Supinator

Superficial extensors (cut)

The expression "righty-tighty, lefty loosey" is a mental reminder for which direction to turn a screw, but it also applies for which hand to hold the screwdriver. We have more power to supinate than pronate and the world is dominated by right-handed individuals, so screws have been designed to be tightened by supination of the right forearm. This, of course, leaves those who are "south paws" to tighten with weak pronators or the uncoordinated supinators of the right forearm.

Location - Deep to forearm extensors
BLM - Lateral epicondyle of humerus, shaft of radius
Action - Resisted supination of forearm

supinate   **su**-pi-nate   L. bent backward

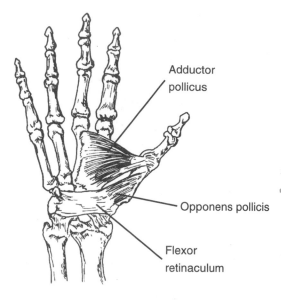

(3.55) Palmar surface of right hand showing muscles of the thenar eminence

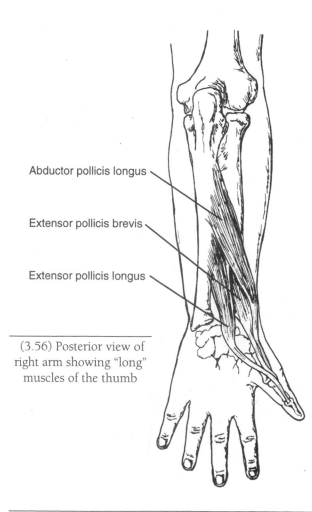

(3.56) Posterior view of right arm showing "long" muscles of the thumb

Muscles of the Thumb

Opponens Pollicis
Adductor Pollicis
Extensor Pollicis Longus and Brevis
Flexor Pollicis Longus
Abductor Pollicis Longus

The muscles which act upon the thumb can be divided into two groups: short muscles located at the thenar eminence (the fleshy mass at the thumb's base) and long, tendonous muscles which attach along the radius or ulna.

The thenar eminence is composed of several muscles. They include the opponens pollicis, which creates the critical movement of opposition, and adductor pollicis, which is responsible for bringing the thumb and index finger together. (3.55)

The long muscles of the thumb include the extensor pollicis longus and brevis, flexor pollicis longus, and abductor pollicis longus. (3.56, 3.57) Their bellies are located on the posterior surface of the forearm and are deep to the extensors.

When the thumb is extended, the tendons of the extensor pollicis and abductor pollicis form the "anatomical snuffbox." Used historically as a platform for inhaling a variety of substances, this small cavity is located along the dorsal surface of the hand, just distal to the styloid process of the radius.

Opponens Pollicis

A - Oppose the thumb

O - Flexor retinaculum and tubercle of the trapezium

I - Entire length of first metacarpal bone, radial side

N - Median

Adductor Pollicis

A - Adduct
Flex the thumb

O - Capitate, second and third metacarpals

I - Base of proximal phalange of thumb

N - Ulnar

opponens	o-**po**-nens	L. against	thenar	**the**-nar	Gr. palm
pollicus	**pol**-li-cus	L. strong			

Extensor Pollicis Longus and Brevis

A - Extend the thumb

O - Posterior surface of radius and ulna, deep to extensors

I - Proximal (brevis) and distal (longus) phalange of thumb

N - Radial

Flexor Pollicis Longus

A - Flex the thumb

O - Anterior surface of radius, deep to flexors

I - Distal phalange of thumb

N - Median

Abductor Pollicis Longus

A - Abduct
Extend the thumb

O - Posterior surface of radius and ulna, deep to extensors

I - Base of first metacarpal

N - Radial

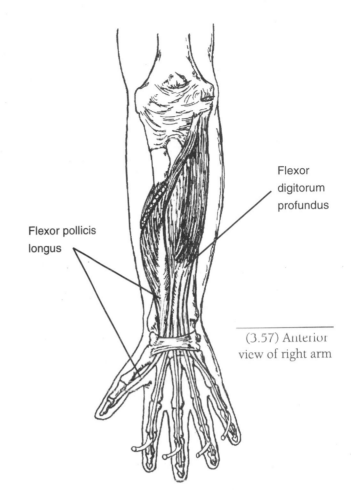

Flexor digitorum profundus

Flexor pollicis longus

(3.57) Anterior view of right arm

👁 Short muscles of the thenar eminence

1) Locate the base of the thumb. Explore all sides of the eminence's thick, movable tissue.
2) Palpate around the surface of the first metatarsal bone, as well as the webbing between the thumb and finger. (3.58)
3) Ask your partner to squeeze her thumb and pinkie fingerpad together. With contraction, note how the eminence becomes dense and compact.

(3.58)

The human thumb has several unique qualities which distinguish it from the thumbs of other primates. One element that is *not* distinctly human is the saddle joint of the first carpometacarpal joint. The joint's shape allows for opposition of the thumb and fingers, a skill shared by many higher primates including chimpanzees, orangutans and gorillas.

One reason for the dexterity of the human thumb is the separation between the flexor pollicis longus and flexor digitorum profundus muscles. In other primates these muscles are united, restricting the ability

Orangutan

Chimpanzee

to move the fingers and thumb independently.

Humans are also able to execute a strong, precise grip between the thumb and fingertips - such as when tightening the lid on a jar. A human's thumb is quite long in proportion to the fingers; thumbs of many primates are shorter than the fingers. Also, the muscles of a human's thenar eminence are larger, while the thenar pad on many primates is typically flat and lacks dense musculature. Chimpanzees, gorillas and other primates can grasp with terrific force by curling their digits around an object, yet to oppose the thumb and finger for a specific, detailed task is something only humans can do.

(3.59)

"Anatomical snuffbox"

Anatomical snuffbox and long thumb muscles

1) With the wrist in a neutral position, ask your partner to extend her thumb ("bring your thumb nail toward your elbow").

2) Just distal to the styloid process of the radius will be a small trough formed by the surrounding tendons. If not seen immediately, change the angle of the thumb. (3.59)

3) Follow the tendons which form the snuffbox (extensor pollicis longus, brevis and abductor pollicis) proximally as they slide over the posterior surface of the radius. Lay your fingers along the posterior surface of the radius as your partner circumducts her thumb to feel a portion of these muscles contract. (3.60)

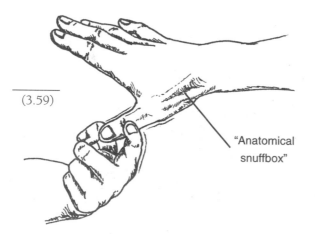

Extensor digitorums and extensor carpi ulnaris (cut)

(3.60) Exploring the bellies of the thumb muscles, deep to the extensors

Ligaments, Nerves and Vessels

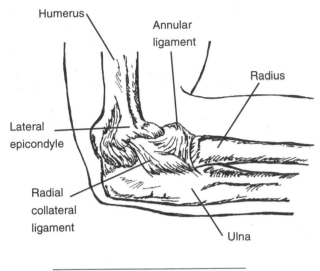

(3.61) Lateral view of right elbow

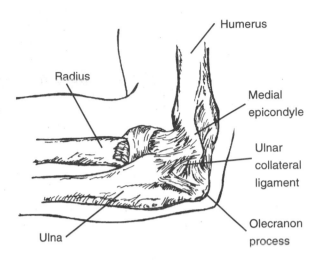

(3.62) Medial view of right elbow

Radial Collateral Ligament

The radial collateral ligament is a cord-like band that stretches from the lateral epicondyle of the humerus to the annular ligament and olecranon process. (3.61) The ligament is deep to the extensors and supinator of the forearm.

👁 1) Shaking hands, locate the lateral epicondyle of the humerus and head of the radius.
2) Between these landmarks will be a slight ditch. Place your fingertip in this space. Visualize the ligament spanning across the ditch and gently roll your finger across the ligament's slender surface. It may feel like a thin strip of duct tape. (3.63)

✔ *Are you between the head of the radius and lateral epicondyle? With the elbow flexed, do the fibers of the ligament run parallel with the forearm?*

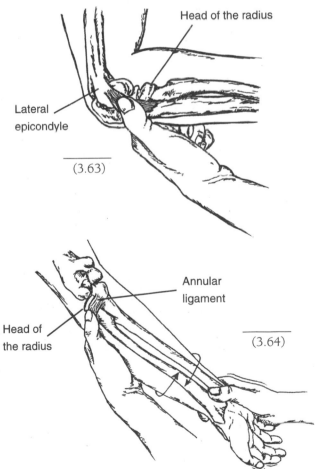

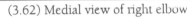

(3.63)

(3.64)

Annular Ligament

The annular ligament wraps around the radius' head and neck, stabilizing the proximal radius against the ulna during pronation and supination. (3.61) It lies deep to the supinator and forearm's extensor muscles. Although the annular ligament cannot be specifically palpated, its location can be isolated.

annular	**an**-u-ler	L. ring
collateral	ko-**lat**-er-al	L. together, pertaining to the side

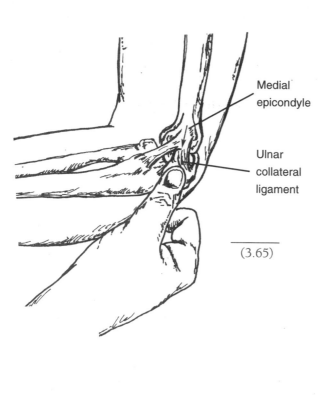

Medial epicondyle

Ulnar collateral ligament

(3.65)

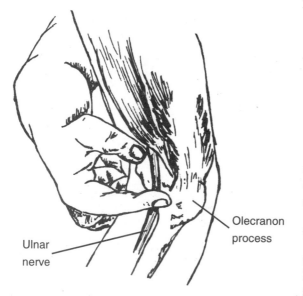

Ulnar nerve

Olecranon process

(3.66) Posterior/medial view of right elbow

While palpating be sure not to press too hard and impinge the ulnar nerve which can create tingling or numbness in the forearm or hand.

👁 1) With the elbow flexed, place your thumb pad on the head of the radius.

2) While passively supinating and pronating the forearm, allow the head and neck of the radius to pivot under your thumb. (3.64) You may not feel the annular ligament specifically, but visualize it stabilizing the head of the radius to ulna.

Ulnar Collateral Ligament

The ulnar collateral ligament is a strong, triangular-shaped ligament that originates on the humerus' medial epicondyle. (3.62) Its fibers spread out and attach to the coronoid process of the ulna and olecranon process. The collateral ligament is deep to the common flexor tendon, but superficial to the ulnar nerve.

👁 1) With the elbow flexed, locate the humerus' medial epicondyle and the medial aspect of the olecranon process.

2) Place your finger between these landmarks.

3) Palpating through the overlying muscle tissue, explore the ligament's thin fibers which run transversely to the muscle fibers. (3.65) You may not feel anything distinct, but if you are between the stated landmarks you are in the right location.

Ulnar Nerve

The ulnar nerve passes between the medial epicondyle and olecranon process as it extends down the arm. Between these two landmarks the nerve is superficial and easily accessible. Bumping your elbow may aggravate the ulnar nerve, creating the annoying "funny bone" sensation down the arm.

👁 1) With the elbow flexed, locate the medial epicondyle and olecranon process.

2) Using gentle pressure, slide your finger into the space between these landmarks and palpate for the tube-shaped nerve. (3.66)

3) Explore its location in relation to the triceps tendon and common flexor tendon (p. 102).

✔ *Is the structure you feel soft and moveable? Are you palpating the ulnar nerve or the triceps brachii's fibrous tendon? Ask your partner to extend her arm and the tendon will tighten and the nerve will "disappear" into the tissue.*

Retinacula of the Wrist and Palmar Aponeurosis

The *flexor retinaculum* is located on the anterior surface of the wrist just distal to the flexor crease. (3.2) It has transverse fibers which lie deep to the palmaris longus tendon (p. 102) and superficial to the other flexor tendons and median nerve. The flexor retinaculum and carpal bones form the carpal tunnel (or canal) in which the flexor tendons and median nerve pass. (3.67)

Isolating the thin flexor retinaculum can be difficult, but its transverse fibers (which are perpendicular to the deeper tendons) can be a helpful distinction. Also, if the retinaculum is "tight" the tissue of the anterior wrist may have an inflexible feel.

The thick *palmar aponeurosis* is a continuation of the antebrachial fascia that stretches superficially across the palm of the hand and is an attachment for the palmaris longus tendon. It is shaped in a similar way to the plantar aponeurosis (p. 280) on the sole of the foot. It may not be specifically palpable but, like the flexor retinaculum, its tensile quality can be felt.

The *extensor retinaculum* is superficial and located on the posterior wrist. Like the flexor retinaculum, it is a thickening of fascia that has transverse fibers which stretch across the wrist to attach to underlying bones. It stabilizes the wrist and thumb extensors. It is roughly three-quarters of an inch wide and located distal to the head of the ulna and the styloid process of the radius.

Flexor retinaculum and palmar aponeurosis

1) Cradle your partner's hand so your thumb pads are on the flexor crease of the wrist. Slide half an inch distal to the crease and sink into the thick tissues of the "heel" of the hand. (3.68)
2) As you explore the carpal space, visualize the retinaculum spanning across the carpals. Passively flex and extend the wrist and feel how the tension of the retinaculum changes.
3) Slide distally onto the palm of the hand and palpate for the thick, superficial palmar aponeurosis.

✔ *When palpating the flexor retinaculum, are you distal to the level of the pisiform (p. 89)?*
To highlight the palmar aponeurosis, ask your partner to tighten his hand as if he were "palming a basketball." (3.69) Note how this action also brings the palmaris longus tendon into view.

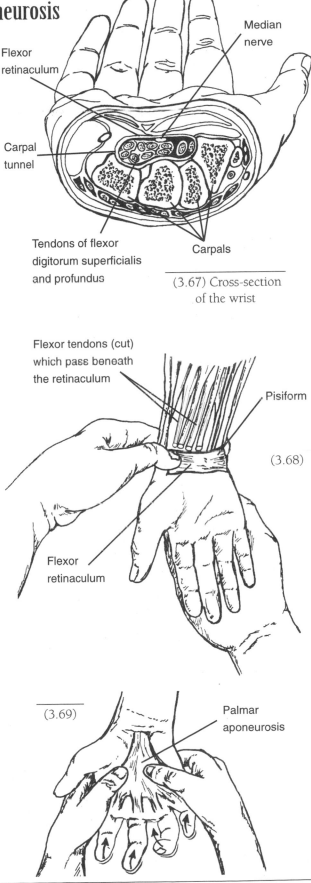

Median nerve

Flexor retinaculum

Carpal tunnel

Tendons of flexor digitorum superficialis and profundus

Carpals

(3.67) Cross-section of the wrist

Flexor tendons (cut) which pass beneath the retinaculum

Pisiform

(3.68)

Flexor retinaculum

(3.69)

Palmar aponeurosis

retinaculum   **ret**-i-**nak**-u-lum   L. halter
aponeurosis   **ap**-o-nu-**ro**-sis   Gr. *apo*, from + *neuron*, nerve or tendon

✋ Extensor retinaculum

1) Ask your partner to extend her fingers and wrist. The pressure from the bulging extensor tendons will make the retinaculum more distinct.
2) Locate the head of the ulna and the styloid process of the radius.
3) Palpate just distal to these landmarks by sliding across the transverse fibers of the thin retinaculum. (3.70)

✔ *Are you distal to the head of the ulna and the styloid process of the radius? Can you distinguish superficial, transverse fibers?*

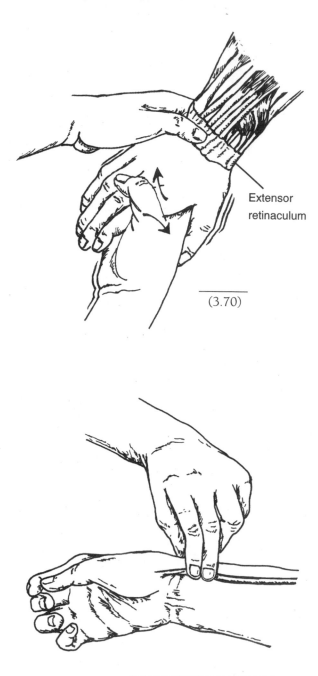

Extensor retinaculum

(3.70)

Radial and Ulnar Arteries

The radial and ulnar arteries branch off the brachial artery and travel down the forearm to the hand. The radial artery is often used to take the pulse. It is palpable on the anterior wrist between the tendon of flexor carpi radialis and the shaft of the radius.

The ulnar artery is found proximal to the pisiform and medial to palmaris longus tendon. Its pulse may not be as readily accessible as the radial pulse.

✋ 1) Palpate the radial pulse by placing two fingerpads on the flexor side of the wrist. Move laterally and gently press to feel the pulse. (3.71)
2) Locate the ulnar pulse by moving your fingerpads to the medial side of the flexor surface. (3.72)

(3.71) Feeling the pulse
of the radial artery

(3.72) Feeling the pulse
of the ulnar artery

Chapter
Spine & Thorax 4

Topographical Views

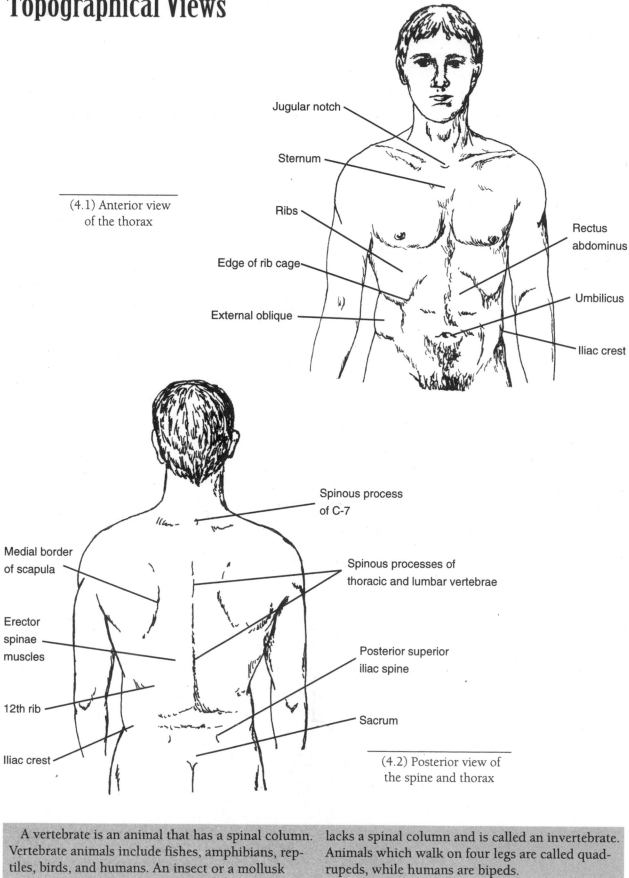

Jugular notch

Sternum

(4.1) Anterior view
of the thorax

Ribs

Rectus
abdominus

Edge of rib cage

Umbilicus

External oblique

Iliac crest

Spinous process
of C-7

Medial border
of scapula

Spinous processes of
thoracic and lumbar vertebrae

Erector
spinae
muscles

Posterior superior
iliac spine

12th rib

Sacrum

Iliac crest

(4.2) Posterior view of
the spine and thorax

A vertebrate is an animal that has a spinal column. Vertebrate animals include fishes, amphibians, reptiles, birds, and humans. An insect or a mollusk lacks a spinal column and is called an invertebrate. Animals which walk on four legs are called quadrupeds, while humans are bipeds.

Bones of the Spine and Thorax

The *vertebral column* (or spine) consists of twenty-four vertebrae: seven *cervical* in the neck, twelve *thoracic* of the thorax and five *lumbar* in the lower back. (4.3) The sacrum and coccyx, both composed of fused vertebrae, are also considered part of the vertebral column. For palpation and clarity purposes, the sacrum and coccyx are included in Chapter 6 - Pelvis and Thigh.

The *cervical vertebrae* are the most mobile and accessible vertebrae of the spine. The twelve *thoracic vertebrae* articulate with the twelve sets of ribs. Designed for minimal movement, the thoracic vertebrae help to stabilize the thoracic area and protect the internal organs. The large, stocky *lumbar vertebrae* support the weight of the upper body and are located between the twelfth rib and the posterior iliac crest.

When palpating along the back, all twenty-four vertebrae are deep to layers of muscle tissue. However, the spinous and transverse processes, which project out from each vertebrae, are helpful location guides.

The *thorax* includes the sternum and rib cage. The superficial *sternum* ("breast bone") is located along the midline of the chest. The *rib cage* consists of costal cartilage and twelve sets of ribs. The cartilage is shaped identically to the ribs and serves as a bridge between the ribs and sternum. Ribs 1 through 7 are known as "true ribs," since they attach directly to the sternum. Ribs 8 through 12 attach indirectly to the sternum by way of the costal cartilage and are therefore referred to as "false ribs." The eleventh and twelfth ribs, which do not attach to the sternum or cartilage, are also considered "floating ribs."

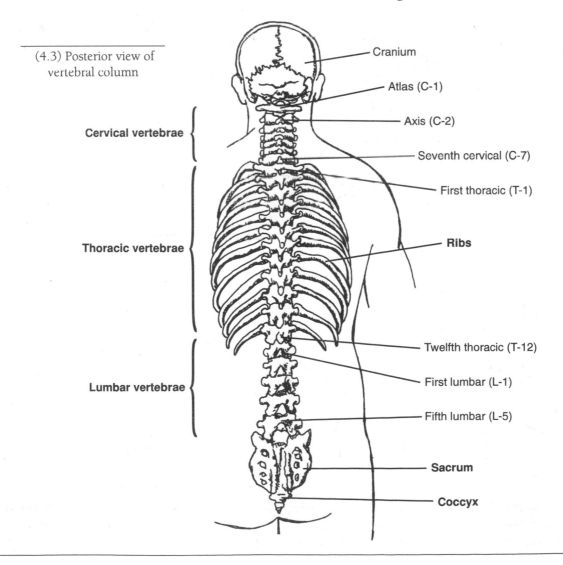

(4.3) Posterior view of vertebral column

Cervical vertebrae

Thoracic vertebrae

Lumbar vertebrae

Cranium

Atlas (C-1)

Axis (C-2)

Seventh cervical (C-7)

First thoracic (T-1)

Ribs

Twelfth thoracic (T-12)

First lumbar (L-1)

Fifth lumbar (L-5)

Sacrum

Coccyx

chest   AS. box
spine   L. thorn

Bony Landmarks

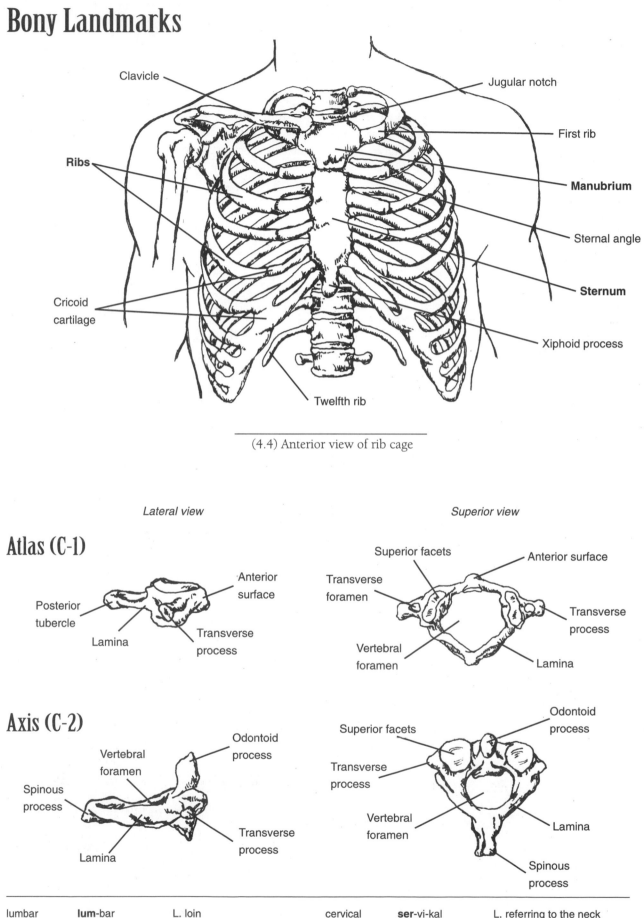

(4.4) Anterior view of rib cage

- Clavicle
- Jugular notch
- First rib
- **Ribs**
- **Manubrium**
- Sternal angle
- **Sternum**
- Cricoid cartilage
- Xiphoid process
- Twelfth rib

Atlas (C-1)

Lateral view

- Posterior tubercle
- Lamina
- Anterior surface
- Transverse process

Superior view

- Superior facets
- Anterior surface
- Transverse foramen
- Transverse process
- Vertebral foramen
- Lamina

Axis (C-2)

Lateral view

- Vertebral foramen
- Odontoid process
- Spinous process
- Transverse process
- Lamina

Superior view

- Odontoid process
- Superior facets
- Transverse process
- Vertebral foramen
- Lamina
- Spinous process

lumbar	**lum**-bar	L. loin	cervical	**ser**-vi-kal	L. referring to the neck
vertebra	**ver**-ta-bra	L. joint	thoracic	tho-**ras**-ik	Gr. chest

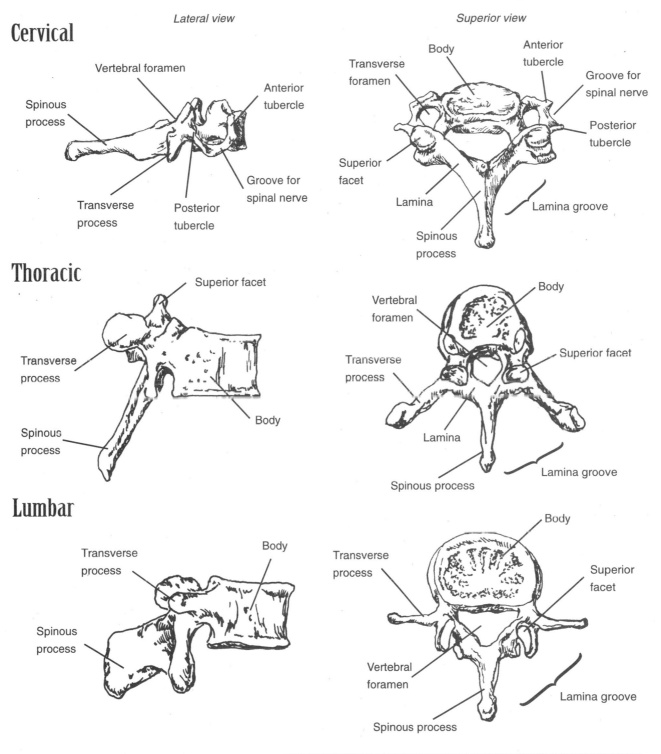

Cervical

Lateral view

Vertebral foramen

Spinous process

Anterior tubercle

Transverse process

Posterior tubercle

Groove for spinal nerve

Superior view

Body

Transverse foramen

Anterior tubercle

Groove for spinal nerve

Posterior tubercle

Superior facet

Lamina

Spinous process

Lamina groove

Thoracic

Superior facet

Transverse process

Spinous process

Body

Vertebral foramen

Body

Transverse process

Superior facet

Lamina

Spinous process

Lamina groove

Lumbar

Transverse process

Body

Spinous process

Body

Transverse process

Superior facet

Vertebral foramen

Spinous process

Lamina groove

When standing, the entire weight of the trunk, head and arms is transferred through the bodies of the vertebrae. The pressure is greatest on the lumbar vertebrae at the base of the spine.

Cushioning some of this shock are the intervertebral disks. Located between the bodies of each vertebrae, the disks are composed of a tough outer layer, the annulus fibrosis, and a liquid center called the nucleus pulposus.

Together these structures act as a shock absorber. When weight is placed on the disk, the annulus fibrosis supports the nucleus pulposus to compress and distribute the pressure.

The nucleus is mostly water and in the course of a day some of its gets squeezed out. During sleep, when pressure is off the spine, the disk fully regains itself and in the morning you wake up half an inch taller than you were the night before.

facet	**fas**-et	Fr. small face	odontoid	o-**don**-toyd	L. toothlike
foramen	for-**a**-men	L. a passage or opening			

Bony Landmark Trails of the Spine and Thorax

Trail 1 - "Midline Ridge" explores the spinous processes of the vertebrae, and the spaces between them, as they run down the middle of the back.

Trail 2 - "Crossing Paths" describes surrounding bony landmarks which intersect with specific spinous processes. (4.5)

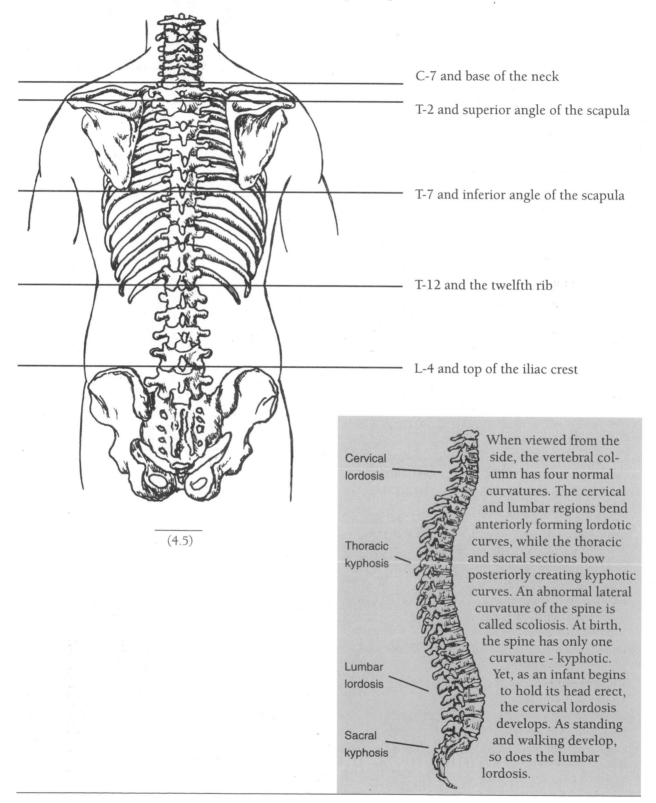

C-7 and base of the neck

T-2 and superior angle of the scapula

T-7 and inferior angle of the scapula

T-12 and the twelfth rib

L-4 and top of the iliac crest

(4.5)

Cervical lordosis

Thoracic kyphosis

Lumbar lordosis

Sacral kyphosis

When viewed from the side, the vertebral column has four normal curvatures. The cervical and lumbar regions bend anteriorly forming lordotic curves, while the thoracic and sacral sections bow posteriorly creating kyphotic curves. An abnormal lateral curvature of the spine is called scoliosis. At birth, the spine has only one curvature - kyphotic. Yet, as an infant begins to hold its head erect, the cervical lordosis develops. As standing and walking develop, so does the lumbar lordosis.

Trail 3 - "Nape Lane" locates the landmarks of the cervical vertebrae. (4.6)

   a) Spinous processes of cervicals
   b) Transverse processes of cervicals
   c) Lamina groove of cervicals

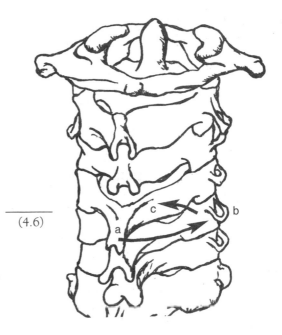

(4.6)

Trail 4 - "Buried Boulevard" delves into the middle and low back to locate landmarks of the thoracic and lumbar vertebrae. (4.7)

   a) Spinous processes
   b) Transverse processes
   c) Lamina groove

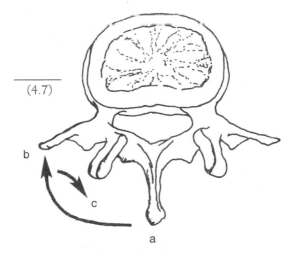

(4.7)

Trail 5 - "Breastbone Ridge" explores the sternum and its landmarks. (4.8)

   a) Jugular notch
   b) Manubrium
   c) Body of sternum
   d) Xiphoid process

Trail 6 - "One Bumpy Road" accesses the rib cage.

   Ribs and costal cartilage
   First rib
   Eleventh and twelfth ribs

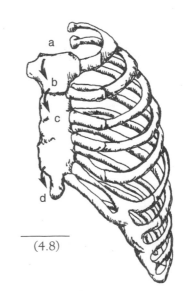

(4.8)

Trail 1 - "Midline Ridge"
Spinous Processes of the Vertebrae

A spinous process is a vertebrae's posterior projection. As a group, the processes form the visible line of bumps down the center of the back. They are designed as attachment sites for layers of muscles, ligaments and fascia.

The lumbar, thoracic and cervical spinous processes vary in several respects. The *spinous processes of the lumbar vertebrae* are much larger than the thoracic or cervical processes. Tall and stocky, the tips of the lumbar processes may feel more like short strips than points. The bodies of the lumbar vertebrae are quite massive and tall; the processes may have a finger width of space between them. The *thoracic spinous processes* angle downward and are smaller and closer together than the lumbar processes.

The *cervical spinous processes* are shorter and smaller than the thoracic processes. Because of the lordotic curve in the cervical spine and the overlying ligamentum nuchae, the cervical spinous processes are deeper than the thoracic and lumbar vertebrae. The first cervical (C-1) is the only vertebrae that does not have a spinous process.

1) Seated with the trunk and neck flexed (this will stretch the overlying tissues and allow the processes to be accessed easily). Place your fingers along the centerline of the back and locate the long line of processes. (4.9)

2) Slide your fingers slowly up and down the spine, palpating the different sizes, prominences and spaces between them. Some processes may present themselves immediately, others may be more difficult to find. Ask your partner to slowly flex and extend his spine and note the movements of the processes.

3) Try this same method with your partner prone.

✔ *Can you sculpt out the sides of the process you are palpating? Is there a dip superior and inferior to the point you feel? Can you line up three to four fingers on a series of processes or the spaces between the processes?*

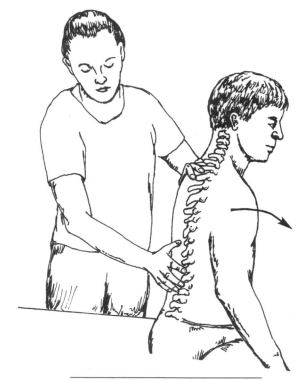

(4.9) Palpating the spinous processes and spaces between them

Prone or seated. Palpate the entire spinal column and count the spinous processes. How many can you feel? All vertebrae (except C-1) have spinous processes, making a total of twenty-three. Use the intersecting spinous processes such as C-7, T-12, L-4 to ensure your accuracy.

Trail 2 - "Crossing Paths"

Several spinous processes can be identified with the help of intersecting bony landmarks. For example, a line drawn between the tops of the iliac crests will cross the spinous process of L-4, which will lead you to the surrounding processes. Because each body is unique, these intersecting landmarks are not definitive and may be best used as helpful guides.

Top of the Iliac Crest and L-4

1) Prone or standing. Locate the lateral aspect of the iliac crests (p. 198).
2) With your index fingers along the top of the crests, slide your thumbs medial, meeting at the spine. (4.10)
3) Isolate the large knob of L-4. Explore superiorly and inferiorly for the surrounding lumbar processes.

✔ *Are you at the level of the iliac crests? Can you feel a firm protuberance at the centerline of the body?*

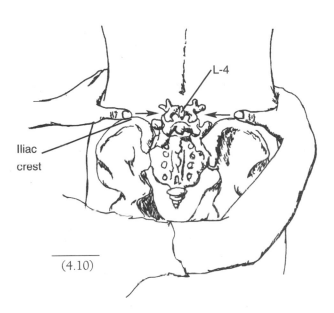

(4.10)

Twelfth Rib and T-12

The eleventh and twelfth ribs do not attach to the anterior costal cartilage and are considered "floating ribs." The twelfth rib has a slender, spear-like shape and angles inferiorly. The twelfth rib may vary between three to six inches in length and can be used as an indicator for the spinous process of T-12.

1) Prone. The strategy is to locate the tip of the twelfth rib and follow its shaft to the spinous process. Reaching across to the opposite side of the body, place your hand along the lateral side of the ribs.
2) Slide inferiorly to the bottom of the rib cage and explore for the tip of the twelfth rib. (4.11)
3) With the tip isolated, gently follow the shaft of the rib medially, noting how it lies at an angle. The rib lies deep to the erector spinae muscles, so you may lose contact with its most medial portion. Continue to slide your fingers where the shaft was headed and palpate for the spinous process.

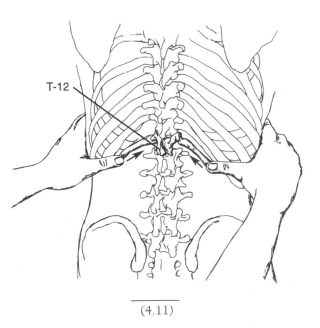

(4.11)

✔ *If you have located L-4, can you count the processes superiorly to T-12?*

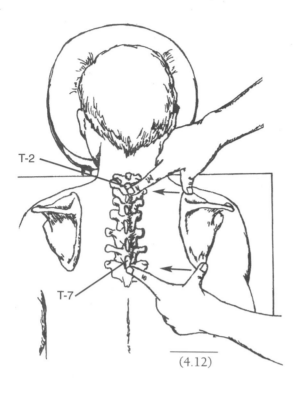

T-2

T-7

(4.12)

Inferior Angle and T-7
Superior Angle and T-2

Body type, muscular contraction and other factors can affect the position of the scapulas. The inferior angle of the scapula will lie generally at the level of the spinous process of T-7, while the superior angle is at the level of T-2.

👁✋ 1) Prone or standing. Locate the inferior angle. (p. 41) Keep one hand at the angle, while the other slides medially to the vertebral column.
2) Locate the superior angle. Keep one hand at the angle, while the other slides medially to the vertebrae. (4.12)

✔ *From T-7, can you count the processes inferiorly to T-12? Can you count superiorly to T-2? From T-2, can you count down to T-7?*

Base of the Neck and C-7

The spinous process of C-7 is located at the base of the neck. It protrudes further than C-6, C-5 and C-4, a helpful distinction when locating structures of the upper back and neck.

👁✋ 1) Prone or supine. Place your thumb pad superior to the base of the neck along the midline of the body.
2) Slide inferior. At the base of the neck, your thumb will bump into the process of C-7. (4.13)
3) Explore its edges and surrounding processes.

✔ *Are you at the base of the neck? Is the process superior to your finger smaller than the process you are palpating? Is there an equally protruding process (T-1) immediately inferior?*

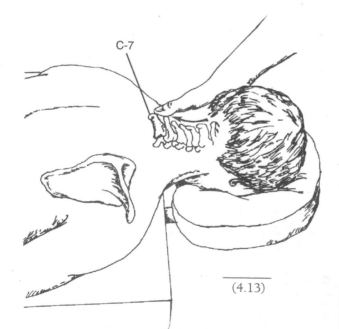

C-7

(4.13)

When flexing the neck, the spinous process of C-7 shifts superiorly. T-1, however, is buckled in by the first ribs and does not move.

👁✋ With your partner seated, place a finger on the spinous process C-7 and T-1. Have your partner slowly flex his neck and observe the movements at C-7 and T-1. Does C-7 tilt superiorly, while T-1 is stationary?

Trail 3 - "Nape Lane"
Spinous Processes of Cervicals

The spinous processes of C-6, C-5, C-4 and C-3 project posteriorly at approximately the same distance. C-2, however, has a larger, more distinct process. The tips of the cervical spinous processes are deep to the ligamentum nuchae (p. 153), a flat ligament attaching to the processes and running superiorly to the occiput (p. 162).

👁 1) Supine. Locate the spinous process of C-7.
2) Using gentle pressure, explore the tip and sides of the other cervical processes. (4.14) Strum transversely across the dense ligamentum nuchae that spans across the tips of the spinous processes.
3) Continue superiorly until you reach the prominence of C-2. Passively flex, extend and rotate the neck as you explore the spinous processes.

✔ *Can you feel the subtle ridge formed by the processes along the back of the neck? When exploring the spinous process of C-2, are you inferior to the level of the ear lobes? Is the process of C-2 larger and more pronounced than the other cervicals?*

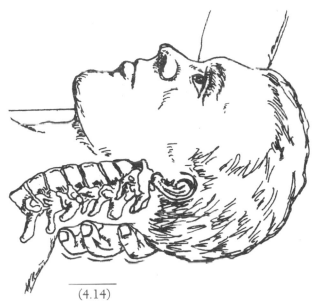

(4.14)

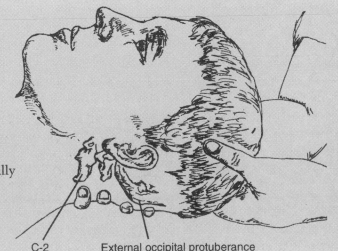

👁 Differentiating between the spinous process of C-2 and the external occipital protuberance (p. 163) can be helpful to navigate the posterior neck. Begin by laying your fingers horizontally along the base of your partner's head.

Place your ring finger at the external occipital protuberance, while your index finger locates the spinous process of C-2. Your middle finger will lie between these structures at the level of C-1. Explore the distance between these two prominent landmarks.

C-2   External occipital protuberance

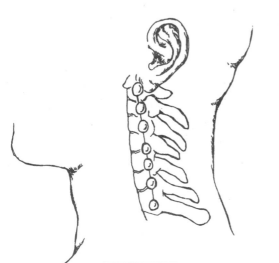

(4.15) The location of the cervical TVPs is similar to that of a long dangling earring

Compression or impingement of the brachial plexus (p. 188) or one of its nerves can create a sharp, shooting sensation down the arm. If this occurs, immediately release and adjust your position posteriorly. Also, ask your partner for feedback.

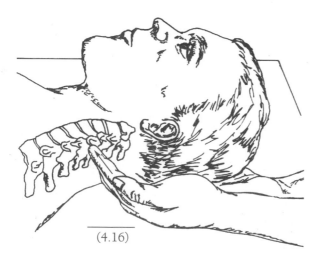

(4.16)

Transverse Processes of Cervicals

The transverse processes (TVPs) of the cervical vertebrae are located on the lateral side of the neck. (4.15) Old Hollywood films put Frankenstein's neck bolts into his TVPs!

The TVPs extend inferiorly from the mastoid process and many are deep to the sternocleido-mastoid muscle. All of the TVPs have the same width, except the TVPs of C-1, which are much wider. The TVPs of C-1 are located just distal and posterior to the tip of the mastoid process and are relatively accessible.

All of the TVPs serve as attachment sites for numerous muscles including the scalenes and levator scapula. The brachial plexus, a large group of nerves which innervate the arm, exit between the TVPs. When first palpating the TVPs, use the flat of your fingerpads. As your palpation skills improve, explore the TVPs' surfaces more specifically.

1) Supine or prone. Place your fingers below the ear lobes on the lateral side of the neck.
2) Using your flat fingerpads, slide anteriorly and posteriorly to feel the ridge of TVPs. Explore the length of the neck. (4.16)
3) You may not feel the tips of individual processes, but instead the ridge formed by the TVPs beneath the overlying tissue.

Are you palpating inferior to the ear lobe? Do you feel a subtle ridge running down the side of the neck? If you passively flex, laterally flex or rotate the neck, can you feel the TVPs move individually?

1) Supine. Rotate the head 45˚ to the left.
2) With the head in this position, the TVPs form a line from the mastoid process to the center of the shaft of the clavicle. Draw an imaginary line from these two landmarks, and visualize and palpate the TVPs along the path.

C-6 has a large anterior tubercle called the carotid tubercle. Its name refers to the carotid artery (p. 185) that passes immediately lateral to it. Not that you would want to do this, but by placing your finger lateral to the cricoid cartilage and pressing directly posterior, the carotid artery can be occluded against the carotid tubercle. Long ago such a dramatic maneuver was used in emergency rooms as a last ditch effort to stop a hemorrhage in the head.

👁 Transverse processes of C-1

1) Seated or supine. Locate the mastoid process of the temporal bone. (p. 164)
2) Using your broad fingerpads, slide half an inch inferiorly and half an inch posteriorly from the mastoid process. Explore deep to the sternocleido-mastoid muscle for the solid bump of the transverse process of C-1. (4.17) Pressing gently on these points may be uncomfortable for your partner.
3) For reference, locate both transverse processes.

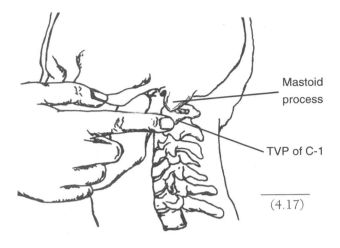

Mastoid process

TVP of C-1

(4.17)

Cervical Lamina Groove

The lamina groove is the trough-like space between the spinous and transverse processes of the vertebrae. Although it is a sizable channel on a skeleton, your partner's groove is filled with layers of muscles, rendering it mostly inaccessible. The lamina groove is best thought of as a helpful region for locating muscle bellies.

👁 1) Supine. Scoop the head with one hand, and with your other hand locate the cervical TVPs.
2) Slide posteriorly off the TVPs. Explore the space between the transverse and spinous processes - the lamina groove of the cervical vertebrae. (4.18) Again, since the groove is filled with muscles, the bone which forms the groove is impalpable.

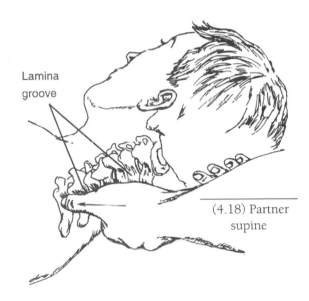

Lamina groove

(4.18) Partner supine

Some bony landmarks serve as attachment sites for numerous tendons and connective tissues. Whether palpating with your fingers or dissecting with a scalpel, these tissues are often very difficult to distinguish. The TVPs of the cervical vertebrae are a case in point: tendons come from several directions to attach to the TVPs and, to complicate matters, nerves pass between the tendons.

To coordinate the tendons and the spinal nerves of the cervical and brachial plexus,

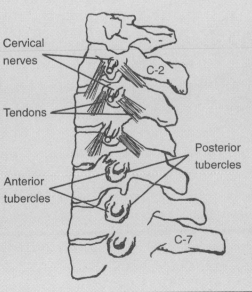

Cervical nerves

C-2

Tendons

Posterior tubercles

Anterior tubercles

C-7

the TVPs of C-2 through C-7 have anterior and posterior tubercles (right). The tubercles are small tips situated on either side of the sulcus (or canal) that carries the cervical nerves.

The anterior tubercle is an attachment site for the scalenes and other anterior muscles. Levator scapula, splenius cervicis and other posterior muscles attach to the posterior tubercles. It can be initially difficult to palpate the specific tubercles, but with experience the tubercles become detectable.

lamina   **lam**-i-na   L. thin plate

Trail 4 - "Buried Boulevard"
Transverse Processes of Thoracic and Lumbar Vertebrae

The TVPs of the thoracic vertebrae are shorter and do not extend as far laterally as the TVPs of the lumbar vertebrae. They are palpable deep to the erector spinae muscles and superficial to the connecting aspect of the ribs.

The TVPs of the lumbar vertebrae are located deep to the erector spinae muscles. Extending an inch or two laterally, their solid presence can be felt beneath the overlying muscles.

Thoracic transverse processes

1) Prone. Locate a portion of the thoracic spinous processes.
2) Move roughly one inch laterally and sink your fingers through the thick erector muscles.
3) Roll your fingers superiorly and inferiorly, palpating and exploring for their subtle, knobby shape. (4.19)

✔ *Slide further laterally to the thoracic transverse processes onto the posterior ribs. Can you determine where the ribs and transverse processes meet? Can you feel the short processes beneath the erector spinae fibers?*

Lumbar transverse processes

1) Prone. Locate the lumbar spinous processes.
2) Slide roughly two inches laterally, to avoid the thick mound of the erector spinae (p. 138).
3) Slowly sink your fingers through the muscle tissue. Direct your pressure at a medial angle (as if toward the navel) and explore for the tips of the TVPs. (4.20) Because of the thick overlying tissue, the individual processes may not be directly palpable, but try to sense the solid ridge they form.

✔ *Ask your partner to raise her feet slightly and determine if you are lateral to the erector spinae muscles. Can you feel the hard surface of the processes running horizontally?*

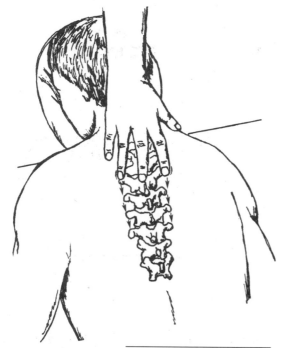

(4.19) Palpating the TVPs of thoracic vertebrae

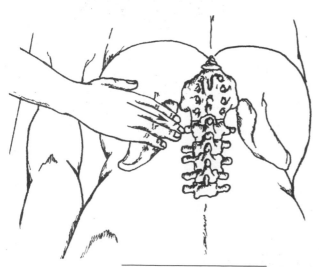

(4.20) Palpating the TVPs of lumbar vertebrae

Lamina Groove of the Thoracic and Lumbar Vertebrae

The lamina groove of the thoracic and lumbar vertebrae is located between their spinous and transverse processes. Shaped like a long, vertical trough, the lamina groove expands in depth and width as it progresses down the spine. In the thoracic and lumbar vertebrae the lamina groove is filled with layers of the transversospinalis and erector spinae muscles (4.22). Because of this overlying tissue, the lamina groove is difficult to access directly, but its borders (the spinous and transverse processes) are palpable.

👁 1) Prone. Locate the spinous processes of the thoracic vertebrae. With the other hand, locate the TVPs of the thoracic vertebrae.
2) Using firm pressure, explore between these landmarks into the lamina groove. (4.21) Note the thick muscle tissue lying in the groove.
3) Try this same method in the lumbar region. (4.23) Observe how the lamina groove widens and deepens and the muscle tissue thickens compared to the thoracic region.

✔ *Are you between the transverse and spinous processes of the vertebrae? Can you slide your fingers between the muscle fibers and sink into the lamina groove?*

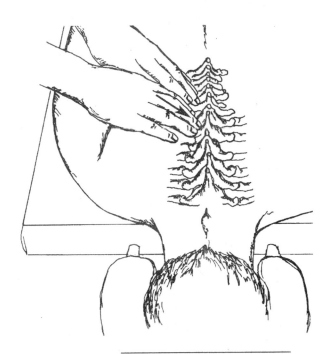

(4.21) Palpating the lamina groove of thoracic vertebrae

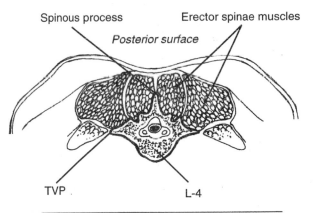

Spinous process   Erector spinae muscles

Posterior surface

TVP   L-4

(4.22) Cross-section of spine at level of L-4

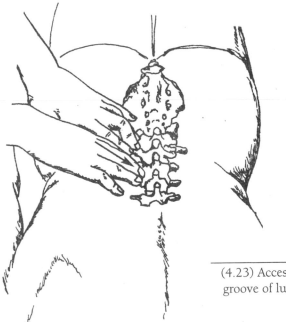

(4.23) Accessing the lamina groove of lumbar vertebrae

The jugular notch, sternal angle and xiphoid process can be helpful guideposts to the vertebral column. The jugular notch lies on the same transverse plane as the spinous process of T-2 (a). The sternal angle lines up with the spinous process of T-4 (b), while the xiphoid process is directly across from the body of T-10 (c). Of course, many factors such as posture and body type will effect the placement of the ribs, so use this as a guide only.

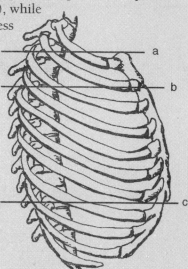

👁 1) Partner standing. Stand beside your partner. Palpate the jugular notch with one hand while the other locates the spinous process of T-2. 2) Note if you can see or feel a difference in the level of these landmarks. Follow the same procedure for T-4 and T-12.

Trail 5 - "Breastbone Ridge"
Sternum

The sternum has several landmarks. (4.4) Located at the top of the sternum, the *jugular notch* is between the sternal heads of the clavicles. It may be flat or bowl-shaped, and although no muscles attach to it directly, the sternocleidomastoids pass superficially and the infrahyoids attach deep to it.

The *manubrium*, the superior portion of the sternum, articulates with the clavicles, the first rib and second rib. The *body* of the sternum is located inferior to the manubrium and forms the majority of the sternum. The junction between the manubrium and body of the sternum is the *sternal angle*.

Extending off the bottom of the sternum, the *xiphoid process* can be an inch in length or completely absent. It is an attachment for the abdominal aponeurosis.

The manubrium, body and xiphoid process of the sternum are superficial, covered only by fascia and pectoralis major tendon.

👁 Jugular notch, manubrium and sternum

1) Supine. Place your finger at the center of your partner's chest, upon the sternum.
2) Slide superiorly until you reach the jugular notch at the top of the sternum. Explore the notch and its location next to the sternoclavicular joints.
3) Move your fingers inferiorly off the jugular notch onto the manubrium and body of the sternum. Explore any crevices or hills upon this "flat" bone. Also, palpate laterally to its attachments with the costal cartilage.

👁 Xiphoid process

Slide your fingers inferiorly until they drop off the sternum and fall into the muscles of the abdomen. Backtrack to the most inferior tip of the sternum - the xiphoid process. (4.24) Gently sculpt its tip.

✔ *Are you at the most inferior point of the sternum?*

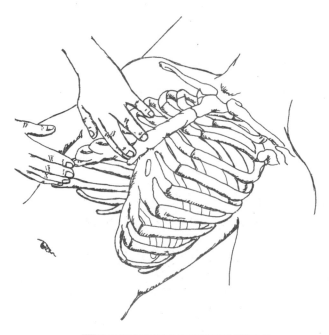

(4.24) Palpating the xiphoid process and the edge of the ribs

| abdomen | **ab**-do-men | L. belly | | cricoid | **kri**-koyd | Gr. ring-shaped |
| costal | **kos**-tal | L. rib | | jugular | **jug**-u-lar | L. throat |

Trail 6 - "One Bumpy Road"
Ribs and Costal Cartilage

The ribs articulate posteriorly with the thoracic vertebrae and then curve around the thorax to the anterior chest. (4.25) Extending off the ribs is the costal cartilage which attaches to the sternum. There are six or seven costal branches, all of which are identical in shape and feel to the ribs. Together, the ribs and costal cartilage run at a variety of angles around the trunk.

The entire rib cage is deep to muscle tissue. However, the ribs along the sides of the trunk are easily accessed. The spaces between the ribs and cartilage are filled with thin intercostal muscles which can be easily palpated.

As you explore the thorax, avoid accessing mammary (breast) tissue. Ask your partner, male or female, if you may palpate the surrounding areas.

👁✋ 1) Supine. Slide laterally from the sternum onto the costal cartilage. Use your fingertips to locate one costal branch and palpate its rounded surface.
2) Roll off the cartilage into the space between the branches. Explore this groove as it extends laterally.
3) Continue along the length of the sternum, locating and exploring each branch of cartilage and rib and the spaces between them. (4.26)

✔ *Can you determine how the angle of the ribs changes as you move around the body? Can you differentiate between the round shaft of the ribs and the ditch-like spaces between them? Ask you partner to breathe deeply and note the spatial changes you feel between the ribs.*

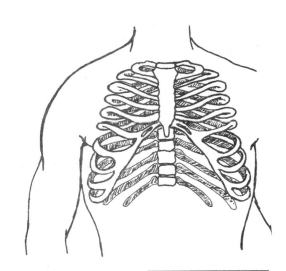

(4.25) Position of the ribs during inhalation and exhalation

👁✋ The sternal angle is the junction point between the manubrium and sternum. Stretching horizontally, it may feel like a small speed bump or a dip. Locate the jugular notch and glide inferiorly along the surface of the manubrium. Within an inch or two, palpate for a ridge or ditch stretching horizontally across the sternum.

✔ *The second rib attaches to the sternum at the level of the sternal angle. Slide your fingers laterally off the angle. Can you feel the round surface of the second rib?*

(4.26) Exploring the ribs

manubrium	ma-**nu**-bre-um	L. handle
sternum	**ster**-num	Gr. chest

thorax	**tho**-raks	Gr. chest
xiphoid	**zif**-oyd	Gr. sword-shaped

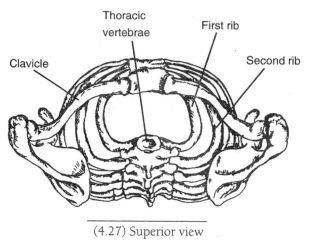

(4.27) Superior view

Clavicle
Thoracic vertebrae
First rib
Second rib

Compression or impingement of the brachial plexus or one of its nerves can create a sharp, shooting sensation down the arm. If this occurs, adjust your position to one side. Also, ask your partner for feedback.

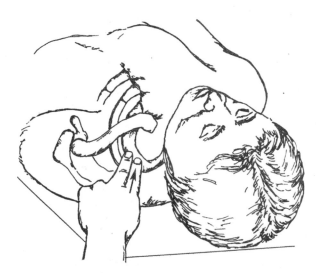

(4.28) Partner supine, palpating the first rib

First Rib

Unlike its cohorts, the first rib is difficult to isolate along the anterior chest. It lies directly beneath the clavicle and then quickly curves posteriorly toward the back. (4.27) It can, however, be accessed in the posterior triangle (p. 158) of the neck which is formed by the clavicle and the sternocleidomastoid and trapezius muscles.

The scalene muscles (p. 173) fan across the posterior triangle and attach to the first and second ribs. To access the first rib, you must palpate through the scalenes. The brachial plexus and subclavian artery (p. 188) pass between the first rib and the clavicle.

1) Supine. (Soften the overlying tissue by passively elevating your partner's shoulder.)
2) Locate the clavicle and upper flap of the trapezius to identify the posterior triangle. Place two fingerpads between these structures.
3) Slowly sink into the tissue of the scalene muscles, directing your fingers straight inferior toward the feet. (4.28) You will meet the solid resistance of the first rib's shaft as your fingers sink into the tissue.

✔ *Differentiate between the ribs and the superior angle of the scapula by locating where you believe the rib to be. Passively elevate and depress the scapula. If the structure your are palpating moves, it is the superior angle. Also, ask you partner to take a slow, deep breath into the upper chest. Do you feel the rib rise?*

Discrepancies in the number of ribs is not unusual. There are ordinarily twelve pairs of ribs, but eleven or thirteen pairs can be found on some people. If an extra rib is present, it may be bilateral or unilateral and found either in the cervical or lumbar areas.

A cervical rib often articulates with C-7 and can be felt in the posterior triangle region of the neck at the level of the clavicle. An extra rib in the lumbar will extend off L-1.

Eleventh and Twelfth Ribs

The eleventh and twelfth ribs are called "floating ribs" because they do not attach to the anterior costal cartilage. Both ribs have a slender, spear-like shape and lie at roughly a 45° angle. Their medial portions lie deep to the thick erector spinae muscles; however, their lateral aspects and tips are palpable.

The eleventh rib is six to eight inches in length and extends halfway around the body. The twelfth rib may vary from three to six inches in length. Since abnormalities as to the length and number of ribs are not uncommon, your partner's ribs may not match this description.

1) Prone. Reaching across to the body's opposite side, place your hand along the lateral side of the ribs.
2) Slide inferiorly to the bottom of the rib cage, allowing your hand to slide into the soft abdominal tissue.
3) Compressing your fingerpads into side of the thorax, explore this region for the tips of the eleventh and twelfth ribs. (4.29)
4) With the tips isolated, gently follow the shaft of the rib medially, noting how they run at an angle.

Can you feel two tips - one more lateral to the other? Ask your partner to take a slow, deep breath and note if the tips or bodies of the ribs press into your hand.

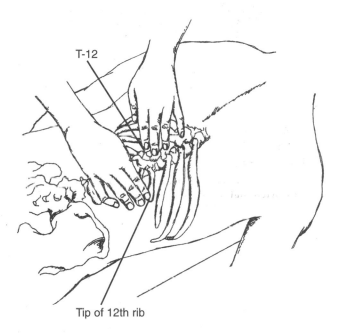

T-12

Tip of 12th rib

(4.29) Partner prone

Ideally, the ribs are designed to expand in three dimensions during inhalation: anterior/posterior, lateral and superior. Yet, for reasons ranging from posture to emotional trauma, few people truly breathe in this manner. Often the breath becomes limited to a portion of the thorax and the ribs move in only one or two dimensions.

With this in mind, ask your partner to stand and breathe normally. Observe any changes in the shape or movement of the thorax, shoulders and abdomen. Then lay your hands on all sides of the rib cage and feel for any activity of the thorax. Do the ribs move in all three directions? Do some ribs move individually?

Ask your partner to inhale deeply and exhale fully. Explore the ribs and anterior neck muscles (scalenes and sternocleidomastoid) during inhalation. These muscles will tighten to elevate the upper ribs. Try these exercises with your partner seated, prone and supine.

Muscles of the Spine and Thorax

The following muscles are situated along the back and abdominal regions and create movement of the vertebral column and rib cage. (4.30-4.35)

The muscles of the spine have a unique arrangement. Unlike the limb muscles, which can often be individually distinguished, the spine muscles are composed of numerous bands of muscle fibers which have a dense, interwoven design. For this reason, it can be challenging to isolate a specific portion of these muscles.

The muscles of the spine may be divided into small, individual sections or combined into a few major groups. For our purposes, the muscles of the spine will be presented in four groups:

1) The large *erector spinae group* is the most superficial of the spinal muscles and has three major branches.

2) The smaller *transversospinalis group* also has three branches and lies deep to the erectors.

3) The two *splenius* muscles are located along the posterior neck.

4) The eight short *suboccipitals* are the deepest muscles. They are located at the base of the head.

Other muscles which affect the thorax, notably the scalenes and sternocleidomastoid, are included in Chapter 5 - Head, Neck and Face.

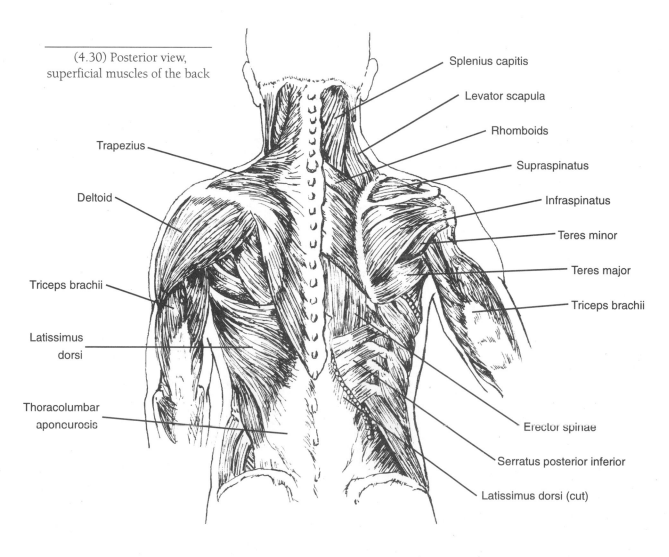

(4.30) Posterior view, superficial muscles of the back

Splenius capitis

Levator scapula

Rhomboids

Supraspinatus

Infraspinatus

Teres minor

Teres major

Triceps brachii

Trapezius

Deltoid

Triceps brachii

Latissimus dorsi

Thoracolumbar aponeurosis

Erector spinae

Serratus posterior inferior

Latissimus dorsi (cut)

Latissimus dorsi, trapezius and deltoid are removed on right side

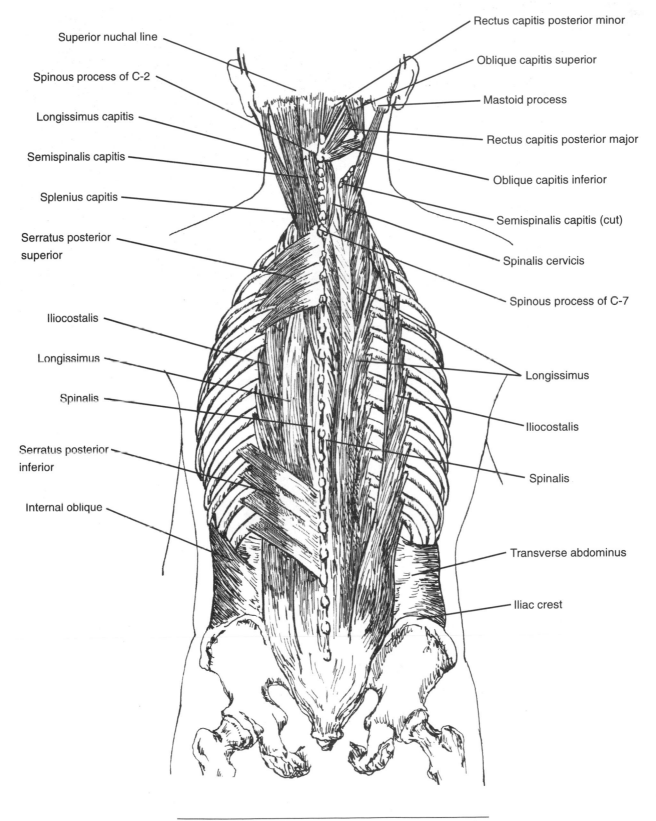

Superior nuchal line

Spinous process of C-2

Longissimus capitis

Semispinalis capitis

Splenius capitis

Serratus posterior
superior

Iliocostalis

Longissimus

Spinalis

Serratus posterior
inferior

Internal oblique

Rectus capitis posterior minor

Oblique capitis superior

Mastoid process

Rectus capitis posterior major

Oblique capitis inferior

Semispinalis capitis (cut)

Spinalis cervicis

Spinous process of C-7

Longissimus

Iliocostalis

Spinalis

Transverse abdominus

Iliac crest

(4.31) Posterior view, intermediate muscles of the back

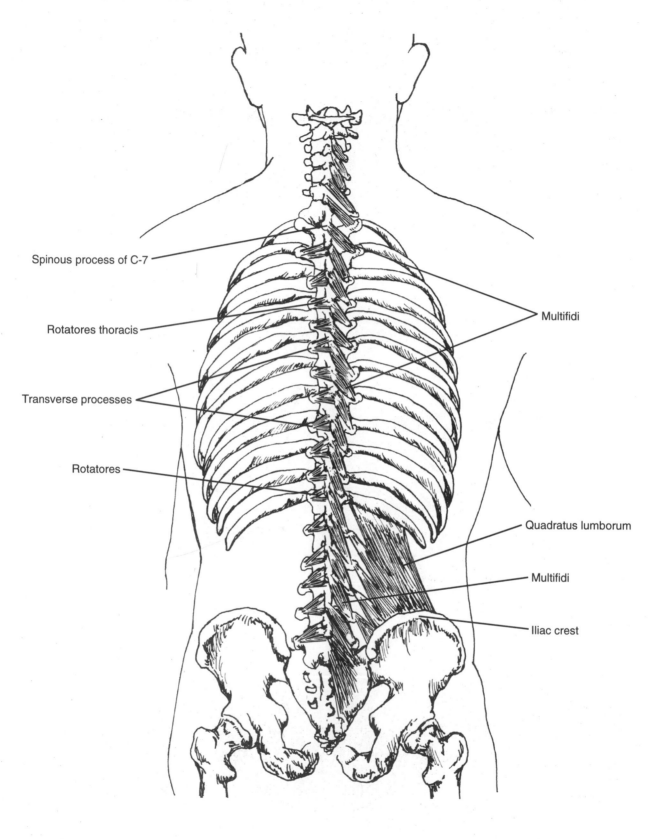

Spinous process of C-7

Rotatores thoracis

Transverse processes

Rotatores

Multifidi

Quadratus lumborum

Multifidi

Iliac crest

(4.32) Posterior view, deep muscles of the back

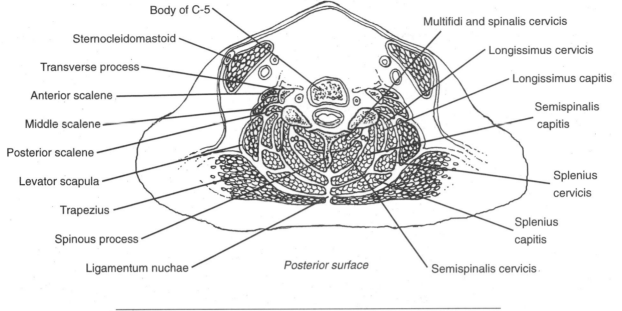

Body of C-5

Sternocleidomastoid

Transverse process

Anterior scalene

Middle scalene

Posterior scalene

Levator scapula

Trapezius

Spinous process

Ligamentum nuchae

Multifidi and spinalis cervicis

Longissimus cervicis

Longissimus capitis

Semispinalis capitis

Splenius cervicis

Splenius capitis

Semispinalis cervicis

Posterior surface

(4.33) Cross-section of the neck at the level of fifth cervical vertebrae

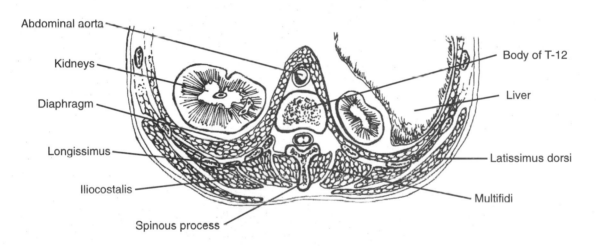

Abdominal aorta

Kidneys

Diaphragm

Longissimus

Iliocostalis

Spinous process

Body of T-12

Liver

Latissimus dorsi

Multifidi

(4.34) Cross-section of the thorax at the level of twelfth thoracic vertebrae

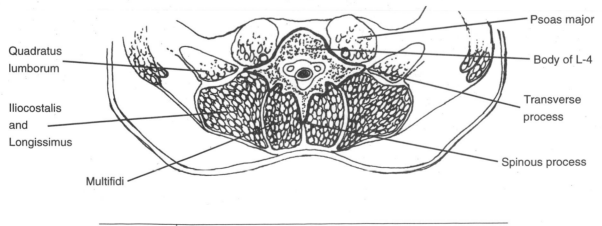

Quadratus lumborum

Iliocostalis and Longissimus

Multifidi

Psoas major

Body of L-4

Transverse process

Spinous process

(4.35) Cross-section of the thorax at the level of fourth lumbar vertebrae

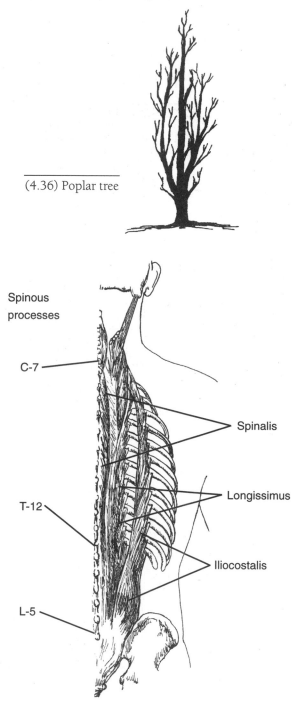

(4.36) Poplar tree

Spinous
processes

C-7

Spinalis

Longissimus

T-12

Iliocostalis

L-5

(4.37) Posterior view of right side
showing erector spinae group

Erector Spinae Group

Spinalis
Longissimus
Iliocostalis

The erector spinae group spans from the sacrum to the occiput along the posterior aspect of the vertebral column. It has an immense, layered arrangement which can be difficult to comprehend. Simply stated, the erector spinae is like a tall poplar tree (4.36) with three main branches, the spinalis, longissimus and iliocostalis. (4.37) These branches can be divided into numerous, smaller branches such as longissimus thoracis, longissimus cervicis and longissimus capitis.

The *spinalis* is the smallest of the three and lies closest to the spine in the lamina groove. (4.38) The thick *longissimus* and lateral *iliocostalis* form a visible mound alongside the lumbar and thoracic spine. (4.39, 4.40) The long tendons of iliocostalis extend laterally beneath the scapula.

In the lumbar region, the erectors lie deep to the thin but dense thoracolumbar aponeurosis. In the thoracic and cervical areas, they are deep to the trapezius, the rhomboids, and the serratus posterior superior and serratus posterior inferior. As a group, the erectors are easily palpated along the entire back and neck. However, specifically locating a particular branch of the erectors can be challenging.

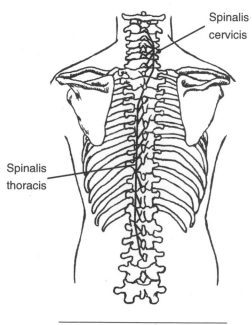

Spinalis
cervicis

Spinalis
thoracis

(4.38) Branches of the spinalis

Erector Spinae Group

A -
Bilaterally:
  Extend the vertebral column
Unilaterally:
  Laterally flex vertebral column to the same side

O - Common tendon (thoracolumbar aponeurosis) that attaches to the posterior surface of sacrum, iliac crest, spinous processes of the lumbar and last two thoracic vertebrae

I - Various attachments at the posterior ribs, spinous and transverse processes of thoracic and cervical vertebrae, mastoid process of temporal bone

N - Dorsal primary divisions of spinal nerves

Branches of the Erector Spinae Group

Spinalis

O - Spinous processes of the upper lumbar and lower thoracic vertebrae, ligamentum nuchae and spinous process of C-7

I - Spinous processes of upper thoracic and cervicals, except C-1

Longissimus

O - Common tendon, transverse processes of upper five thoracic vertebrae

I - Lower nine ribs, transverse processes of thoracic and cervical vertebrae, mastoid process of temporal bone

Iliocostalis

O - Common tendon, posterior surface of ribs 1-12

I - Posterior surface of ribs 1-12

The upper fibers of longissimus and iliocostalis (longissimus cervicis and capitis, iliocostalis cervicis) assist in lateral flexion and rotation of the neck and head.

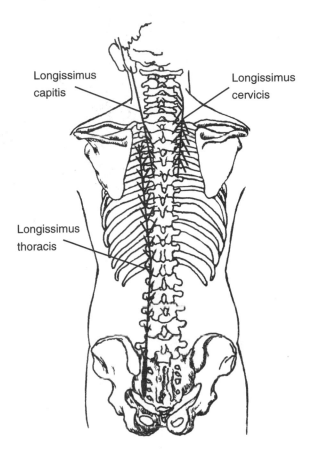

(4.39) Branches of the longissimus

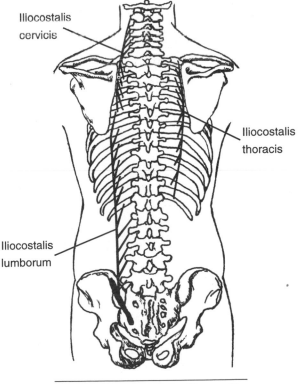

(4.40) Branches of the iliocostalis

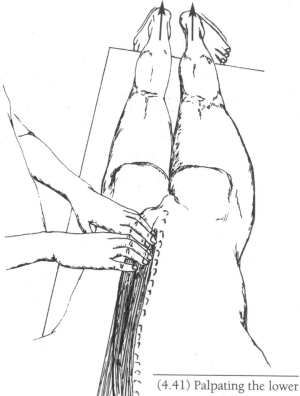

(4.41) Palpating the lower erectors while partner raises feet

Location - Alongside the spine
BLMs- Spinous processes of all vertebrae
Action - "Extend your spine" or
"raise your feet slightly"

👁 Erector spinae group

1) Prone. Lay both hands along either side of the lumbar vertebrae. Locate the region of the lower erectors by asking your partner to alternately raise and lower his feet slightly. The erectors, of course, do not raise the feet, but contract to stabilize the pelvis.
2) Notice the strong, rounded erector fibers tighten and relax with this action. (4.41)
3) As your partner continues this action, palpate inferiorly onto the sacrum, and then superiorly along the thoracic vertebrae. Your partner could raise his feet or extend his spine and neck slightly to contract the erectors in the thoracic region. (4.42)
4) Follow the ropy fibers of the erectors between the scapulas and along the back of the neck. These fibers are smallest in the cervical region, and are situated primarily lateral to the lamina groove.
5) With your partner relaxed, sink your fingers into the erector fibers and feel their ropy texture and vertical fiber direction.

✔ *Is the tissue you are palpating directly beside the spinous processes of the vertebrae? Is the fiber direction parallel with the spine?*
When the muscles are contracted, can you locate the lateral edge of the erector group? Between the scapulas, can you distinguish between the fiber direction of the middle trapezius, rhomboids and erectors?

(4.42) Palpating the upper erectors while partner arches his back

Transversospinalis Group

Multifidi
Rotatores
Semispinalis

Deep to the erector spinae muscle group is the transversospinalis muscle group. The transversospinalis is composed of three branches - multifidi, rotatores and semispinalis - and extends the length of the vertebral column. Unlike the long, vertical erector fibers, the branches of the transversospinalis have many short, diagonal fibers. These fibers form an intricate stitch-like design to link together the vertebrae. The name "transversospinalis" refers to its fibers which extend at varying lengths from transverse and spinous processes of the vertebrae.

The *multifidi* are surprisingly thick, and are directly accessible in the lumbar spine. They are the only muscles whose fibers lie across the posterior surface of the sacrum. The short, smaller *rotatores* lie deep to the multifidi. (4.43) The *semispinalis* are located along the thoracic and cervical vertebrae and ultimately reach the cranium. (4.44)

It can be difficult to isolate the individual bellies of the transversospinalis since they are crowded together and interwoven. However, as a group, its mass or density can be felt along the lamina groove of the thoracic and lumbar vertebrae.

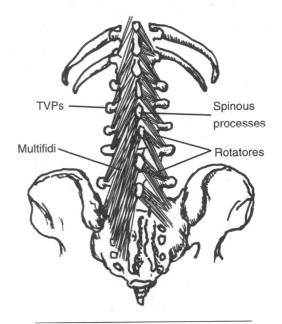

(4.43) Posterior view of the multifidi and rotatores in lumbar and sacral area

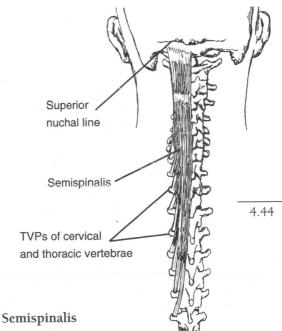

Multifidi and Rotatores

A-
Extend
Rotate the vertebral column to the opposite side

O - Sacrum and transverse processes of lumbar through cervical vertebrae

I - Spinous processes of lumbar vertebrae through second cervical vertebrae
(Multifidi span two to four vertebrae)
(Rotatores span one to two vertebrae)

N - Dorsal primary divisions of spinal nerves

Semispinalis

A -
Extend vertebral column and head

O - Transverse processes of thoracic vertebrae, articular processes of lower cervicals

I - Spinous processes of upper thoracic, cervicals (except C-1) and superior nuchal line of occiput

N - Dorsal primary divisions of spinal nerves

Location - Deep to erectors, alongside spine
BLMs - Spinous and transverse processes
Action - "Extend and/or rotate your spine"

---

multifidi	mul-**tif**-i-di	L. *fidi*, to split	semispinalis **sem**-e-spi-**na**-lis
rotatores	**ro**-ta-**tor**-ays	pl. for rotators	

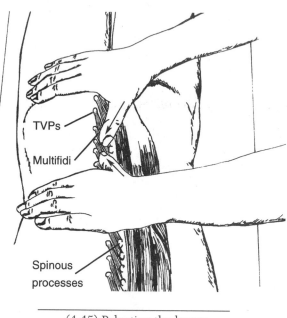

TVPs

Multifidi

Spinous
processes

(4.45) Palpating the lower
multifidi

1) Prone. Locate the spinous processes of the lumbar vertebrae.
2) Slide your fingers laterally off the spinous processes, sinking between them and the erector spinae fibers.
3) Pushing the erectors out of the way laterally, explore deeply for the density and diagonal fibers of the multifidi. (4.45) Progress inferiorly to the sacrum. Explore the multifidi's thin fibers as they lie on the surface of the sacrum.
4) Move superiorly, exploring the lamina groove of the thoracic and cervical areas. Turn your partner supine and palpate the cervical region.

✔ *Are you between the spinous and transverse processes? Can you get a sense of these smaller, deeper fibers stretching at an oblique angle?*

Splenius Capitis and Cervicis

The long splenius capitis and splenius cervicis muscles are located along the upper back and posterior neck. (4.46, 4.47) In contrast to the other back muscles which run parallel to the spine, the splenii fibers run obliquely. The *splenius capitis* is deep to the trapezius and rhomboids. Its fibers angle toward the mastoid process and are superficial between the trapezius and SCM.

The *splenius cervicis* is deep to the splenius capitis and not as easily isolated. However, its general location can be outlined in the lamina groove of the upper thoracic and cervical spine.

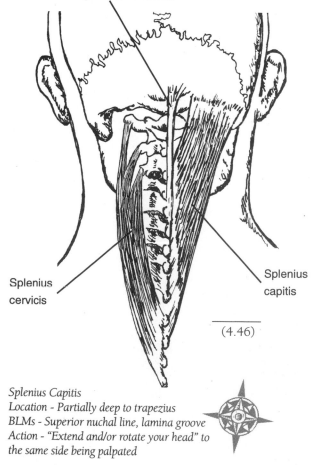

Ligamentum nuchae

Splenius
cervicis

Splenius
capitis

(4.46)

Splenius Capitis
Location - Partially deep to trapezius
BLMs - Superior nuchal line, lamina groove
Action - "Extend and/or rotate your head" to the same side being palpated

A -
Unilateral:
    Rotate the head to the same side
    Laterally flex the head and neck
Bilateral:
    Extend the head

O -
Capitis: Ligamentum nuchae, spinous processes of C-7 to T-3
Cervicis: Spinous processes of T-3 to T-6

I - Capitis: Mastoid process, lateral nuchal line
Cervicis: Transverse processes of the upper cervical vertebrae

N - Branches of dorsal division of cervical

👁 Splenius capitis

1) Prone. Locate the upper fibers of the trapezius.
2) Isolate the lateral edge of the trapezius by having your partner extend his head slightly.
3) Ask your partner to relax. Palpate just lateral to the trapezius for the splenius capitis' oblique fibers, following it up to the mastoid process and inferiorly through the trapezius. (4.48)

✔ *Do the fibers you feel lead toward the mastoid process? Distinguish between the trapezius fibers and the splenius capitis fibers by asking your partner to rotate his head slightly toward the side you are palpating. Do you feel these oblique fibers contract as the trapezius lies passively?*

◈ Locate the mastoid process and slide medially and inferiorly onto the superficial capitis fibers.

👁 Both splenii muscles

1) Supine. Cradle the head with one hand, while the other hand locates the lamina groove of the upper thoracic and cervical vertebrae.
2) Passively extend the neck slightly to shorten the tissue and palpate through the overlying trapezius fibers. (4.49) These bellies will not be particularly distinct or clear. However, the density of both splenii will be felt in the lamina groove.

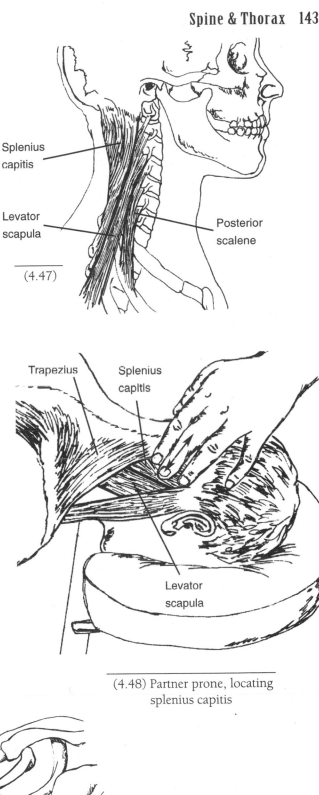

(4.47)

Splenius capitis

Levator scapula

Posterior scalene

Trapezius

Splenius capitis

Levator scapula

(4.48) Partner prone, locating splenius capitis

(4.49) Partner supine

Spinous process of T-3

Splenius capitis

| capitis | **kap**-i-tis | L. referring to the head | splenius | **sple**-ne-us | Gr. bandage |
| cervicis | ser-**vi**-sis | L. neck | | | |

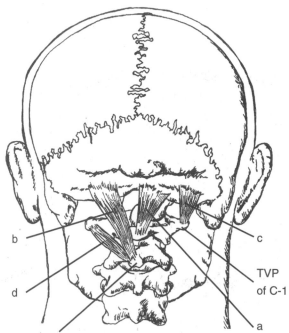

Spinous process of C-2

b

d

c

TVP
of C-1

a

(4.50) Posterior view

In 1995, researchers discovered that the rectus capitis posterior minor not only attaches to the occiput but also to the dura mater, the connective tissue which surrounds the spinal cord and brain. It is suspected that this connection could cause headache pain by affecting cerebrospinal fluid fluctuations and normal functioning of the vertebral artery and suboccipital nerve.

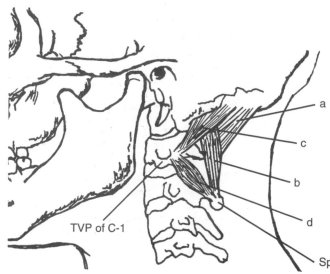

TVP of C-1

a

c

b

d

Spinous process of C-2

(4.51) Lateral view

Suboccipital Muscles

a) Rectus Capitis Posterior Minor
b) Rectus Capitis Posterior Major
c) Oblique Capitis Superior
d) Oblique Capitis Inferior

The eight small suboccipitals are the deepest muscles of the upper, posterior neck. (4.50) They are involved in stabilizing the axis and atlas, as well as creating intrinsic movements such as rocking and tilting of the head. To outline the suboccipitals' location, find the spinous process of C-2, the transverse process of C-1, and the space between the superior nuchal line of the occiput and C-2. (4.51)

The upper fibers of the trapezius can be a valuable marker as well; its lateral edge is the same width as the suboccipitals. The density of the suboccipital bellies can certainly be felt, but accessing specific muscle bellies can be challenging.

All Suboccipitals:

A -

All, except Oblique Inferior:
    Rock and tilt head back into extension
Rectus Capitis Major and Oblique Inferior:
    Rotate the head to the same side

Rectus Capitis Posterior Major

O - Spinous process of the axis

I - Inferior nuchal line of the occiput

N - Suboccipital

Rectus Capitis Posterior Minor

O - Tubercle of the posterior arch
    of the atlas

I - Inferior nuchal line of the occiput

N - Suboccipital

dura mater   **dyoo**-ra **ma**-ter   L. tough mother
occiput   **ok**-si-put   L. the back of the skull

Oblique Capitis Superior

O - Transverse process of the atlas

I - Between the nuchal lines of the occiput

N - Suboccipital

Oblique Capitis Inferior

O - Spinous process of the axis

I - Transverse process of the atlas

N - Suboccipital

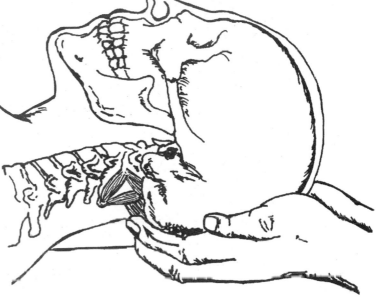

(4.52) Partner supine, curling the fingers under the occiput

1) Supine. Cradle the head in both hands. Passively extending the neck slightly will soften the overlying tissue. Locate the superior nuchal line of the occiput and the spinous process of C-2. Stretching between these landmarks are the suboccipitals.

2) Cradle the head with one hand, while two fingertips of the other hand slowly palpate through the trapezius, splenius capitis and semispinalis capitis fibers. (4.52)

3) Roll your fingers across the suboccipitals' small, short bellies. Again, initially you may not feel any specific bellies.

✔ *Are you between the C-2 and the superior nuchal line of the occiput? If you ask your partner to "tilt your head back ever-so-slightly," do you feel any contraction in the deepest layer of tissue?*

Prone. Locate the lateral edge of the trapezius' upper fibers. (4.53) Palpating beside the level of C-1, place one finger at the lateral edge of the trapezius. Slowly sink medially into the suboccipitals.

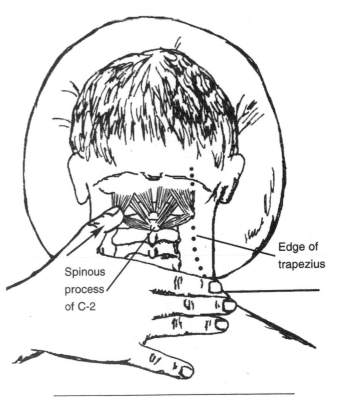

Spinous process of C-2

Edge of trapezius

(4.53) Partner prone, sinking your thumb medially, just lateral to the trapezius

Suboccipitals
Location - Deepest muscles in posterior neck
BLMs - Spinous process of C-2, superior nuchal line
Action - "Extend your head ever-so-slightly"

Quadratus Lumborum

The quadratus lumborum is considered an abdominal muscle, but is actually the deepest muscle of the low back. Stretching from the posterior ilium to the transverse processes of the lumbar vertebrae and twelfth rib, the quadratus lumborum is sometimes known as the "hip hiker" for its ability to elevate the hip.

The medial portion of the quadratus lumborum is buried beneath the thoracolumbar aponeurosis, the thick erector spinae and the transversospinalis muscles. (4.54) The lateral edge of the quadratus, however, is accessible when approached from the lateral side of the body.

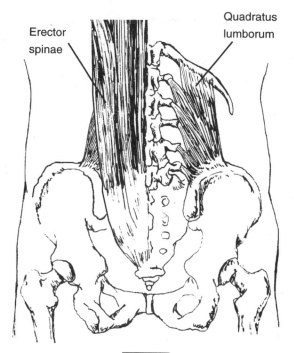

Erector spinae

Quadratus lumborum

(4.54)

A -
Unilateral:
    Elevate the hip
    Laterally flex
    Assists to extend the vertebral column
Bilaterally:
    Fixes the last rib during respiration

O - Posterior iliac crest, iliolumbar ligament

I - Last rib, transverse processes of first through fourth lumbar vertebrae

N - Branches of first lumber and twelfth thoracic

The external oblique fibers are superficial to the quadratus and assist in hiking the hip. When you are palpating the quadratus, be certain you are accessing beyond the superficial muscle layers.

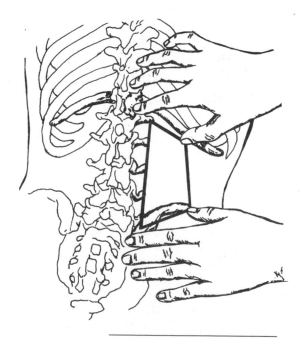

(4.55) Laying the fingers along the corners of the QL

Location - Deep to erectors, low back
BLMs - Twelfth rib and posterior iliac crest
Action - "Hike your hip toward your shoulder"

quadratus lumborum   **kwod**-rait-us lum-**bor**-um   L. four-sided muscle of the lumbar region

1) Prone. Isolate the borders of the quadratus by locating the twelfth rib, posterior iliac crest and transverse processes of the lumbar vertebrae.
2) Lay your fingers along these landmarks to outline the edges of the quadratus. (4.55)
3) Lay your fingerpads along the lateral edge of this square. Using slow firm pressure, sink your fingerpads medially toward the lumbar vertebrae and into the edge of the quadratus. (4.56)
4) Ask your partner to elevate his hip toward his shoulder to feel its solid contraction. The hip should remain on the table.

✔ *As you palpate, be sure you are accessing the deeper tissue in the low back and not the superficial external oblique fibers. When your partner hikes his hip, do you feel the lateral edge of the quadratus contract? Can you distinguish between the edge of the erector spinae (p. 138) and the quadratus?*

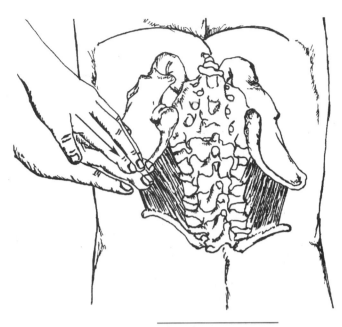

(4.56) Partner prone

Abdominals

Rectus Abdominus
External Oblique
Internal Oblique
Transverse Abdominus

The four abdominal muscles expand far beyond the "stomach" region. In fact, they form a muscular girdle that reaches around the sides of the thorax to the thoracolumbar aponeurosis, superiorly to the middle ribs, and inferiorly to the inguinal ligament. The immense span of these muscles, as well as their overlapping arrangement and unique fiber direction, help to stabilize the entire abdominal region.

The revered "washboard belly" is formed by the multiple, superficial bellies of the *rectus abdominus*. (4.57) Lateral to the rectus abdominus is the *external oblique*. (4.58) Unlike the round bellies of the rectus, the external oblique is a broad, superficial muscle best palpated at its attachments to the lower ribs.

The thin *internal oblique* fibers are deep and perpendicular to the external oblique and difficult to distinguish. (4.59) The *transverse abdominus*, the deepest muscle of the group, plays a major role in forced exhalation, yet cannot be palpated specifically.

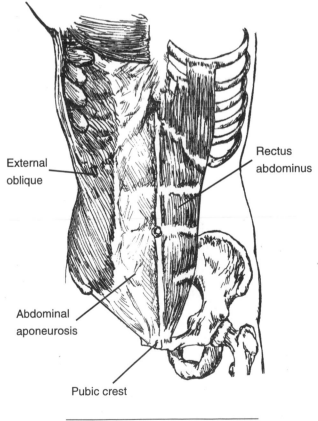

(4.57) Anterior view of abdomen

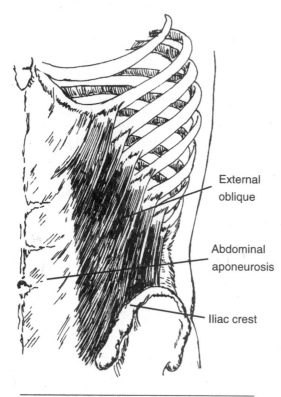

(4.58) Anterior/lateral view of abdomen

Internal Oblique

A -
Bilateral:
    Flex the thorax
    Compress abdominal contents
Unilateral:
    Laterally flex
    Rotate vertebral column to same side

O - Lateral inguinal ligament, iliac crest, thoracolumbar fascia

I - Cartilage of bottom three ribs, abdominal aponeurosis to linea alba

N - Branches of intercostals

Transverse Abdominus

A -
Compress abdominal contents

O - Lateral inguinal ligament, iliac crest, thoracolumbar fascia, cartilage of lower six ribs

I - Abdominal aponeurosis to linea alba

N - Branches of intercostals

Rectus Abdominus

A -
Flex the vertebral column

O - Crest of pubis, pubic symphysis

I - Cartilage of fifth, sixth and seventh ribs and xiphoid process

N - Branches of intercostals

External Oblique

A -
Bilateral:
    Flex the thorax
    Compress abdominal contents
Unilateral:
    Laterally flex
    Rotate vertebral column to opposite side

O - Lower eight ribs

I - Anterior part of the iliac crest, abdominal aponeurosis to linea alba

N - Branches of intercostals

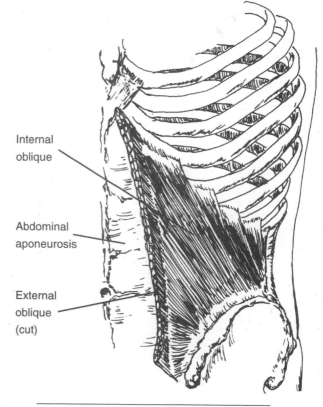

(4.59) Anterior/lateral view of abdomen

| oblique | o-**bleek** | L. diagonal |
| rectus | **rek**-tus | L. straight |

👁✋ Rectus abdominus

1) Supine with the knees bent. Locate the xiphoid process and the ribs just lateral to the xiphoid. Also locate the pubic crest. (p. 206)
2) Place your hand between these landmarks and ask your partner to alternately flex and relax his trunk slightly (a small sit-up).
3) Explore the entire length of the rectus and sculpt between its rectangular muscle bellies. (4.60)

✔ *While your partner flexes his trunk, can you palpate the lateral edges of the rectus abdominus?*

👁✋ External oblique (left side)

1) Supine with the knees bent. Lay your hand on the left side of the abdomen and lower ribs. Ask your partner to raise his left shoulder toward his right hip (rotating his trunk).
2) Palpate across the superficial fibers of the external obliques, noting their diagonal direction. (4.61)
3) With the trunk still rotated, follow the fibers superiorly to where they interdigitate with the serratus anterior, inferior to the abdominal aponeurosis and lateral to the iliac crest.

✔ *Are you palpating lateral to the edge of rectus abdominus? Are the fibers superficial and running in an oblique direction?*
Palpate lateral to the rectus abdominus with the abdomen relaxed. Can you distinguish between the external oblique and the deeper internal oblique fibers? Their fiber directions should be virtually perpendicular to each other.

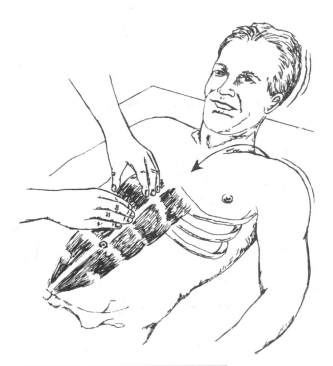

(4.60) Palpating rectus abdominus while partner flexes trunk

Abdominals
Location - Superficial, between ribs and pelvis
BLMs - Iliac crest, bottom edge of ribs
Action - "Flex your trunk"

(4.61) Accessing external oblique while partner rotates trunk

interdigitate   **in**-ter-**dij**-i-tate   to interlock, as fingers of clasped hands

An involuntary contraction of the diaphragm causes air to rush into the lungs and the vocal cords to snap shut. The audible result is a hiccup.

Diaphragm

The diaphragm is the primary muscle of respiration and unique in both its design and abilities. Its broad, umbrella-like shape separates the upper and lower thoracic cavities. The diaphragm's muscle fibers attach to the inner surface of the ribs and the lumbar vertebrae and converge at a central tendon. (4.63)

The diaphragm creates inspiration by the following process: the diaphragm's muscle fibers contract and pull the central tendon inferiorly. Because the central tendon is attached to the connective tissue which surrounds the lungs, a vacuum is created in the upper thoracic cavity and air is pulled into the lungs. For exhalation, the muscle fibers of the diaphragm relax, allowing the central tendon to elevate and the lungs to deflate. (4.62)

Although only a small portion of the diaphragm is accessible, the muscle's effect on the thorax and breathing can be easily felt.

A -
Draw central tendon down
Increase volume of thoracic cavity

O - Sternal part: Inner part of xiphoid process
Costal part: Inner surface of lower six ribs
Lumbar part: Upper two or three lumbar vertebrae

I - Central tendon

N - Phrenic

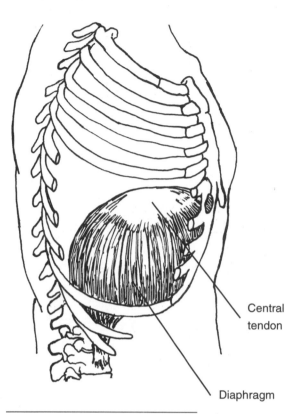

(4.62) Lateral view of thorax, diaphragm in position of exhalation

Central tendon

Diaphragm

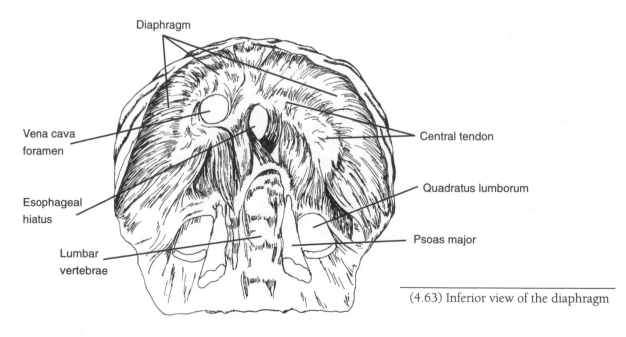

Diaphragm

Vena cava foramen

Esophageal hiatus

Lumbar vertebrae

Central tendon

Quadratus lumborum

Psoas major

(4.63) Inferior view of the diaphragm

diaphragm   **di-a-fram**   Gr. a partition

1) Supine, with knees bolstered. Locate the inferior edge of the rib cage, lateral to the xiphoid process.
2) Lay your thumbpads just inferior to the ribs on the abdomen and ask your partner to take slow, deep breaths.
3) Moving only when your partner exhales, slowly press and curl your thumbpads underneath the edge of the ribs. (4.64) During inhalation, you may not feel the tissue of the diaphragm, but will most likely feel the diaphragm pushing other tissues into your fingertips.

Try the above procedure with your partner sidelying and the trunk flexed slightly. This position allows the abdominal contents to shift away from where you are accessing and the overlying abdominal muscles to soften. (4.65)

✔ *Are your thumbs curling under the ribs rather than sinking into the abdominal organs? Ask your partner to "breathe into your belly" and notice how the abdominal region expands as the diaphragm contracts.*

Location - Underside of the ribs
BLMs   Bottom edge of rib cage
Action - "Inhale into your belly"

Move slowly and communicate with your partner when palpating. If, at any point, he does not feel safe or comfortable, gently remove your hands.

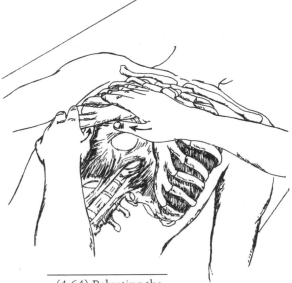

(4.64) Palpating the diaphragm supine

(4.65) Fingers curling around the ribs to access the diaphragm

The heart is also affected by the movements of the diaphragm. The heart's fibrous pericardium is attached to the diaphragm's central tendon by way of ligaments. The heart literally rides up and down on the diaphragm as you breathe. The yogis were right - breathing can massage the heart!

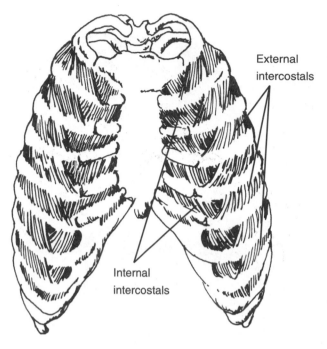

External
intercostals

Internal
intercostals

(4.66) Anterior view of rib cage

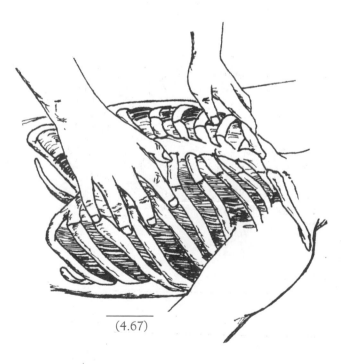

(4.67)

Location - *Spaces between the ribs*
BLMs - *Body of the ribs*
Action - *"Inhale and exhale"*

Intercostals

Better known to carnivores as the meat on spare ribs, the intercostals are small, slender muscles located between the ribs. They are divided into two groups: the external and internal intercostals. (4.66) These two groups have fibers perpendicular to each other and can be thought of as extensions of the external and internal oblique muscles (p. 147).

The function of the intercostals is to stabilize the framework of the rib cage and assist in respiration, although their specific role is debated. The entire rib cage is deep to one or numerous layers of muscle, but portions of the intercostals can be easily palpated. The internal and external inter-costals cannot be separately distinguished. The ribs and the spaces between them can be sensitive areas to access, so use slow, firm hand movements.

A -
External Intercostals:
> Draw the ventral part of the ribs upward, increasing the space of the thoracic cavity
Internal Intercostals:
> Draw the ventral part of the ribs downward, decreasing the space of the thoracic cavity

0 - Inferior border of the rib above

I - Superior border of the rib below

N - Intercostal

1) Supine. Begin just inferior to the pectoralis major on the side of the rib cage. Working across the body, position your fingers in the spaces between the ribs.
2) With one fingerpad, specifically palpate the tissue between two ribs. Roll your finger along the rib space and palpate the short, dense intercostals bridging across the ribs. (4.67)
3) Ask your partner to take several slow, deep breaths. Note any expansion or collapsing you feel in the spaces between the ribs. Turn your partner prone or sidelying and continue to explore the intercostals.

✔ *Are you between the rib spaces or too superficial? Can you specifically roll your fingers across the small intercostal fibers? Can you sink your fingers through the pectoralis major, latissimus dorsi or external obliques and isolate the intercostals?*

Ligaments and Vessels of the Spine and Thorax

Ligamentum Nuchae

The ligamentum nuchae is a finlike sheet of connective tissue that runs along the sagittal plane from the external occipital protuberance to the spinous process of C-7. (4.68)

Its chief function is to help stabilize the neck and head, but it also plays an important role as an attachment site for the superficial muscles of the posterior neck. The cervical spinous processes do not extend far enough posteriorly to serve as an attachment site for all the layers of muscle at the back of the neck. So muscles like trapezius and splenius capitis utilize the ligamentum nuchae as an attachment site.

In terms of palpation, the posterior edge of the ligamentum nuchae is superficial, yet can be difficult to discern from the surrounding muscle tissue.

1) Supine. Locate the external occipital protuberance (p. 163) and the spinous process of C-7.
2) Palpate between these landmarks along the midline of the neck. Be certain you are superficial to the spinous processes. It might help you to detect the ligamentum nuchae by rolling your fingertips across its fiber direction and explore for what may feel like a flap of soft rubber. (4.69)
3) Passively flex and extend the head slowly, rolling your fingers across the fibers of the ligamentum nuchae. Note how this changes the tension of the ligament.

Ask your partner to sit and flex her head and neck as far as comfortably possible. In this position, the ligamentum nuchae will stretch and rise to the surface and feel like a long, thin "speed bump" along the posterior neck.

✔ *Are you palpating superficial to the spinous processes of the vertebrae?*

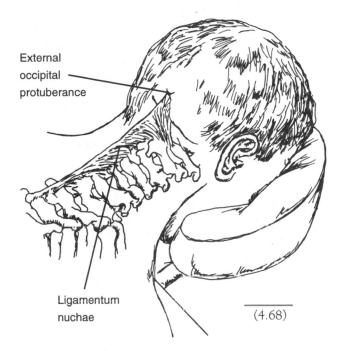

External occipital protuberance

Ligamentum nuchae

(4.68)

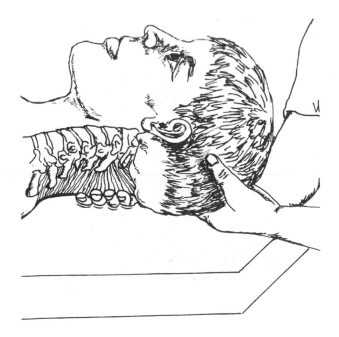

(4.69) Palpating the ligamentum nuchae

nuchae   **nu**-kay   L. nape of neck

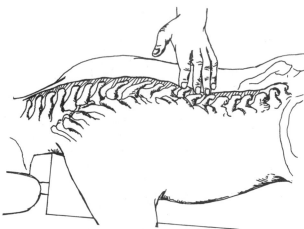

(4.70) Partner prone, exploring the supraspinous ligament

Supraspinous Ligament

The long, thin supraspinous ligament extends inferiorly from the ligamentum nuchae. It continues down the spine, attaching to the spinous processes of the thoracic and lumbar vertebrae. It is superficial and easily located in the spaces between the spinous processes.

👁 1) Prone. Locate several thoracic or lumbar spinous processes. (4.70)
2) Palpate between the spinous processes. Feel the slender shape and vertical fiber direction of the ligament by rolling your fingertip across its surface.

✔ *Ask your partner to sit in a chair and then slowly flex and extend his trunk. As he flexes forward, can you feel any changes in the tension or prominence of the ligament?*

Abdominal Aorta

Measuring nearly an inch in diameter, the abdominal aorta is the chief artery that carries blood to the abdominal organs and lower appendages. It lies on the anterior surface of the vertebrae, deep to the small intestines. The psoas major (p. 231) is lateral to the aorta.

1) Supine. Locate the umbilicus.
2) Place your fingerpads two inches superior to the umbilicus. Access the pulse of the abdominal aorta by slowly, but firmly, pressing straight down into the abdomen. Its strong pulse should be easily detected. (4.71)

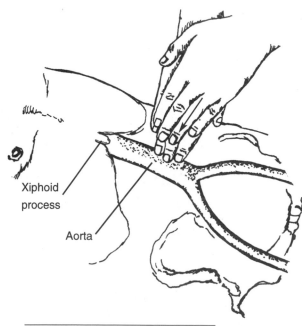

Xiphoid process

Aorta

(4.71) Partner supine, feeling the pulse of the abdominal aorta

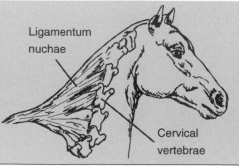

A horse's head and neck can weigh up to 300 pounds, requiring a broad ligamentum nuchae. Because the head and neck are held in a normal resting posture without any muscular effort, the ligamentum

Ligamentum nuchae

Cervical vertebrae

nuchae acts primarily as an antigravity device. A small contraction brings the head down to the ground, which simultaneously stretches the ligament. When the muscles relax, the ligament recoils and the head raises.

Thoracolumbar Aponeurosis

Despite its long name, the thoracolumbar aponeurosis is just what it says it is: a broad, flat tendon that stretches across the thorax and lumbar regions. Specifically, it is a thick, diamond-shaped tendon that lies superficially across the posterior thorax, stretching across the sacrum to the posterior iliac crest and up to the lower thoracic vertebrae. (4.72)

The aponeurosis serves as an anchor for several muscles in the thorax and hips, including the latissimus dorsi and the erector spinae group. It has a flat, dense texture that can be challenging to distinguish from the deeper muscles.

👁 1) Prone. Draw out the diamond shape of the aponeurosis by locating the posterior iliac crest, the surface of the sacrum, and the lower thoracic vertebrae. (4.73)
2) Lay the flat of your hands upon this diamond and use gentle pressure to palpate the thick layer of connective tissue underneath the skin yet superficial to the erector spinae muscles.

✔ *Ask your partner to alternately raise her elbows slightly and relax (this will contract the latissimus dorsi and tighten the aponeurosis). Do you feel any change in the superficial tissue?*
Move your hands laterally off the 'diamond' and onto the latissimus dorsi muscle belly. Do you notice any textural differences between the tissues?

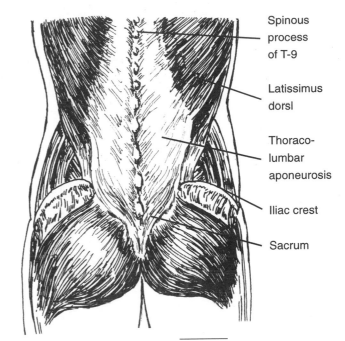

Spinous process of T-9

Latissimus dorsl

Thoraco-lumbar aponeurosis

Iliac crest

Sacrum

(4.72)

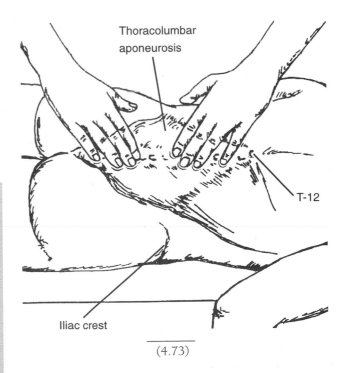

Thoracolumbar aponeurosis

T-12

Iliac crest

(4.73)

A giraffe's neck is more than five feet in length, yet still contains only seven cervical vertebrae. The atlas and axis are relatively short, while the five cervicals can measure eleven inches each. The neck and head are stabilized by a massive ligamentum nuchae and an array of short muscles which weave along the posterior surface of the neck. The anterior surface of the cervicals is covered by the retractor muscle. It extends from the giraffe's sternum all the way up to its hyoid bone and serves to draw back the tongue.

aponeurosis	**ap**-o-nu-**ro**-sis	Gr. *apo*, from + *neuron*, nerve or tendon
thoracolumbar	**tho**-rak-o-**lum**-bar	

Head, Neck & Face

Chapter **5**

Topographical Views

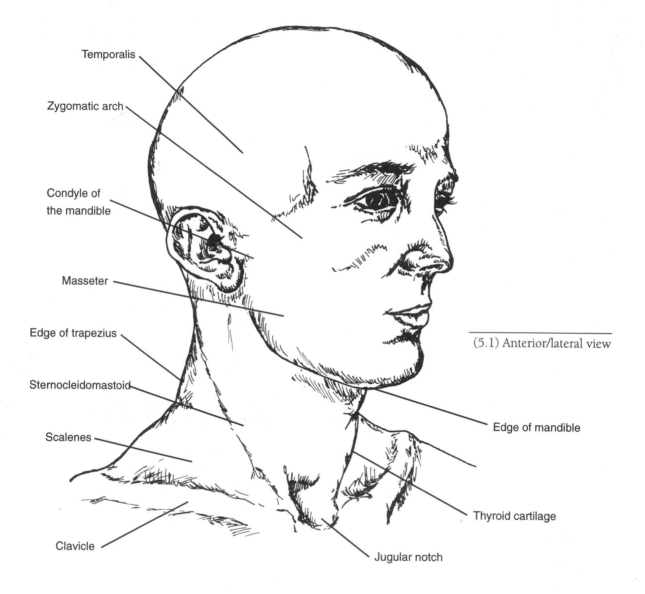

Temporalis

Zygomatic arch

Condyle of
the mandible

Masseter

Edge of trapezius

Sternocleidomastoid

Scalenes

Clavicle

Edge of mandible

Thyroid cartilage

Jugular notch

(5.1) Anterior/lateral view

Edge of
mandible

Trachea

SCM

a

b

Trapezius

Clavicle

The anterior and lateral sides of the neck
can be divided into two triangular regions.
The anterior triangle (a) is bordered by the
sternocleidomastoid muscle, the base of the
mandible and the trachea. The hyoid bone,
thyroid gland, carotid artery, submandibular
salivary gland and styloid process of the
mandible are some of the structures within
the anterior triangle.

The posterior triangle (b) is formed by the
sternocleidomastoid, clavicle and trapezius
and contains, among other structures, the
brachial plexus and the external jugular vein.

Bones and Bony Landmarks

The *skull* is composed of twenty-two bones: eight in the cranium and fourteen in the facial region. Seven of the eight *cranial* bones are directly accessible. The eighth, the ethmoid, is accessible only by way of the nasal cavity. Most of the cranial bones are superficial. Seven of the fourteen *facial* bones are palpable, including numerous bony landmarks of the *mandible* (jaw). (5.2 - 5.4)

The articulations between the cranial bones are different than the articulations of the appendages. The joints of the arms and legs have a synovial (mobile) joint structure. The cranial bones, instead, have fibrous joints which weave together to form tight-fitting sutures.

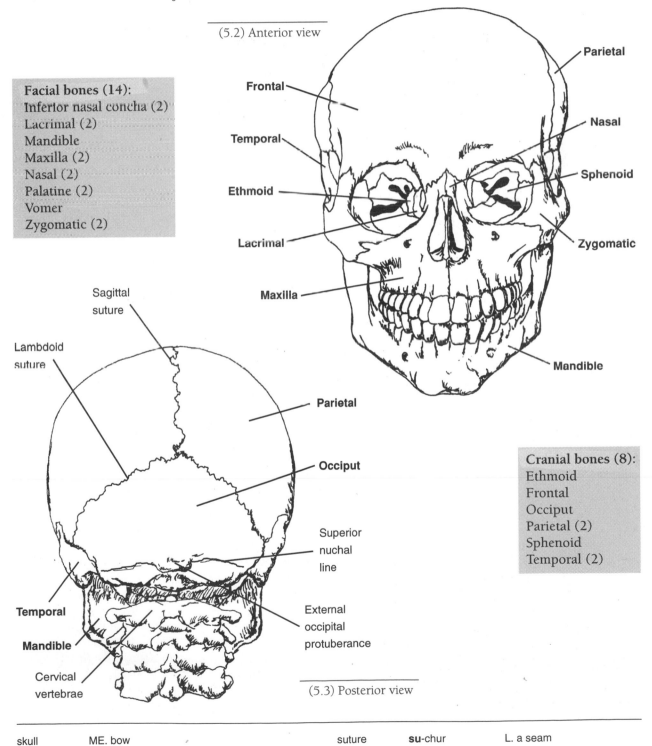

Facial bones (14):
Inferior nasal concha (2)
Lacrimal (2)
Mandible
Maxilla (2)
Nasal (2)
Palatine (2)
Vomer
Zygomatic (2)

(5.2) Anterior view

- Parietal
- Frontal
- Temporal
- Nasal
- Ethmoid
- Sphenoid
- Lacrimal
- Zygomatic
- Maxilla
- Mandible

Sagittal suture
Lambdoid suture
Parietal
Occiput
Superior nuchal line
External occipital protuberance
Temporal
Mandible
Cervical vertebrae

(5.3) Posterior view

Cranial bones (8):
Ethmoid
Frontal
Occiput
Parietal (2)
Sphenoid
Temporal (2)

skull	ME. bow	suture	**su**-chur	L. a seam
cranio-	L. skull			

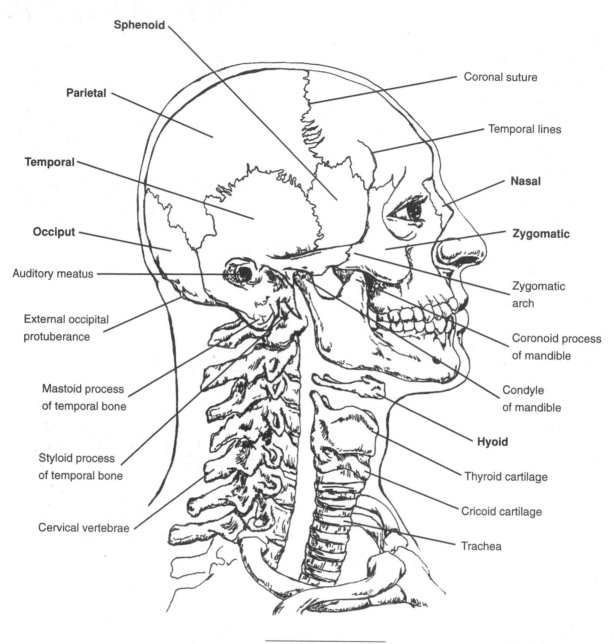

Sphenoid

Coronal suture

Parietal

Temporal lines

Temporal

Nasal

Occiput

Zygomatic

Auditory meatus

Zygomatic arch

External occipital protuberance

Coronoid process of mandible

Mastoid process of temporal bone

Condyle of mandible

Styloid process of temporal bone

Hyoid

Thyroid cartilage

Cricoid cartilage

Cervical vertebrae

Trachea

(5.4) Lateral view

It was long believed among the medical community that the cranial bones did not move. Initial observation would support this theory, since the bones are designed to protect the brain and the joints consist of tightly-woven sutures.

However, in the 1920s, a young osteopath named William Sutherland was determined to prove the presence of an infinitesimal, yet palpable, motion or rhythm of the cranial bones. Using himself as a guinea pig, Sutherland tested and proved his theory by applying a variety of homemade contraptions to his head including a football helmet with screws

drilled through it. While Sutherland monitored his cranial rhythm, his wife quietly detailed his dramatic personality and appearance changes.

Sutherland's research and perseverance among his colleagues helped cranial osteopathy to be accepted by the medical establishment.

Bony Landmark Trails

Trail 1 - "Around the Globe" palpates the bones and bony landmarks of the cranium and face. (5.5)

   a) Occiput
      External occipital protuberance
      Superior nuchal line
   b) Parietal
   c) Temporal
      Mastoid process
      Zygomatic arch
      Styloid process
   d) Frontal
   e) Sphenoid
   f) Zygomatic, maxilla and nasal

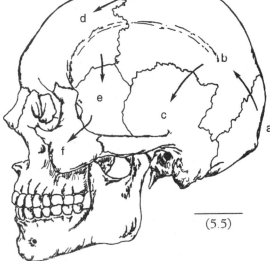

(5.5)

Trail 2 - "Jaw Jaunt" explores the mandible. (5.6)

   a) Body
   b) Submandibular fossa
   c) Angle
   d) Ramus
   e) Condyle
   f) Coronoid process

(5.6)

Trail 3 - "Horseshoe Trek" locates the cartilaginous structures of the anterior neck and the horseshoe-shaped hyoid bone. (5.7)

   a) Trachea
   b) Thyroid cartilage
   c) Cricoid cartilage
   d) Hyoid bone

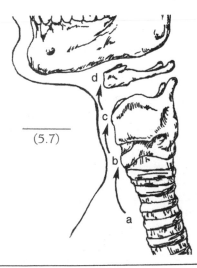

(5.7)

As evolution progressed, the more advanced creatures had fewer and fewer skull bones. For example, some fish have more than one hundred bones of the skull, reptiles may have seventy and primitive mammals forty. A human skull contains twenty-two bones, eight of which form the cranium. From a design perspective, this makes good sense: fewer bones mean fewer sutures, and fewer seams creates greater protection.

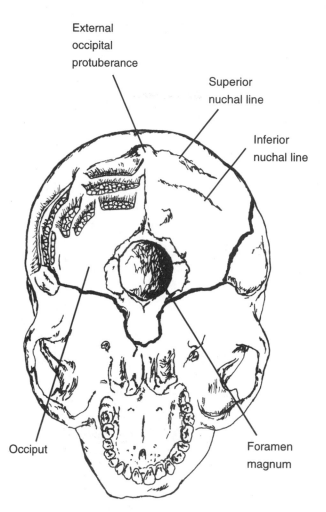

External
occipital
protuberance

Superior
nuchal line

Inferior
nuchal line

Occiput

Foramen
magnum

(5.8) Inferior view of cranium,
muscle attachment sites on left

Trail 1 - "Around the Globe"
Occiput

External Occipital Protuberance
Superior Nuchal Line

The occiput is located at the posterior and inferior aspects of the cranium. It extends superiorly from the external occipital protuberance and laterally to the mastoid processes of the temporal bone. The portion of the occiput superior to the occipital protuberance is superficial and easily palpable. The portion inferior to the protuberance curves in and under the head and is covered by layers of tendon and muscle. (5.8)

Sometimes called the "bump of knowledge," the *external occipital protuberance* is a small, superficial point located along the back of the head at the center of the occiput. It lies between the attachment sites of both trapezius muscles and is the superior attachment for the ligamentum nuchae. Regardless of intelligence, it varies in size.

Located along either side of the occipital protuberance, the *superior nuchal lines* are faint, sometimes bumpy ridges which extend laterally to the ears and the mastoid processes. The nuchal lines are attachment sites for the trapezius and splenius capitis muscles.

👁 General location of the occiput

1) Prone. Lay your hand on the back of the head between the ears.
2) Explore its surface by sliding your fingers
- superiorly from the occipital protuberance two or three inches;
- inferiorly where the occiput curves and sinks into the muscles of the neck;
- laterally to the mastoid processes behind the ears.

The superior nuchal line is an attachment site for several muscles. Metaphorically, it represents the "shoreline" between the dry land of the cranium and the sea of neck muscles.

nuchal   **nu**-kal   L. back of neck
occiput   **ok**-si-put   L. the back of the skull

☝ External occipital protuberance

1) Prone or supine. Place your fingers along the back of the neck at the body's midline. (5.9)
2) Slide superiorly onto the bony surface of the cranium. At the "shoreline" between the neck muscles and the cranium will be the protuberance.

✔ *Are you at the level of the top of the ears? If your partner is prone, ask her to extend her head slightly. Is the bump you feel just superior to where the muscles tighten?*

☝ Superior nuchal lines

1) Prone or supine. Stand at the head of the table and place both index fingers at the external occipital protuberance.
2) Allow the other fingers to fall in place beside them. Glide your fingertips up and down and palpate the edge of the superior nuchal lines.
3) Follow these ridges laterally as they extend toward the ear and mastoid processes.

✔ *Are you just lateral to the occipital protuberance? Do the ridges lead toward the back of the ear? Can you find them from a prone position? Are you on the cranium as opposed to the muscles of the neck?*

Parietal Bone

The large, rectangular parietal bones form the top and sides of the cranium. Positioned between the frontal, occipital and temporal bones, the parietals are saucer-shaped and extend anteriorly to the level of the ear canal. They merge at the body's midline to form the sagittal suture, forming a slight crest that can often be felt.

☝1) Prone, supine or seated. To locate the general location of the parietals, place both hands on top of the cranium.
2) Palpate the sagittal suture between the parietals. If you cannot feel its crest, visualize it along the top of the skull.
3) Follow it anteriorly to the level of the ear canal and posteriorly to the occiput. (5.10)

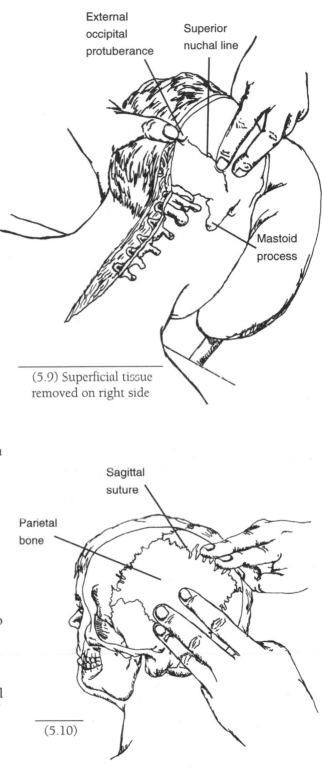

External occipital protuberance

Superior nuchal line

Mastoid process

(5.9) Superficial tissue removed on right side

Sagittal suture

Parietal bone

(5.10)

At birth, the cranial bones are not fully developed and joined. Usually there are six unossified gaps in the skull called fontanels. The name (L. little fountain) perhaps came from the pulse of blood vessels felt under the skin reminding physicians of the spurting of a fountain. The fontanels close between two and twenty-four months.

parietal   puh-**ri**-e-tul   L. wall

(5.11) Accessing the
mastoid process

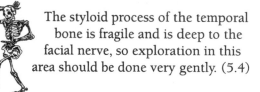

The styloid process of the temporal
bone is fragile and is deep to the
facial nerve, so exploration in this
area should be done very gently. (5.4)

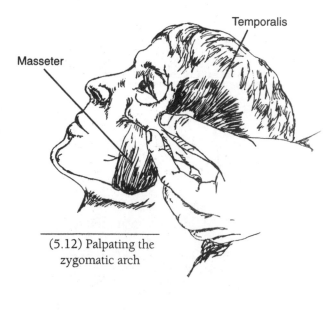

(5.12) Palpating the
zygomatic arch

Temporal Bone

Mastoid Process
Zygomatic Arch
Styloid Process

The temporal bone is located on the side of the
head, encompassing the area around the ear. It has
three important bony landmarks: the mastoid
process, the zygomatic arch and the styloid process.
(5.4) The temporal bone is superficial, except for
its superior aspect deep to the temporalis muscle.

The *mastoid process* forms a large, superficial
bump directly behind the earlobe. It is an
attachment site for the sternocleidomastoid and
other muscles.

The superficial *zygomatic arch* (cheek bone) is
formed by the temporal and zygomatic bones. It is
an attachment site for the masseter muscle. The
space between the zygomatic arch and the cranium
is filled by the thick temporalis muscle.

The *styloid process* is located behind the earlobe
between the mastoid process and the posterior edge
of the mandible. Its fang-like shape serves as an
attachment site for several ligaments and muscles.
The styloid process is deep to overlying muscles
and tissue and is not directly palpable; however, its
location can be accessed.

Locate the mastoid process by placing your
finger behind the earlobe. Sculpt around its edges,
exploring its entire surface. (5.11)

*Are you behind the ear lobe? Is the bone you feel
round and superficial? Can you palpate posteriorly
onto the superior nuchal line?*

Explore the zygomatic arch by placing your
finger anterior to the ear canal. Move anteriorly
along the arch rolling your finger across its surface.
Follow it anteriorly as it merges with the orbit of
the eye. (5.12)

*Does the ridge run horizontally? Is it level with
the ear canal?*

mastoid	**mas**-toyd	Gr. breast-shaped	styloid	**sti**-loyd	Gr. a pillar
zygomatic	**zi**-go-**mat**-ik	Gr. cheekbone			

Frontal Bone

Located on the anterior aspect of the cranium, the broad frontal bone forms the forehead and upper rim of the eye sockets. It articulates with the parietal bones to form the coronal sutures, which are deep to the occipitofrontalis and lateral edge of the temporalis muscles (p. 177).

Supine. Explore the forehead region superiorly to the coronal sutures, inferiorly to the brow, and laterally to the anterior edge of the temporalis muscle.

Sphenoid Bone

The sphenoid bone is located inside the cranium and has major articulations with fourteen bones of the skull. Located behind the eyeballs and superior to the zygomatic arches, the sphenoid is shaped like a butterfly and its lateral portions are called the greater wings. (5.13) The temporalis lies on top of these flat wings, making them inaccessible.

1) Supine. Place your fingers at the middle of the zygomatic arch (cheek bone) to locate the greater wings of the sphenoid.
2) Slide your fingers superiorly one inch onto the temporalis muscle belly. Deep to the thick temporalis is the location of the greater wing of the sphenoid.

Facial Bones

Nasal
Located at the bridge of the nose, the small nasal bones are positioned between the frontal and maxilla bones. They are virtually indistinguishable from the surrounding bones.

Zygomatic
Better known as the cheek bone, the zygomatic forms the anterior aspect of the zygomatic arch and the lateral portion of the orbit of the eye. (5.14) It serves as an attachment for the masseter muscle.

Maxilla
The maxilla bones form the center of the face, the inferior portion of the orbit of the eye, and the surface around the nose and the upper jaw in which the upper row of teeth articulate.

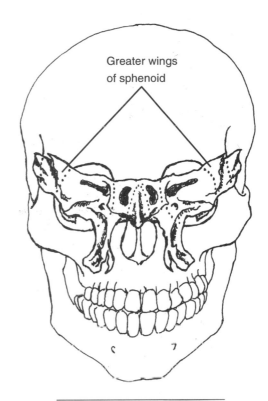

Greater wings
of sphenoid

(5.13) Anterior view showing location of the sphenoid

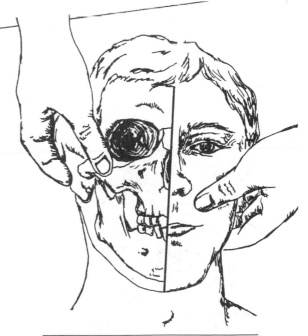

(5.14) Exploring the facial bones, superficial tissue removed on left side

---

maxilla	**max**-il-a	L. jawbone		sphenoid	**sfe**-noyd	G. wedge-shaped
nasal	**na**-zl	L. nose				

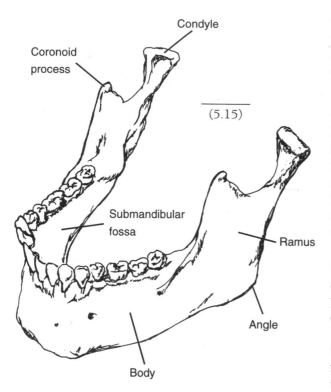

Coronoid process

Condyle

(5.15)

Submandibular fossa

Ramus

Angle

Body

Exploring in the submandibular fossa can be uncomfortable for your partner because of neighboring glands and nerves. Move slowly and check in with your partner.

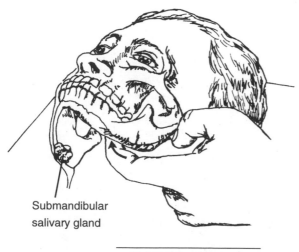

Submandibular salivary gland

(5.16) Palpating in the submandibular fossa

Trail 2 - "Jaw Jaunt"
Mandible

Body, Submandibular Fossa, Angle, Ramus, Condyle and Coronoid Process

The *mandible* or "jaw" has numerous landmarks which are superficial and accessible. (5.15) The *body* is the flat surface of the mandible located inferior to the lower teeth. The base or "jaw line" is the edge of the body and an attachment site for the thin platysma muscle. The *submandibular fossa* is located on the underside of the mandible and is an attachment site for the suprahyoid muscles and the anterior belly of the digastric muscle (p. 178).

The superficial *angle of the mandible* is located at the posterior end of the base. It forms part of the attachment for the masseter. The flat *ramus* is the posterior, vertical portion of the mandible and is deep to the masseter.

The mandible articulates with the cranium at two temporomandibular joints. The superficial *condyle* is located just anterior to the ear canal and inferior to the zygomatic arch. The deeper, inaccessible head of the condyle forms the articulating surface of the mandible at the temporomandibular joint. (5.1)

The *coronoid process* is located an inch anterior to the condyle of the mandible and the attachment of the temporalis muscle. When the jaw is closed, the coronoid process lies underneath the zygomatic arch and is unreachable. Opening the mouth fully, however, brings the coronoid process out from under the arch and allows the process to be accessed.

✋ Body, base and submandibular fossa

1) Supine. Place your fingers inferior to the bottom teeth and explore the superficial surface of the body.
2) Move inferiorly and palpate the base or edge of the mandible. Explore its entire length from the chin to the angle of the mandible.
3) With one hand stabilizing the mandible, slowly curl a fingertip underneath its edge and into the submandibular fossa. (5.16)

mandible	**man**-di-ble	L. lower jawbone	
condyle	kon-**dil**	Gr. knuckle	

ramus	**ray**-mus	L. branch	
coronoid	**kor**-o-noyd	Gr. crown-shaped	

👁 Angle and ramus of mandible

1) Slide posteriorly along the base of the mandible to the angle. Clarify your location by asking your partner to open her mouth and noting how the angle moves.

2) Slide superiorly from the angle onto the ramus, which is deep to the masseter muscle.

👁 Condyle of mandible

1) Place your fingerpad anterior to the ear canal and below the zygomatic arch.

2) Ask your partner to open her mouth fully. With this action the condyle will become more palpable as it slides anteriorly and inferiorly. (5.17)

3) As the jaw closes, follow the condyle as it returns to its original position.

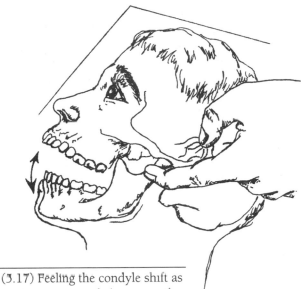

(5.17) Feeling the condyle shift as partner opens and closes mouth

✔ *Are you anterior to the ear canal, below the zygomatic arch? While your partner opens her mouth, can you palpate both condyles simultaneously?*

👁 Coronoid process of mandible

1) Place your fingerpad on the middle aspect of the zygomatic arch.

2) Drop half an inch inferiorly and ask your partner to open her mouth fully. As the jaw drops, the large process will press into your finger. (5.18)

3) With the mouth still open, explore the surfaces of the process.

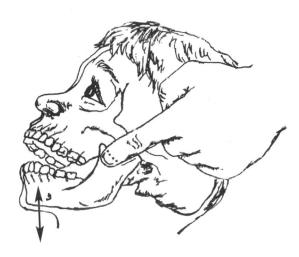

(5.18) Accessing the coronoid process

✔ *Are you inferior to the zygomatic arch? When the mouth is open, can you feel the anterior edge of the process?*

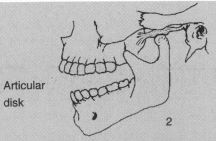

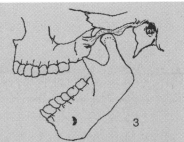

Articular disk

The temporomandibular joint is the most frequently used joint in the body, moving 2000-3000 times a day. Compounding this strain is the incongruity of its joint surfaces, the mandibular condyle and fossa. Luckily, the TM joint is equipped with an articular disk (1). Shaped like a lifesaver, the disk lies on top of the condyle and creates a more congruent joint surface and reduces bone deterioration.

When the mandible depresses, the condyle and disk move in tandem, pivoting anteriorly and inferiorly (2, 3). The reverse occurs as the mandible elevates.

jugular   **jug**-u-lar   L. throat

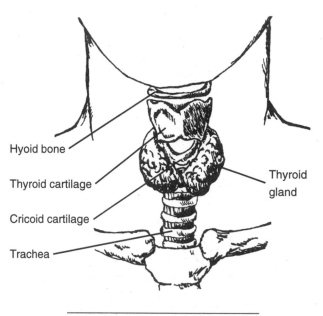

Hyoid bone

Thyroid cartilage

Cricoid cartilage

Trachea

Thyroid gland

(5.19) Anterior view of the neck

The thyroid cartilage is sometimes referred to as the "Adam's Apple." The name comes from a folk legend which told of the biblical Adam's first bite of apple getting stuck partway down his throat. The male descendants, as the legend goes, have carried on this trait.

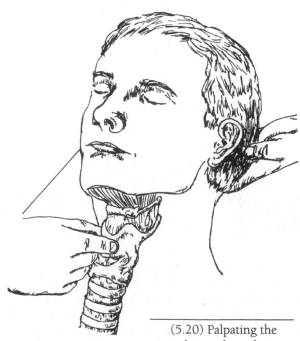

(5.20) Palpating the
trachea and cartilages

Trail 3 - "Horseshoe Trek"
Trachea

Cricoid Cartilage, Thyroid Cartilage, Hyoid Bone

The *trachea* (windpipe) is located at the center of the anterior neck. (5.19) It is a ribbed, cartilaginous tube that is roughly an inch in diameter and deep to the thyroid gland. The *cricoid cartilage* is a slightly larger ring of the trachea superior to the thyroid gland. The *thyroid cartilage* (Adam's Apple) is superior to the cricoid cartilage below the level of the chin. Present on both sexes, the thyroid cartilage is larger and visibly protruding on adult males. These three structures are partially deep to the slender infrahyoid muscles, yet clearly palpable.

The horseshoe-shaped *hyoid bone* is located superior to the thyroid cartilage. It is roughly an inch in diameter and lies parallel to the base of the mandible (jaw line) and the third cervical vertebrae. The hyoid bone serves as an attachment site for the supra- and infrahyoid muscles. It is accessible and elevates upon swallowing.

👁 Trachea and cartilages

1) Supine or seated. Using a fingerpad and thumbpad to palpate, gently explore the anterior surface of the neck for the tubular trachea.
2) Slide your finger up and down to feel the trachea's ribbed surface and slowly and gently shift it side to side noting its pliability.
3) The cricoid ring can be isolated by sliding your finger and thumb superiorly along the trachea to just below the thyroid cartilage. Explore for its large, ringed surface. (5.20)
4) Slide superiorly from the cricoid ring onto the thyroid cartilage. Palpate its sides and central tip.

✔ *Are you at the midline of the neck? Can you distinguish any rings along the trachea's surface? Is the trachea roughly an inch in diameter? With your fingerpad on the thyroid cartilage, ask your partner to swallow. Do you feel it move up and down?*

| hyoid | **hi**-oid | Gr. U-shaped |
| thyroid | **thi**-royd | Gr. shield |

| trachea | **tray**-ke-a | Gr. rough |
| cricoid | **kri**-koyd | Gr. ring-shaped |

✋ Hyoid bone

1) Supine or seated. Place your index finger upon the thyroid cartilage.

2) Roll your fingerpad superiorly over the thyroid cartilage, onto the hyoid.

3) While the index finger maintains this position, gently palpate the sides of the hyoid with your thumb and middle finger. (5.21) The hyoid will be wider than the trachea.

4) Using gentle pressure, explore the surface and small side-to-side movements of the hyoid. If it is difficult to access the hyoid, encourage your partner to relax her tongue and jaw.

✔ *Are you superior to the thyroid cartilage (Adam's Apple)? Can you gently move the hyoid from side to side? With your thumb and middle finger on either side of the hyoid, ask your partner to swallow. Do you feel the hyoid raise up and then return?*

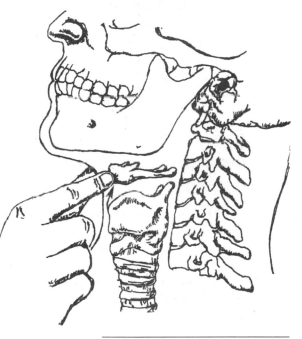

(5.21) Accessing the hyoid bone

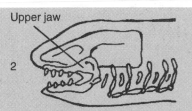

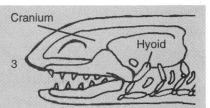

The hyoid bone is an ancestral remnant, descending from the tissue that formed the gills. During the evolution of jaws, the gill arches (the bones around the gills) - (1) gravitated toward the head to hold the upper jaw next to the cranium (2). For fish, who do not have long necks as we do, the position of the hyoid serves as an important link between the jaw and cranium (3).

For humans, the hyoid lost this function and shifted down the neck to become the only bone in the body which does not articulate with other bones. It is supported by the muscles which attach to its surface such as the suprahyoids and infrahyoids.

Muscles of the Head, Neck and Face

The head and face contain over thirty pairs of muscles, many of which are small, thin, and difficult to isolate (p. 182). However, the several muscles which act upon the mandible are accessible on the side of the jaw.

The anterior and lateral neck muscles perform a wide variety of tasks, including moving the head and neck, assistance in swallowing, and raising the ribcage during inhalation. The posterior neck muscles, which act primarily upon the cervical spine and head, are detailed in Chapter 4 - Spine and Thorax.

Before palpating the following muscles on your partner, it is advised that you skip to the back of this chapter and familiarize yourself with the arteries, glands and nerves of the head, neck and face.

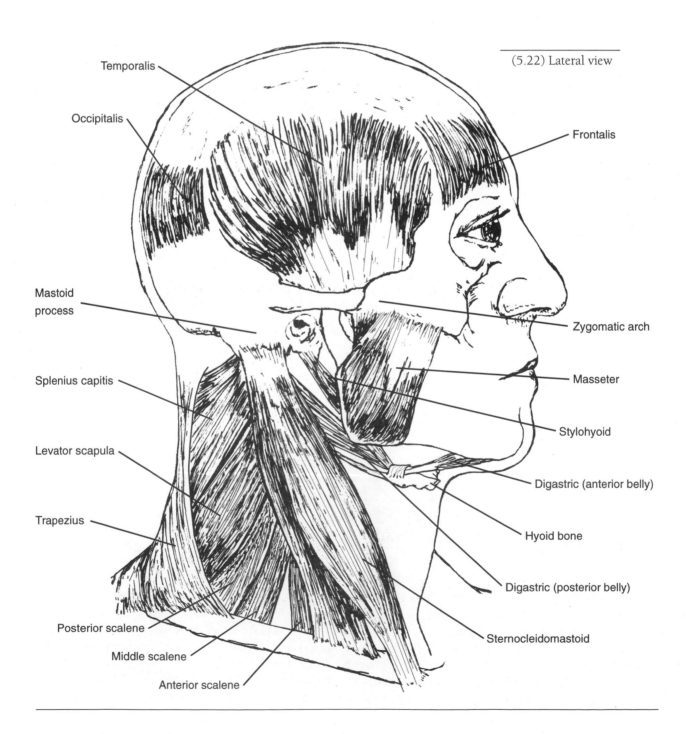

(5.22) Lateral view

Temporalis

Occipitalis

Frontalis

Mastoid process

Splenius capitis

Zygomatic arch

Masseter

Levator scapula

Stylohyoid

Trapezius

Digastric (anterior belly)

Hyoid bone

Posterior scalene

Digastric (posterior belly)

Middle scalene

Sternocleidomastoid

Anterior scalene

The sternocleidomastoid and upper fibers of the trapezius (p. 52) begin as one muscle in the embryo and then split in later development. The location of their attachment sites hints at their initial relationship: they form an almost continuous tendon along the superior nuchal line and mastoid process. Their other attachments are at either end of the clavicle.

Mylohyoid

Submandibular salivary gland

Digastric (anterior belly)

Carotid artery

Thyroid cartilage

Trachea

Omohyoid (cut)

Sternohyoid (cut)

Thyrohyoid

Omohyoid (superior belly)

Sternohyoid

Scalenes

Trapezius

Omohyoid (inferior belly)

Sternocleidomastoid

Sternothyroid

Clavicle

(5.23) Anterior view of neck

The smallest muscle in the human body is located in the middle ear. The stapedius muscle measures less than 1/20 of an inch, thinner than a U.S. dime. It activates the stirrup, one of the tiny bones of the ear, which sends vibrations from the eardrum into the inner ear.

The stapedius, however, may not be the absolute shortest muscle. A minuscule involuntary muscle called the arrector pili (p. 8) is attached to every hair follicle on the human body. These microscopic muscles have a big responsibility. When you are cold or respond to an emotion such as fear, the arrector pili raise the hair and produce goose bumps or help retain body heat. They are also believed to have given our ancestors the hair-raising ability of appearing larger to potential enemies.

stapedius   sta-**pe**-de-us   L. stirrup
arrector pili   **a**-rek-tor **pee**-li   L. *pilosus*, hairy

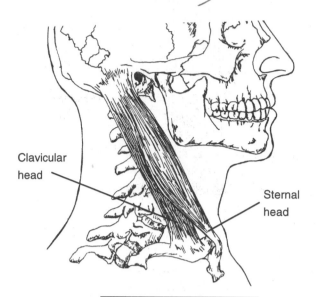

Clavicular head

Sternal head

(5.24) Sternocleidomastoid

(5.25)

Location - *Superficial, anterior and lateral neck*
BLMs - *Mastoid process, medial clavicle*
Action - *"Flex your neck" or "inhale deeply"*

(5.26)

Sternocleidomastoid

The sternocleidomastoid (SCM) is located on the lateral and anterior sides of the neck. It has a large belly which is composed of two heads: a flat, clavicular head and a slender, sternal head. (5.24) Both heads merge together to attach at the mastoid process behind the ear. The carotid artery (p. 185) passes deep and medial to the SCM. The external jugular lies superficial to it.

The SCM is superficial, completely accessible, and often visible when the head is turned to the side in a Lord Byron-like fashion. (5.25)

A -
Unilateral:
    Laterally flex the head to the same side
    Rotate the head to the opposite side
Bilateral:
    Flex the neck
    Assist in inspiration

O - Sternal head: Top of manubrium
Clavicular head: Medial one-third of the clavicle

I - Mastoid process of temporal bone, lateral superior nuchal line of occiput

N - Spinal accessory

1) Supine with practitioner at head of table. Locate the mastoid process of the temporal bone, the medial clavicle and the top of the sternum.
2) Draw a line between these landmarks to delineate the location of the SCM. Note how the two SCMs form a "V" on the front of the neck.
3) Ask your partner to raise her head very slightly off the table as you palpate the SCM. It will usually protrude visibly. (To make the SCM more distinct, rotate the head slightly to the opposite side and then ask her to flex.)
4) Palpate along the borders of the SCM, follow it behind the ear lobe, and then down to the clavicle and sternum. (5.26) Sculpt around the skinny sternal tendon and the wide clavicular tendon.

✔ *With your partner relaxed, can you grasp the SCM between your fingers and outline its thickness and shape? How much space is between the clavicular attachments of the SCM and trapezius? It should be roughly two to three inches.*

Scalenes

Anterior Scalene
Middle Scalene
Posterior Scalene

Located on the anterior, lateral neck, the three scalenes are sandwiched between the sternocleidomastoid and the anterior flap of the trapezius. Their fibers begin at the side of the cervical vertebrae, dive underneath the clavicle and attach to the first and second ribs. (5.27) During quiet inspiration, the scalenes perform the vital action of elevating the upper ribs.

The *anterior scalene* lies partially tucked beneath the sternocleidomastoid. The *middle scalene* is slightly larger and lies lateral to the anterior scalene. Both muscle bellies are directly accessible. The smaller *posterior scalene* is located between the middle scalene and levator scapula (5.29). The posterior scalene is positioned deeper than the other scalenes. Due to its small size and buried location, the posterior can be challenging to distinguish from the surrounding bellies.

The large branches of the brachial plexus and subclavian artery (p. 188) pass through a small gap between the anterior and middle scalenes. Individual nerves of the brachial plexus may penetrate through or in front of the anterior scalenes. (5.28)

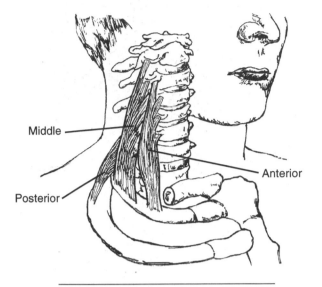

(5.27) Anterior/lateral view of scalenes

Compression or impingement of the brachial plexus or one of its nerves can create a sharp, shooting sensation or numbness down the arm. If this occurs, immediately release and adjust your position posteriorly. Ask your partner for feedback while palpating the scalenes.

A -
Bilateral:
    Elevate the ribs during inhalation (All)
Unilateral:
    With rib fixed, laterally flex the neck (All)
    Rotate head and neck to the opposite side (All)
    Flex the neck (Anterior)

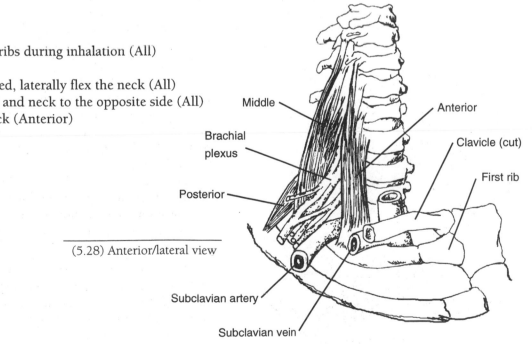

(5.28) Anterior/lateral view

scalene   **skay**-leen   Gr. uneven

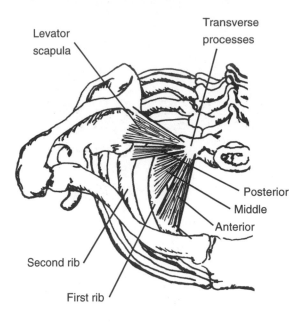

Levator scapula

Transverse processes

Posterior

Middle

Anterior

Second rib

First rib

(5.29) Superior view of scalenes and levator scapula

Location - Lateral neck, between SCM and trapezius
BLMs - Transverse processes of cervicals
Action - "Inhale into your upper chest" or "flex your neck"

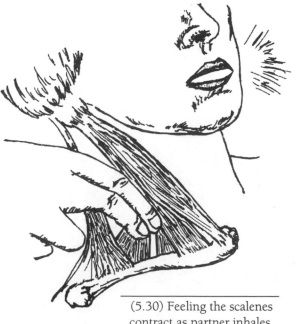

(5.30) Feeling the scalenes contract as partner inhales

> Variations of the scalenes include the presence of a scalene minimus. Present in roughly 40% of the population, the minimus often attaches from the sixth and seventh cervical vertebrae to the first rib or pleural dome of the lung. It lies inferior and deep to the anterior scalene and may be quite a strong muscle.

Anterior Scalene

O -Transverse processes of third to sixth cervical vertebrae

I - First rib

N - Cervical plexus

Middle Scalene

O - Transverse processes of second to seventh cervical vertebrae

I - First rib

N - Cervical plexus

Posterior Scalene

O - Transverse processes of fifth and sixth cervical vertebrae

I - Second rib

N - Cervical plexus

👁 Scalenes as a group

1) Supine, with practitioner at head of table. Cradle the head (passively flexing it) to allow for easier palpation. Lay your fingerpads along the anterior and lateral sides of the neck, between the SCM and trapezius.
2) With the flat of your fingers, use gentle pressure to palpate the stringy, superficial muscle bellies in this triangle.

✔ *Are you between the SCM and trapezius? Ask your partner to inhale deeply into her upper chest. At full inhalation, do you feel the muscles in this triangle contract? (5.30)*

👁 Anterior and middle scalenes

1) Since the anterior scalene lies partially deep to the lateral edge of the SCM, rotate the head slightly to the opposite side to better expose it. Gently palpate under the SCM's lateral edge and roll across the belly of the anterior scalene. (5.31)
2) Follow it inferiorly as it tucks under the clavicle.
3) Move laterally to explore the middle scalene, noting its similarly shaped belly. (5.32)

✔ *Do the muscles you feel have a slender, stringy texture? If you follow them inferiorly, do they sink beneath the clavicle (in the direction of the ribs)? Can you follow them superiorly to the transverse processes of the cervical vertebrae? Ask your partner to flex her head slightly. Do you feel the scalenes contract?*

👁 Posterior scalene

1) Squeezed between the middle scalene and levator scapula (p. 63), the posterior scalene extends laterally off the neck.
2) Locate the middle scalene and the levator scapula. Place a finger between these bellies and sink inferiorly. (5.33)
3) Slowly strum across the thin band of tissue running lateral from the TVPs to the second rib.

✔ *To distinguish between posterior scalene and levator scapula, locate the scalene and ask your partner to slowly elevate her scapula. Since the scalene does not create this action, there should be no contraction of its fibers. However, ask your partner to slowly inhale into the upper chest, and you should feel the posterior scalene contract.*

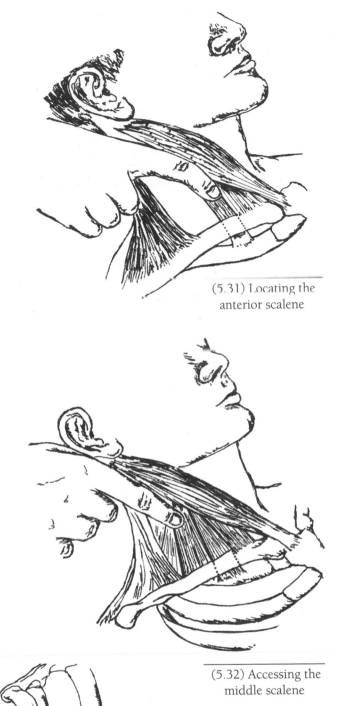

(5.31) Locating the anterior scalene

(5.32) Accessing the middle scalene

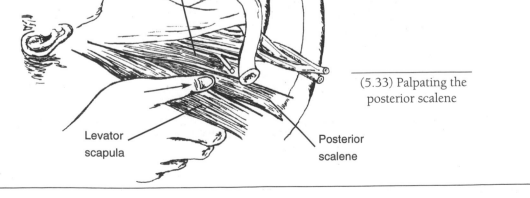

Middle scalene

Levator scapula

Posterior scalene

(5.33) Palpating the posterior scalene

Masseter

In proportion to its size, the masseter is the strongest muscle in the body. Together, the two masseters exert a biting force of nearly one-hundred and fifty pounds of pressure - enough to bite a finger off! It is the major muscle of chewing and is also involved with speech and swallowing.

Located on the side of the mandible, the masseter is square-shaped and comprised of two overlapping bellies. The superficial belly can be accessed from the face, the deep belly is palpable from inside the mouth. The masseter is situated deep to the parotid gland (p. 187), yet still easily palpable. (5.34)

A - Elevate the mandible

O - Zygomatic arch

I - Angle and ramus of mandible

N - Trigeminal

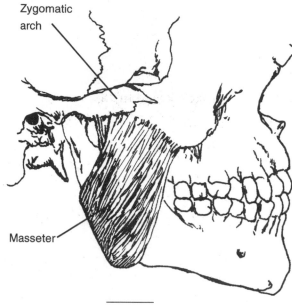

Zygomatic arch

Masseter

(5.34)

 1) Supine. Locate the zygomatic arch and angle of the mandible.
2) Place your fingers between these bony land-marks and palpate the surface of the masseter.
3) Ask your partner to alternately clench and relax her jaw as you sculpt out the square shape of the belly. (5.35)
4) Clarify the masseter's fiber direction by strumming your fingers horizontally across its muscle fibers.

✔ *As your partner clenches, can you outline the anterior edge of the masseter? If your partner opens her jaw as wide as possible, can you feel the tissue lengthen?*

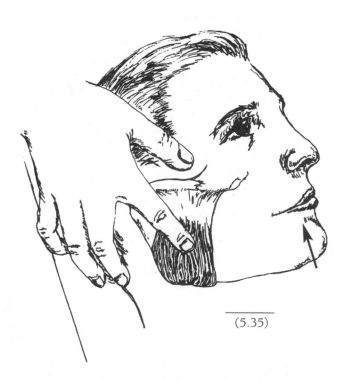

(5.35)

Location - Superficial, side of face
BLMs - Angle of mandible, zygomatic arch
Action - "Clench your jaw"

Temporalis

The temporalis is located on the temporal aspect of the cranium. Its broad origin attaches to the frontal, temporal and parietal bones. (5.36) Its fibers converge in a thick mass, reaching under the zygomatic arch to connect at the coronoid process. Though deep to the temporal fascia and artery, the temporalis is superficial and directly accessible.

A -
Elevate
Retract the mandible

O - Temporal fossa and fascia

I - Coronoid process of the mandible

N - Trigeminal

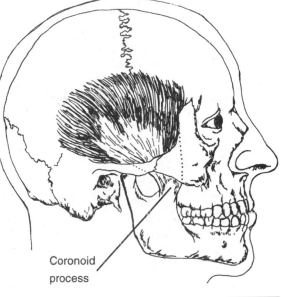

Coronoid
process

(5.36)

 1) Supine, practitioner at head of table. Locate the zygomatic arch.
2) Place your fingerpads one inch superior to the arch and ask your partner to alternately clench and relax her jaw. Do you feel the strong temporalis contracting beneath your fingers? (5.37)
3) To locate the attachment site of the temporalis tendon, ask your partner to open her mouth wide.
4) Locate and explore the coronoid process. Although the coronoid process is easily accessible, you may not be able to specifically isolate the tendon of the temporalis.

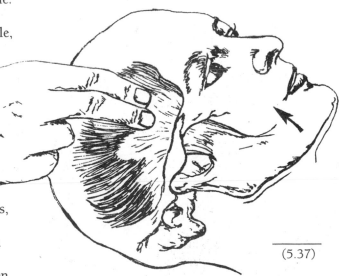

(5.37)

To outline the wide origin of the temporalis, ask your partner to alternately clench and relax her jaw. Lay your fingers in various positions on the side of the head. If your fingers are on the muscle, you will feel the temporalis fibers tighten and soften. If you are off the muscle, nothing will be felt.

✔ *When exploring the muscle belly, are you superior to the zygomatic arch on the side of the head? Can you distinguish the direction and convergence of the muscle fibers?*

Location - Superficial, lateral side of head
BLMs - Zygomatic arch, coronoid process
Action - "Clench your jaw"

temporalis   **tem-po-ra**-lis   L. time, seen by the graying of hairs in this region

Suprahyoids and Digastric

The suprahyoids (*geniohyoid, mylohyoid and stylohyoid*) form a wall of muscle along the underside of the jaw. (5.38) Stretching from the edge of the mandible to the hyoid bone, they lie inferior to the glossus muscles (the muscles of the tongue).

Although each of the three suprahyoids is a small muscle, together they affect the tongue and hyoid bone and are important in chewing, swallowing and speaking. They are partially deep to the digastric muscle, yet accessible. The suprahyoid bellies cannot be distinguished individually.

The long, round *digastric* muscle is comprised of a posterior and an anterior belly. The posterior belly runs from the mastoid process to the hyoid bone and then loops through a tendonous sling on the hyoid's anterior surface. It continues on as the anterior belly to attach at the underside of the chin. (5.39) Both bellies are superficial, yet difficult to distinguish from the deeper suprahyoid muscles.

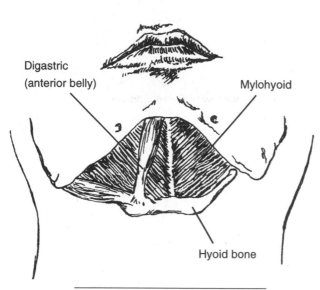

(5.38) Anterior/inferior view, geniohyoid is deep to mylohyoid

Suprahyoids

A -
Elevate hyoid and tongue
Depress mandible

O - Styloid process (Stylohyoid)
Underside of mandible (Mylohyoid, Geniohyoid)

I - Hyoid bone

N - Mylohyoid (Mylohyoid)
Hypoglossal (Geniohyoid)
Facial (Stylohyoid)

Digastric

A -
With hyoid bone fixed, it depresses the mandible
With mandible fixed, it elevates the hyoid
Retracts the jaw

O - Mastoid process (deep to Sternocleidomastoid and Splenius Capitis)

I - Inferior border of the mandible

N - Mylohyoid, Facial

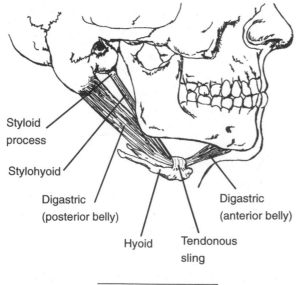

(5.39) Lateral view

Location - Superficial, anterior neck
BLMs - Base (jaw line) of the mandible
Action - "Depress your jaw" or "swallow"

| glossus | **glah**-sis | Gr. tongue | geniohyoid | **je**-ne-o-**hi**-oyd | L. *gena*, cheek |
| stylohyoid | **sti**-lo-**hi**-oyd | | mylohyoid | **my**-lo-**hi**-oyd | L. *myle*, mill |

👁 Suprahyoids

1) Supine or seated. With your partner's jaw closed, place a finger along the underside of the mandible.
2) Contract the suprahyoids by asking your partner to press the tip of the tongue firmly against the roof of the mouth. Note how this action forms a wall of taut muscle along the base of the mandible (jaw line). Follow it as it extends down to the hyoid bone. (5.40)
3) With the tongue relaxed, palpate the flat surface of the suprahyoid tissues, distinguishing them from the lumpy texture of the submandibular salivary glands (p. 187).

✔ *If you place a fingerpad underneath the tip of the chin and ask your partner to gently depress her mandible into your finger, do the suprahyoids contract? Also, palpate the suprahyoids and ask your partner to swallow. Do these tissues contract?*

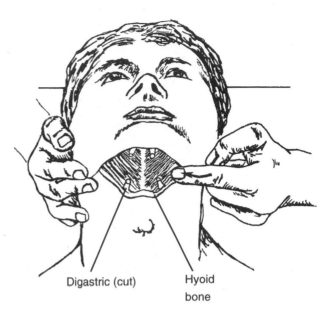

Digastric (cut)   Hyoid bone

(5.40) Palpating the mylohyoid

👁 Digastric

1) Supine, with practitioner at head of table. Locate the mastoid process of the temporal bone and the hyoid bone.
2) Draw an imaginary line between these points. Using your index finger, palpate along this line for the skinny, posterior digastric. (5.41)
3) Draw an imaginary line between the hyoid bone to the underside of the chin and palpate for its anterior belly.
4) To feel the digastric contract, place your finger under the chin and ask your partner to try to open her mouth against your gentle resistance. This contraction will sometimes allow both of the digastric bellies to be located more easily.

✔ *Is the muscle you are palpating superficial and pencil-width? Does it extend from the mastoid process to the hyoid bone to the chin?*

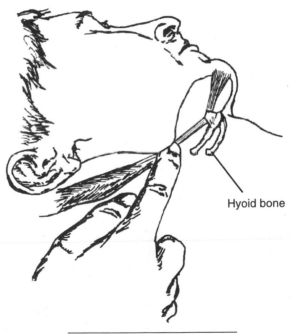

Hyoid bone

(5.41) Locating the digastric

A giraffe's trachea is formed by more than a hundred tracheal rings and creates a unique breathing problem. Because of the windpipe's length, each inhalation contains nearly two gallons of air that never reach the lungs. Comparatively, a resting human takes two gallons of air every minute. To counteract this anatomical dead space, a giraffe is equipped with a massive lung capacity of nearly twelve gallons of air. It has also been suggested that a giraffe's trachea may serve as a cooling device. When the trachea is filled with moist air, it cools the nearby blood vessels travelling up to the brain.

digastric   di-**gas**-trik   Gr. *dis*, twice *gaster*, belly

3

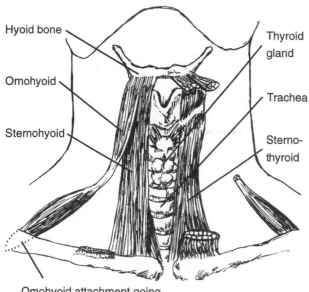

Hyoid bone

Omohyoid

Sternohyoid

Thyroid gland

Trachea

Sterno-thyroid

Omohyoid attachment going to superior angle of the scapula

(5.42) Anterior/inferior view of neck

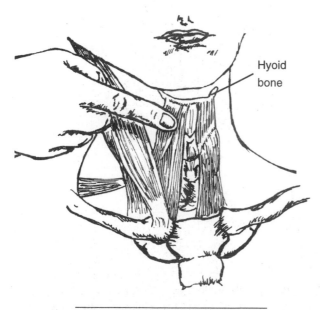

Hyoid bone

(5.43) Palpating the infrahyoids

Infrahyoids

Sternohyoid
Sternothyroid
Omohyoid

The infrahyoids are located on the anterior neck, superficial to the trachea. (5.42) All three muscles are thin and delicate and function as antagonists to the suprahyoids. The superficial sternohyoid and sternothyroid are layered just to the side of the trachea and, although difficult to separately distinguish, are directly accessible.

The omohyoid is perhaps the most bizarre muscle of the body. It has a skinny, ribbonlike belly which begins at the hyoid bone, passes underneath the SCM and scalenes and attaches to the scapula. Aside from depressing the hyoid, the omohyoid tightens the fascia of the neck and dilates the internal jugular vein. Due to its depth and slender belly, most of the omohyoid is inaccessible.

A - Depress the hyoid bone and thyroid cartilage (All)

O - Manubrium (Sternohyoid and Sternothyroid)
Superior border of the scapula (Omohyoid)

I - Hyoid bone (Sternohyoid and Omohyoid)
Thyroid cartilage (Sternothyroid)

N - Upper cervical

👁 Sternohyoid and sternothyroid

1) To avoid irritating the thyroid gland, explore only the superior half of these muscles. Locate the surface of the trachea, just below the thyroid cartilage (Adam's Apple).
2) With one fingerpad, slide to the side of the trachea and gently explore the thin tissue lying superficial to the windpipe. Try to roll your finger across the thin bellies of the infrahyoids. (5.43)
3) Ask your partner to tighten the muscles of the anterior neck. Sometimes this isometric contraction will make the infrahyoids quite solid and easily palpable.

Location - Superficial, anterior surface of neck
BLM - Trachea
Action - Resisted depression of jaw

---

omohyoid   **o-mo-hi**-oyd   Gr. *omo*, shoulder   sternothyroid   **ster-no-thi**-royd
sternohyoid **ster-no-hi**-oyd

Platysma

The platysma is a thin, superficial sheet which spans from the mandible to the chest across the anterior neck. (5.44) The platysma and other facial muscles are integumentary muscles. Instead of connecting to bones, these muscles are invested in the superficial fascia and attach to the skin and overlaying muscle. The platysma's claim to fame is its ability to create the infamous "Creature from the Black Lagoon" expression.

A -
Assist in depressing the mandible
Tighten the fascia of the neck

O - Fascia covering superior part of pectoralis major

I - Edge of mandible, skin of lower part of face

N - Facial

 1) Seated or supine. Ask your partner to jut his head anteriorly and protract his jaw. (5.45) Then ask him to tighten the tissue of the front of the neck.
2) Explore this thin sheet of muscle from the mandible down to the upper chest. Note any "flaps" the platysma forms along the lateral side of the neck.

Location - Superficial, anterior neck
BLMs - Edge of mandible
Action - "Look like a lizard"

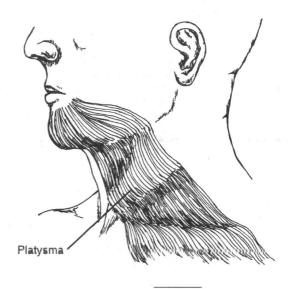

Platysma

(5.44)

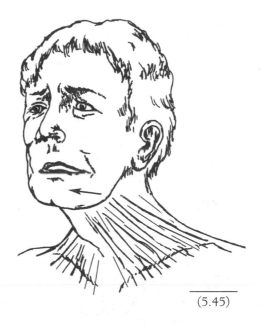

(5.45)

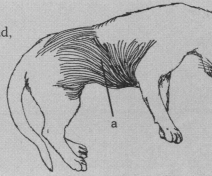

Most mammals have a broad, thin sheet of muscle called the panniculus carnosus (a). It is an integumentary muscle which attaches to the underside of the skin and, on some species, covers the entire thorax.

It enables a horse to shake off flies, an armadillo to roll into a ball and a cat to raise the hair on its back. For humans, the platysma is believed to be all that remains of the panniculus carnosus.

platysma   pla-**tiz**-ma   Gr. plate
panniculus carnosus   pan-**ik**-u-lus car-**no**-sis   L. cutaneous maximus

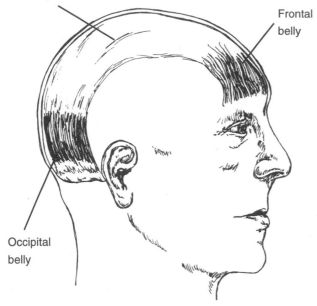

Galea aponeurotica

Frontal belly

Occipital belly

(5.46) Occipitofrontalis

Location- Superficial
BLMs - Forehead and scalp
Action - "Raise your eyebrows"

Occipitofrontalis

Responsible for creating an expression of surprise, the occipitofrontalis is a unique muscle with two, thin bellies - an occipital belly located on the back of the head and a frontal belly on the forehead. The two bellies are joined together by the galea aponeurotica, a broad sheath of connective tissue stretching across the top of the cranium. (5.46) The occipitofrontalis is superficial, yet its thin fibers cannot be specifically isolated.

A -

Frontalis: raise eyebrows and wrinkle forehead,
Occipitalis: anchor and retract the galea posteriorly

O and I - Galea aponeurotica (Both)
Skin over the eyebrows (Frontalis)
Superior nuchal line of the occiput (Occipitalis)

N - Facial

👁 Frontal fibers:
Supine. Place your fingers on the forehead and ask your partner to raise her eyebrows. Do you feel the tissue of the forehead contract?

👁 Occipital fibers:
Supine or prone. Locate the superior nuchal line of the occiput (p. 163) and slide your fingers one inch superiorly and isolate the region of the oval occipital bellies.

Muscles which express emotion are called mimetic muscles. There are thirty pairs of mimetic muscles of the human face - more than any other animal. Together they form an incredible variety of expressions ranging from crinkling the eyebrows in confusion (corrugator supercili), flaring the nostrils in anger (levator labii superioris), puckering the lips for a kiss (orbicularis oris), or raising the chin to pout (mentalis). Smiling is generated by eight muscles while frowning requires up to twenty.

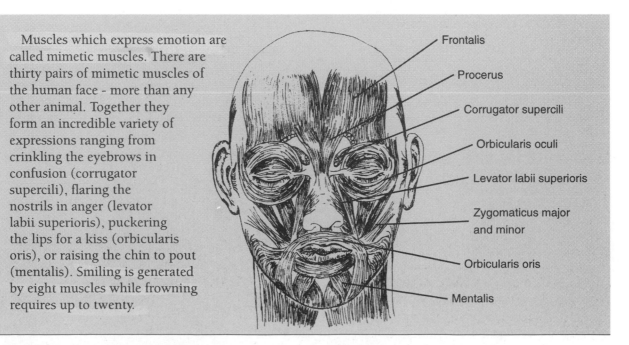

Frontalis

Procerus

Corrugator supercili

Orbicularis oculi

Levator labii superioris

Zygomaticus major and minor

Orbicularis oris

Mentalis

Other Muscles of the Head, Neck and Face

Medial and Lateral Pterygoid

Assisting the masseter and temporalis with movement of the mandible are the medial and lateral pterygoids. The medial pterygoid helps to clench the jaw, while the lateral pterygoids protract the jaw. The *medial pterygoid* is a mirror image of the shape and location of the masseter, but on the medial side of the mandible. (5.47) The *lateral pterygoid* has horizontal fibers which extend from the sphenoid bone to the joint capsule and articular disk of the temporomandibular joint. (5.48)

Portions of the medial and lateral pterygoids are accessible from both inside and outside the mouth.

Longus Capitis and Longus Colli

Tucked between the trachea and the cervical vertebrae are two small muscles - *longus capitis* and *longus colli*. Attaching from the anterior surface of the cervical vertebrae to the occiput and atlas, they primarily flex and laterally flex the head. They also help to straighten and reduce the lordotic curve of the cervical vertebrae. Each muscle has a multi-branch design, similar to the erector spinae. (5.49)

Muscles of the Tongue

There are two groups of muscles which coordinate the tongue. The three glossus muscles attach to the hyoid and other bones and move the tongue during chewing and swallowing. (5.50) Three other intrinsic muscles of the tongue are interwoven with each other and are responsible for changing the tongue's shape during speech. The tongue is basically a bag of water with a constant volume, so the intrinsic muscles mold and twist the tongue in the same way you bend and shape a water balloon.

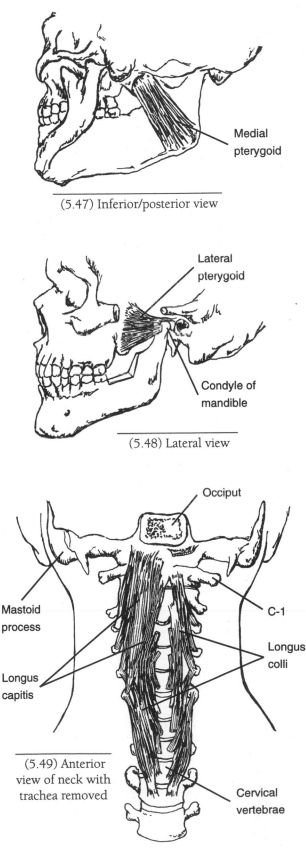

Medial pterygoid

(5.47) Inferior/posterior view

Lateral pterygoid

Condyle of mandible

(5.48) Lateral view

Occiput

Mastoid process

C-1

Longus capitis

Longus colli

Cervical vertebrae

(5.49) Anterior view of neck with trachea removed

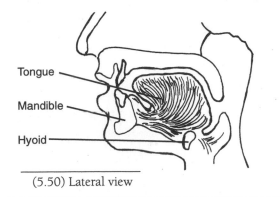

Tongue

Mandible

Hyoid

(5.50) Lateral view

pterygoid   ter-i-**goyd**   Gr. wing-shaped

Glands, Nodes and Vessels of the Head, Neck and Face

There are several accessible arteries, glands and nerves located in the neck and face. (5.51) Many are superficial and delicate and should be palpated gently. It is advised to locate and explore these structures on yourself before palpating your partner.

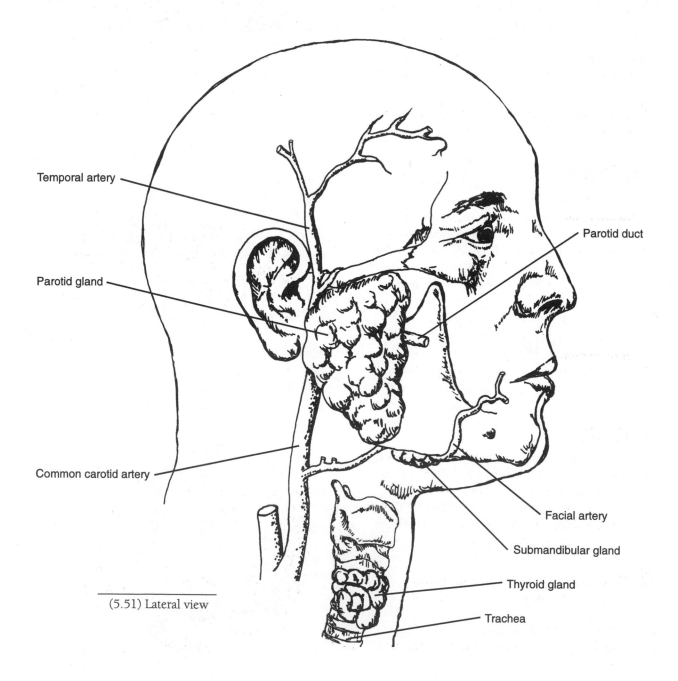

Temporal artery

Parotid gland

Common carotid artery

Parotid duct

Facial artery

Submandibular gland

Thyroid gland

Trachea

(5.51) Lateral view

Common Carotid Artery

The carotid artery is the primary source of blood for the head and neck. It rises up the anterior and lateral sides of the neck and lies deep to the sterno-cleidomastoid (SCM) and infrahyoid muscles. Its pulse can be strongly felt medial to the SCM at the level of the hyoid bone.

👁✋ 1) Supine or seated. Place two finger pads at the angle of the mandible.
2) Slide off the angle in an inferior and medial direction and press gently into the neck. (5.52) The carotid's strong pulse should be quite evident.

✔ *Are you medial to the SCM? Are you under the mandible at the level of the hyoid bone?*

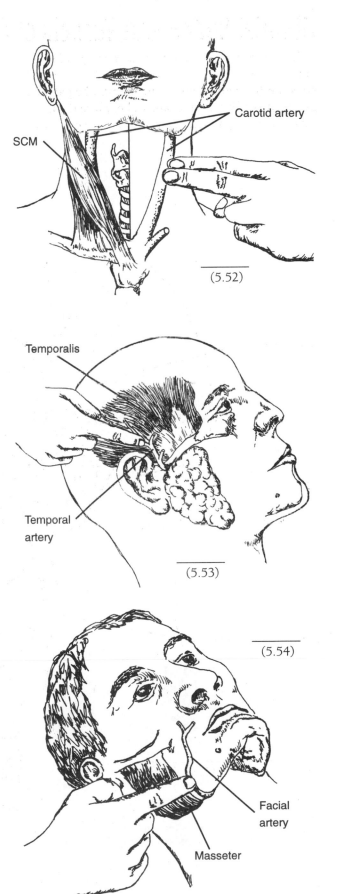

Temporal Artery

The temporal artery branches off the external carotid artery and crosses over the top of the zygomatic arch. It continues superiorly along the side of the cranium, lying superficial to the temporalis muscle. The artery's pulse can be detected in front of the ear along the zygomatic arch.

👁✋ 1) Supine or seated. Place your finger pad in front of the ear at the zygomatic arch. (5.53)
2) Gently explore and palpate for the artery's pulse. If you do not feel it, adjust your finger position and make sure your pressure is not too deep.

Facial Artery

The small, superficial facial artery branches off from the external carotid artery and curves around the base of the mandible (jaw line) toward the mouth and nose. Its pulse may be difficult to detect, but is felt along the base of the mandible at the anterior edge of the masseter.

👁✋ 1) Seated or supine. With your partner clenching her jaw, locate the masseter's anterior edge. (5.54)
2) Position your finger next to the base of the mandible and gently palpate for the artery's pulse.

✔ *Are you at the base of the mandible, along the anterior edge of the masseter?*

---

carotid   ka-**rot**-id   Gr. deep sleep

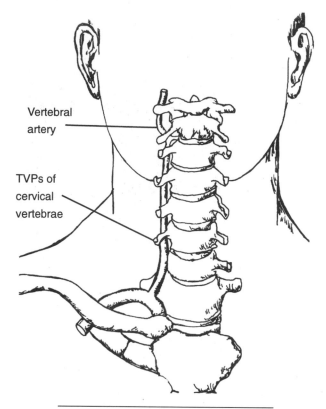

Vertebral artery

TVPs of cervical vertebrae

(5.55) Anterior view of neck and head

Vertebral Artery

The vertebral artery branches off the subclavian artery and supplies blood to the brain and spinal cord. It ascends the neck through the transverse foramen of C-6 to C-1 before passing through the foramen magnum of the occiput. It is inaccessible, of course, but a vital structure to be aware of when palpating and/or passively moving the head and neck. (5.55)

Parotid Gland and Duct
Submandibular Gland

There are three salivary glands in the neck and face: the parotid, submandibular and sublingual. Palpation of the salivary glands may stimulate production of saliva.

The *parotid gland* is located in front of the earlobe, superficial to the masseter muscle. (p. 176) It has a soft, lumpy surface and is penetrated by branches of the facial nerve. The *parotid duct* is a spaghetti-size tube extending anteriorly from the parotid gland. It ducks around the anterior edge of the masseter to funnel saliva to the mouth. (5.56)

As its name describes, the *submandibular gland* is tucked under the base of the mandible. Its round shape can be located anterior to the angle of the mandible.

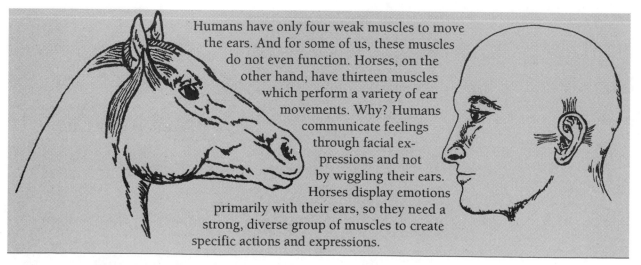

Humans have only four weak muscles to move the ears. And for some of us, these muscles do not even function. Horses, on the other hand, have thirteen muscles which perform a variety of ear movements. Why? Humans communicate feelings through facial expressions and not by wiggling their ears. Horses display emotions primarily with their ears, so they need a strong, diverse group of muscles to create specific actions and expressions.

parotid   pa-**rot**-id   Gr. beside the ear

👁✋ Parotid gland

1) Supine. Place your finger pads in front
of the earlobe on the masseter muscle.
2) Using gentle pressure along the superficial tissue,
palpate between the angle of the mandible and the
zygomatic arch for the gland's gelatinous texture.
3) Press deep to the gland and feel the striated
fibers of the masseter muscle. Compare the
different textures of these structures. (5.56)

✋ Parotid duct

1) Ask your partner to clench her jaw.
2) Place your finger pads below the zygomatic arch,
along the anterior edge of the masseter. Roll your
finger back and forth (in a superior/inferior direction) and palpate for the mobile, horizontal tube.

✔ *Are you along the anterior edge of the masseter?
Is the duct the diameter of a strand of spaghetti, and
does it run horizontally?*

👁✋ Submandibular gland

1) Place a finger along the base of the mandible.
2) Move your fingers medially, underneath the base
to palpate the superficial, marble-size gland. (5.57)

✔ *Can you roll your finger along the surface
of the gland, outlining its shape?*

Thyroid Gland

The left and right lobes of the thyroid gland are
located on the anterior surface of the trachea. The
gland lies deep to the infrahyoid muscles and has a
soft, spongy texture which can be challenging to
distinguish from the surrounding structures.

👁✋ 1) Supine or seated. Using two fingerpads,
locate the surface of the trachea between the
jugular notch and cricoid ring.
2) Palpate for the soft texture of the thyroid gland
lying on top of the trachea. Due to the gland's
delicacy, explore gently and briefly. (5.58)

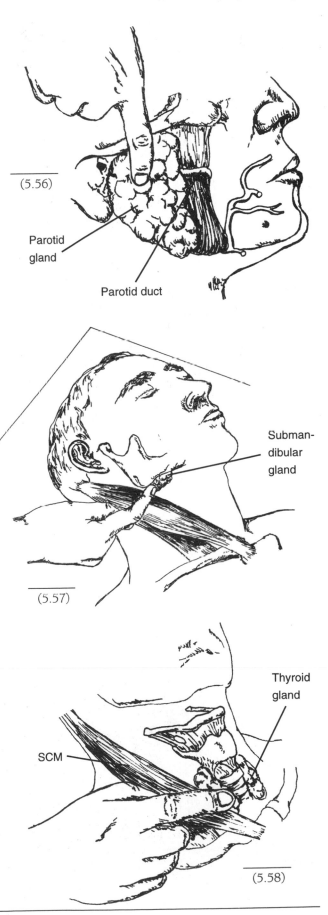

(5.56)

Parotid
gland

Parotid duct

Subman-
dibular
gland

(5.57)

Thyroid
gland

SCM

(5.58)

thyroid   **thi**-royd   Gr. shield

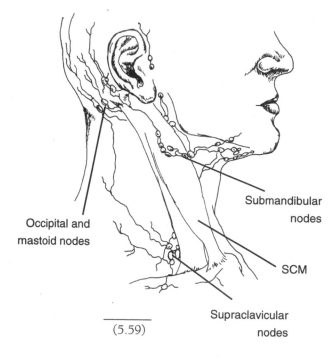

(5.59)

Occipital and
mastoid nodes

Submandibular
nodes

SCM

Supraclavicular
nodes

Cervical Lymph Nodes

The numerous bundles of lymph nodes in the cervical region are divided into two groups: superficial and deep. (5.59) The superficial cervical nodes are located primarily along the underside of the mandible, posterior and inferior to the earlobe, in the posterior triangle (p. 158) between the platysma and the deep fascia. The deep cervical nodes are larger and lie beside several large vessels and glands.

The cervical lymph nodes can often be tender when palpated. They are slightly moveable and the size and texture of soft lentils or moist raisins.

1) Supine or seated. Place your fingers on the lateral side of the neck. Using your broad finger pads, gently palpate under the skin for the superficial cervical nodes.
2) Explore the underside of the mandible and in the posterior triangle. When you have located a node, carefully outline its shape and size.

✔ *Are they slightly moveable and the size and texture of soft lentils?*

Brachial Plexus

The brachial plexus is a large bundle of nerves which innervates the shoulder and upper extremity. Exiting from the transverse processes of C-5 to T-1, it squeezes between the anterior and middle scalenes, continues inferiorly and laterally, and ducks underneath the clavicle to the axillary region. (5.60)

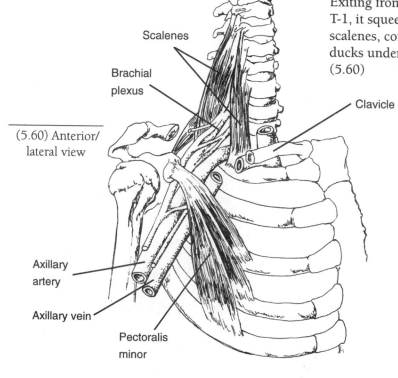

Scalenes

Brachial
plexus

(5.60) Anterior/
lateral view

Clavicle (cut)

Axillary
artery

Axillary vein

Pectoralis
minor

Although the brachial plexus can be accessed, it is best avoided. Compressing or impinging one of its nerves can create a sharp, shooting sensation down the arm.

Chapter
Pelvis & Thigh 6

Topographical Views

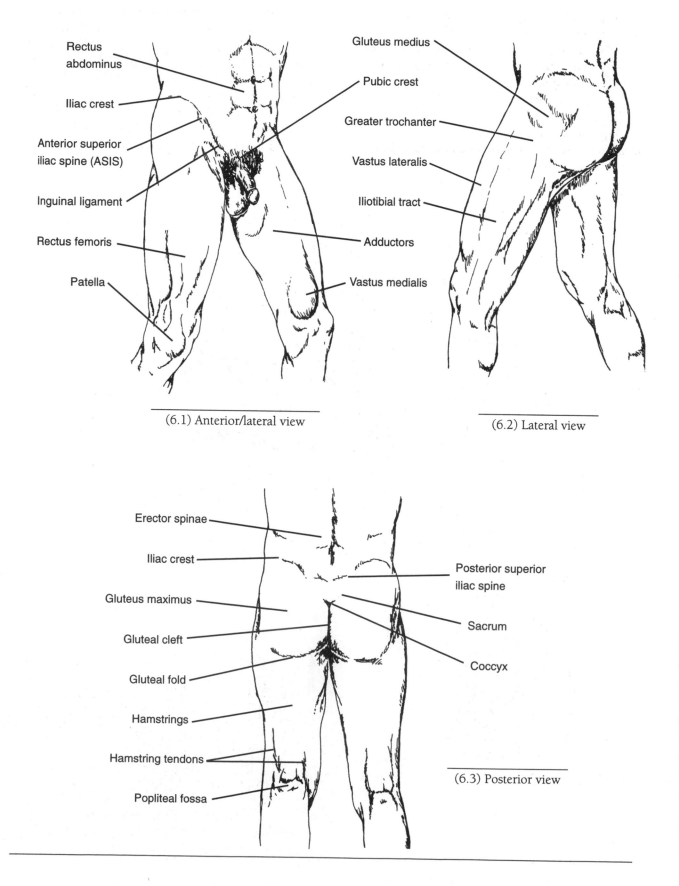

Rectus abdominus

Iliac crest

Anterior superior iliac spine (ASIS)

Inguinal ligament

Rectus femoris

Patella

Gluteus medius

Pubic crest

Greater trochanter

Vastus lateralis

Iliotibial tract

Adductors

Vastus medialis

(6.1) Anterior/lateral view

(6.2) Lateral view

Erector spinae

Iliac crest

Gluteus maximus

Gluteal cleft

Gluteal fold

Hamstrings

Hamstring tendons

Popliteal fossa

Posterior superior iliac spine

Sacrum

Coccyx

(6.3) Posterior view

Bones of the Pelvis and Thigh

The *pelvis* (pelvic girdle) consists of two hip bones and the sacrum and coccyx. (6.4) Each *hip* (coxal) bone is formed by three fused bones: the *ilium, ischium* and *pubis*. (6.5) Although the pelvis is deep to surrounding muscles, organs and adipose tissue, aspects of it are easily palpable.

The superficial *sacrum* lies posteriorly between the hip bones. The small *coccyx* extends inferiorly from the sacrum. The sacrum and coccyx, both composed of fused vertebrae, are considered part of the vertebral column.

The *femur* is the longest, heaviest and strongest bone in the body. Its proximal end articulates with the hip at the acetabulum to form the ball and socket-shaped coxal joint. Portions of the proximal femur are partially accessible. The femoral shaft is surrounded by the thick muscles of the thigh, while the distal end of the femur is superficial.

The femur articulates with the tibia to form the tibiofemoral (knee) joint. The tibiofemoral joint is a modified hinge joint, which means it can flex and extend as well as medially and laterally rotate when the knee is in a flexed position.

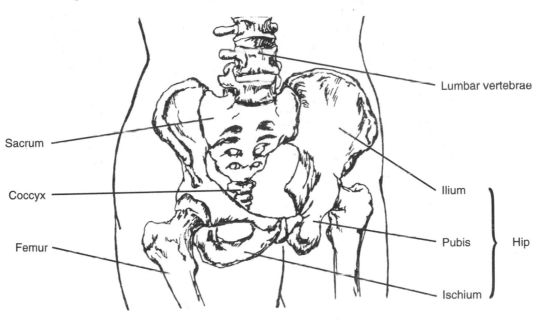

Lumbar vertebrae

Sacrum

Coccyx

Femur

Ilium

Pubis

Ischium

Hip

(6.4) Anterior/lateral view of the pelvis

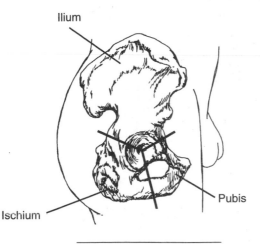

Ilium

Ischium

Pubis

(6.5) Bones of the hip

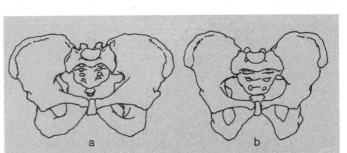

a   b

The shape of the pelvis differs between females (a) and males (b). The female pelvis is broader for carrying and delivering a child; specifically, it has a wider iliac crest, a larger pelvic "bowl" and a greater distance between the ischial tuberosities.

pelvis   **pel**-vis   L. basin

Bony Landmarks of the Pelvis and Thigh

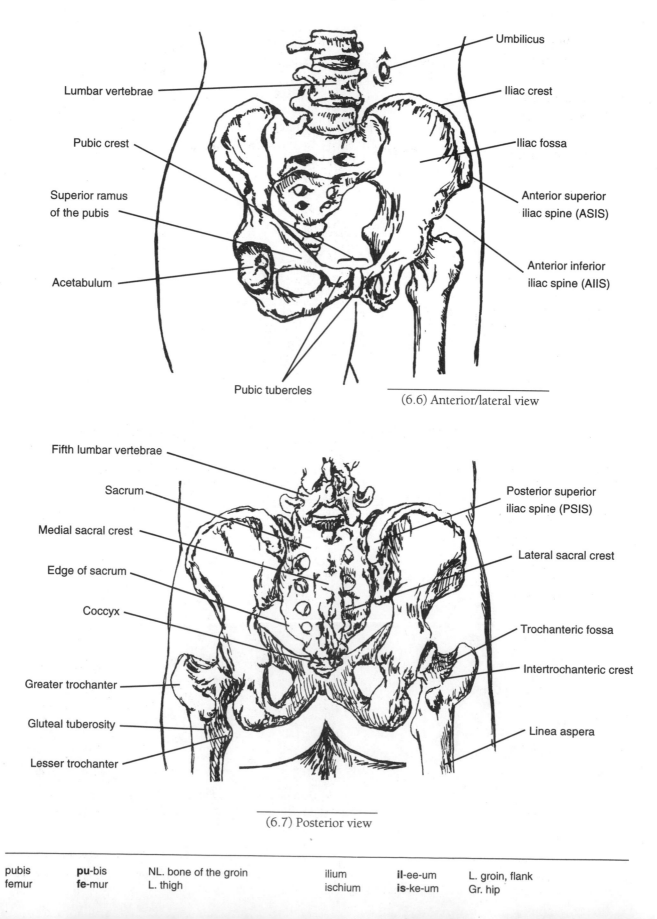

Umbilicus

Iliac crest

Iliac fossa

Anterior superior
iliac spine (ASIS)

Anterior inferior
iliac spine (AIIS)

Lumbar vertebrae

Pubic crest

Superior ramus
of the pubis

Acetabulum

Pubic tubercles

(6.6) Anterior/lateral view

Fifth lumbar vertebrae

Sacrum

Medial sacral crest

Edge of sacrum

Coccyx

Greater trochanter

Gluteal tuberosity

Lesser trochanter

Posterior superior
iliac spine (PSIS)

Lateral sacral crest

Trochanteric fossa

Intertrochanteric crest

Linea aspera

(6.7) Posterior view

| pubis | **pu**-bis | NL. bone of the groin | ilium | **il**-ee-um | L. groin, flank |
| femur | **fe**-mur | L. thigh | ischium | **is**-ke-um | Gr. hip |

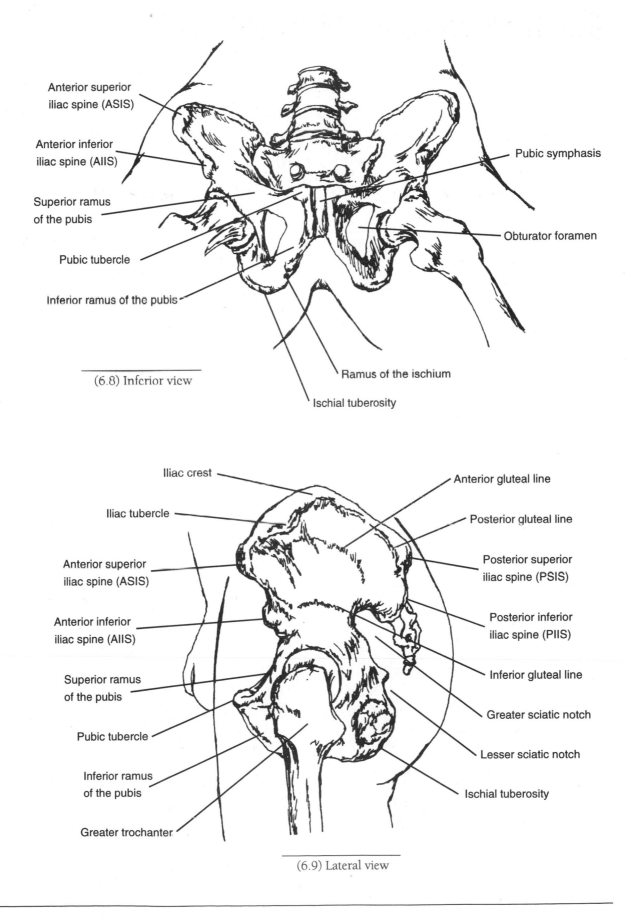

Anterior superior
iliac spine (ASIS)

Anterior inferior
iliac spine (AIIS)

Superior ramus
of the pubis

Pubic tubercle

Inferior ramus of the pubis

Pubic symphasis

Obturator foramen

Ramus of the ischium

Ischial tuberosity

(6.8) Inferior view

Iliac crest

Iliac tubercle

Anterior superior
iliac spine (ASIS)

Anterior inferior
iliac spine (AIIS)

Superior ramus
of the pubis

Pubic tubercle

Inferior ramus
of the pubis

Greater trochanter

Anterior gluteal line

Posterior gluteal line

Posterior superior
iliac spine (PSIS)

Posterior inferior
iliac spine (PIIS)

Inferior gluteal line

Greater sciatic notch

Lesser sciatic notch

Ischial tuberosity

(6.9) Lateral view

| acetabulum | **as**-e-**tab**-u-lum | L. a little saucer for vinegar | linea aspera | **lin**-e-a **as**-per-a | L. rough line |
| foramen | for-**a**-men | L. a passage or opening | | | |

Bony Landmark Trails

Trail 1 - "Solo Pass"
Due to its multifaceted shape and proximity to sensitive areas, palpation of your partner's pelvic region can be initially challenging. Trail 1 is designed to give you an opportunity to first access your own pelvic region. This will generate the confidence necessary to palpate effectively on your partner for the next three trails. (6.10) The six landmarks can be seen as your "base camps" - they are clearly accessible and will lead you to the other landmarks of the pelvis.

a) Anterior superior iliac spine (ASIS)
b) Iliac crest
c) Posterior superior iliac spine (PSIS)
d) Pubic crest
e) Ischial tuberosity
f) Greater trochanter

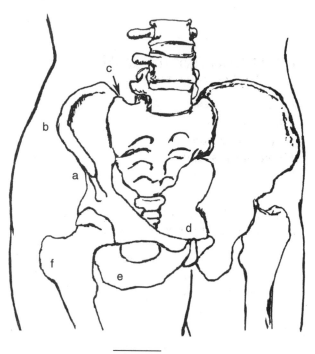

(6.10)

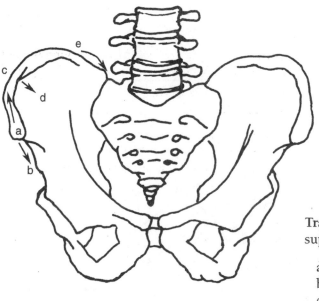

(6.11)

Trail 2 - "Iliac Avenue" travels along the superior aspect of the pelvis on the ilium. (6.11)

a) Anterior superior iliac spine (ASIS)
b) Anterior inferior iliac spine (AIIS)
c) Iliac crest
d) Iliac fossa
e) Posterior superior iliac spine (PSIS)

Trail 3 - "Tailbone Trail" accesses the bones at the base of the spine. (6.12)

a) PSIS
b) Sacrum
c) Medial sacral crest
d) Edge of the sacrum
e) Coccyx

Trail 4 - "Hip Hike" explores the lateral hip and landmarks of the proximal femur. (6.12)

a) Iliac crest
b) Greater trochanter
c) Gluteal tuberosity

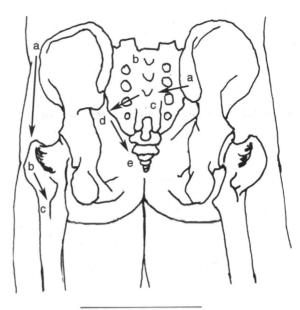

(6.12) Trails 3 and 4

Trail 5 - "The Underpass" follows around the pubic region to access the landmarks of the medial thigh. (6.13)

a) Umbilicus
b) Pubic crest and tubercles
c) Superior ramus of the pubis
d) Inferior ramus of pubis and ramus of ischium
e) Ischial tuberosity

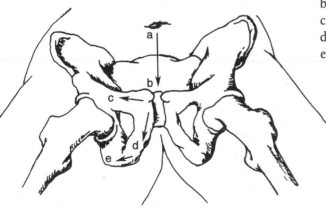

(6.13)

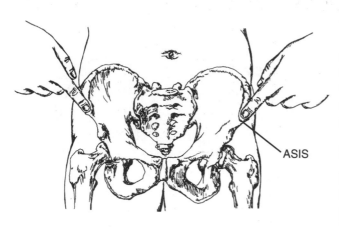

(6.14) Palpating your own ASIS

Trail 1 - "Solo Pass"
Anterior Superior Iliac Spine (ASIS)

As the name suggests, the ASIS is located on the anterior and superior aspect of the ilium. Both ASIS are the superficial tips located below the waistline underneath the pant's front pockets. The ASIS serves as an attachment for the sartorius muscle and the inguinal ligament.

1) Locate both anterior superior iliac spines by placing your hands upon your hips with fingers forward, thumbs behind. Feel for the tip of the pelvis that sticks out anteriorly. (6.14)
2) Explore these points and the surrounding structures of the ilium. Try palpating on yourself from a seated position to soften the overlaying tissue.

✔ *Are the bones you feel just beneath the surface of the skin? Are you inferior to the level of the umbilicus?*

Iliac Crest

The iliac crest is the long, superior edge of the ilium. It begins at the ASIS and extends around the side of the torso to end at the PSIS. Besides helping to keep your pants up, the iliac crest serves as an attachment site for the quadratus lumborum (p. 146) and abdominal muscles. The crest is superficial and easily palpable because the muscles attaching to it do not cross over its edge.

1) Locate the ASIS. Slowly walk your fingers around the sides of your hips, pressing into the wide edge of the crest. Note how the crest rises from the ASIS and may widen laterally soon afterward. (6.15)
2) Follow the crest as it continues around to the posterior side of the body and ends at the PSIS.

✔ *Can you sink your fingers into the flesh of the abdomen just above the iliac crest?*

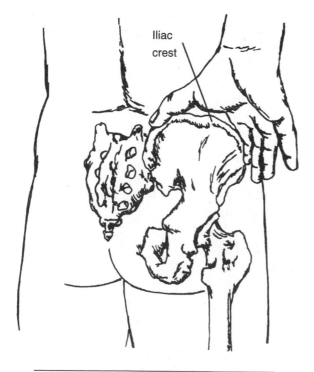

(6.15) Sliding fingers along your iliac crest

Posterior Superior Iliac Spine (PSIS)

The superficial PSIS is located at the posterior end of the iliac crest. In most persons, both PSIS are visibly identified by two small dimples found at the base of the low back. Without the help of a mirror, you may have trouble seeing your own PSIS, but you can still palpate them.

👁🖐 1) Place your thumbs upon your iliac crests. Follow the crests around the posterior hip. Note how they descend as you move medially.
2) The PSIS may feel like shallow humps surrounded by thicker tissues and are not as pronounced as the ASIS. (6.16)

✔ *Are you at the posterior end of the iliac crests? Are the points you feel three to four inches apart from each other?*

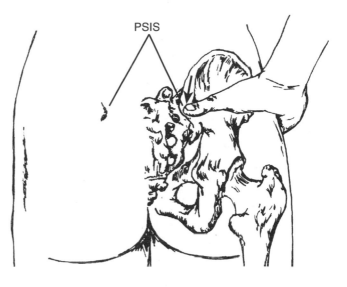

(6.16) Locating your PSIS

Pubic Crest

The pubic crest is located directly inferior to the navel and superior to the genitals. Formed by the superior, medial edge of both pubic bones, the horizontal crest is roughly two inches wide and clearly palpable. It is an attachment site for the rectus abdominus muscle and the abdominal aponeurosis.

👁🖐 1) Position your fingers at your navel.
2) Slowly slide your fingers down the center line of the body toward the pubic region. (6.17) You may travel five to eight inches before you feel the firm ridge of the pubic crest. You will be one to two inches superior to the genitals.

◈ Locate the ASIS. Follow the inguinal ligament (p. 235) inferiorly 45° to the center line of the body until you reach the crest.

✔ *Are you at the midline of the body? Are you inferior to the level of the ASIS? Do you feel a solid, horizontal ridge of bone just above the genital region?*

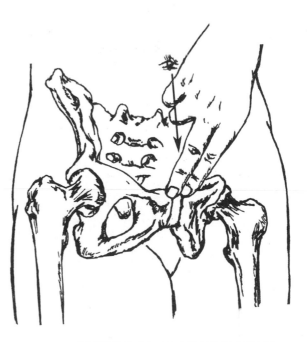

(6.17) Palpating your pubic crest

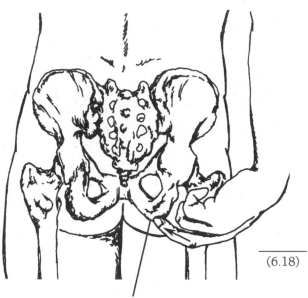

(6.18)

Ischial tuberosity

Ischial Tuberosity

If you have sat through a long music or sporting event on a metal folding chair, the ischial tuberosity is no stranger to you. The "sits bones" are located on the most inferior aspect of the pelvis at the level of the gluteal fold (the horizontal crease between the buttocks and thigh). The ischial tuberosity serves as an attachment site for the hamstrings, adductor magnus and the sacrotuberous ligament.

1) Have a seat on a hard chair or surface and rock side-to-side feeling your "sits bones."
2) Stand up and palpate the bone you were sitting on - your ischial tuberosity. (6.18) Explore in all directions the large surface of the tuberosity.

✔ *Do you feel an identical structure between the other buttock and thigh?*

When exploring the area around the sacrum and posterior iliac crest, it is not uncommon to locate small nodules of fibrofatty tissue. They are situated in the superficial fascia and may vary in size from a pea to a large marble.

Greater Trochanter of the Femur

Located distal to the iliac crest, the greater trochanter is the large, superficial mass located on the side of the hip. It is easily palpable and serves as an attachment site for the gluteus medius, gluteus minimus and deep rotator muscles (p. 228).

1) Locate the middle of the iliac crest.
2) Slide your fingerpads inferiorly four to six inches along the lateral side of the thigh until you reach the superficial mass of the greater trochanter.
3) Explore and sculpt around all sides of its wide hump.

✔ *Medially and laterally rotate your thigh as you palpate the trochanter. Do you feel its wide, knobby surface swivel back and forth under your fingers? (6.19)*

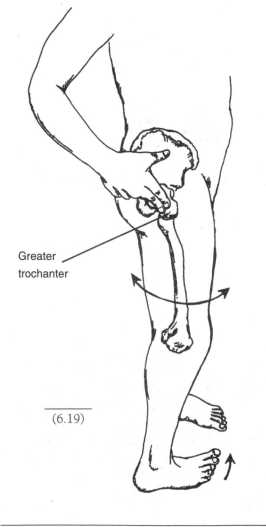

Greater trochanter

(6.19)

trochanter   tro-**kan**-ter   Gr. to run

Trail 2 - "Iliac Avenue"
Anterior Superior Iliac Spine (ASIS)

(Refer to p. 198 for more information.)

👁✋ 1) Supine. Place your hand upon the side of the abdomen, below the level of the umbilicus.
2) Gently compress inferiorly until you feel the superficial tip of the ASIS. (6.20) Palpate and observe the distance between both ASIS's and their relationship to each other.

✔ *Is the bony tip you feel superficial and inferior to the navel? Are you superior to the genital region?*

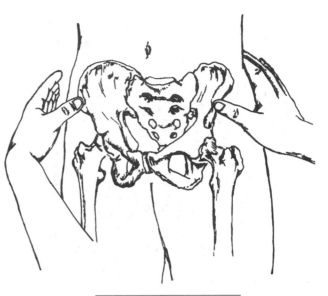

(6.20) Locating both ASIS

Anterior Inferior Iliac Spine (AIIS)

The AIIS is located inferior and medial to the ASIS and is the attachment site for the rectus femoris muscle (p. 211). Smaller and more flat than the ASIS, the AIIS is deep to the sartorius muscle and inguinal ligament. Because of its subtle shape and depth to the sartorius, the AIIS can be challenging to distinguish.

👁✋ 1) Supine. Flex the hip by bolstering under your partner's knee to shorten and soften the overlying tissue.
2) Locate the ASIS. Slide inferiorly and medially approximately one inch.
3) Palpate deep to the overlying tendons and explore for the small mound of the AIIS. (6.21)

✔ *Are you medial and inferior to the ASIS? If your partner flexes his hip slightly, can you feel the tendon of the rectus femoris tighten under your fingers? (The overlying sartorius tendon will also become taut with this action.)*

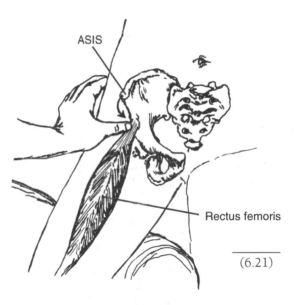

ASIS

Rectus femoris

(6.21)

Iliac tubercle

ASIS

👁✋ Located along the lateral edge of the iliac crest, there is a subtle widening called the iliac tubercle. It designates the boundary between the origin of the tensor fascia latae and gluteus medius muscles.

1) Locate the ASIS.
2) Slide posteriorly along the iliac crest approximately two inches. Explore the lateral edge of the iliac crest where the crest swells slightly. This is the iliac tubercle.

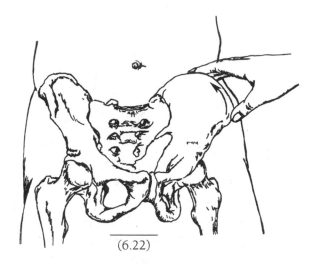

(6.22)

Iliac Crest

(Refer to p. 198 for more information.)

👁 **1)** Sidelying or supine. Locate the ASIS.
2) Slide posteriorly along the iliac crest, observing how it widens and rises up along its path.
3) Follow the crest as it continues around to the posterior side of the body to the PSIS.

✔ *Can you spread the webbing between your finger and thumb along the length of the crest? (6.22)*

Iliac Fossa

The bowl-shaped iliac fossa is located on the medial surface of the ilium and is an attachment site for the iliacus muscle. The presence of the iliacus and the abdominal contents make the majority of the fossa inaccessible. However, you can sink your fingers slowly over the iliac crest and palpate into the fossa.

👁 **1)** Supine. Flex the hip by bolstering under your partner's knee to shorten and soften the overlying tissue.
2) Lay the fingertips of one hand along the iliac crest just posterior to the ASIS.
3) Moving slowly and patiently, curl your fingertips over the lip of the iliac crest into the iliac fossa. (6.23) You may sink in only a small distance depending on the firmness of the tissue.

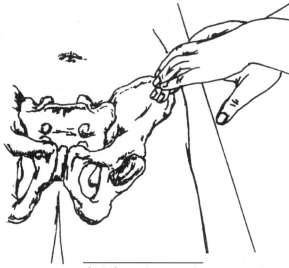

(6.23) Partner supine

Posterior Superior Iliac Spine

(Refer to p. 199 for more information.)

👁 **1)** Prone. Follow both iliac crests posteriorly around the waist.
2) Follow the crests as they descend toward the sacrum and end at either PSIS. The PSIS will feel like a shallow hump surrounded by thicker tissues. It is not as pronounced as the ASIS, but is accessible.
3) If possible, visibly locate the dimples of the low back and explore the surrounding region. (6.24)

✔ *Are you at the posterior end of the iliac crest? Are both landmarks roughly horizontal to each other and three to five inches apart?*

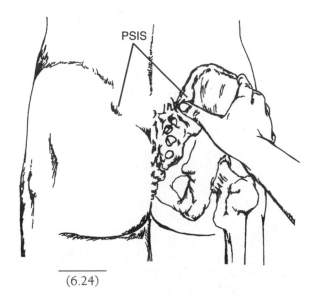

PSIS

(6.24)

fossa   **fos**-a   L. a shallow depression

Trail 3 - "Tailbone Trail"
Sacrum

Medial Sacral Crest
Edge of the Sacrum

The *sacrum* is a large, triangular bone located at the inferior end of the vertebral column. Situated between the overhanging sides of the pelvis, the sacrum is made up of a series of four or five vertebrae that are fused together.

The *medial sacral crest* is composed of three to four points located down the center of the sacrum. On either side of the medial sacral crest is a lateral sacral crest - a smaller series of bony knobs. The *edge of the sacrum* is part of the attachment site for the gluteus maximus and the sacrotuberous ligament.

The sacrum's bumpy surface lies deep to the thoracolumbar aponeurosis and sacroiliac ligaments, yet is easily accessible.

✋👁 1) Prone. Place a thumb and finger upon each PSIS and explore between and below these points for the surface of the sacrum.
2) Locate the midline of the sacrum and explore the points of the sacral crest. (6.25) Palpate superiorly to the level of the PSIS and just above the coccyx.
3) Slide your fingers laterally off the side of the sacrum, pressing your fingertips into its solid edge. (6.26) Follow the lateral edge up toward the PSIS and down to the coccyx.

✔ *How many small tips can you feel along the sacral crest? Can you follow both lateral edges inferiorly where they converge at the coccyx? If you move lateral from the lateral edge of the sacrum, can you feel the mass of the gluteus maximus (p. 217).*

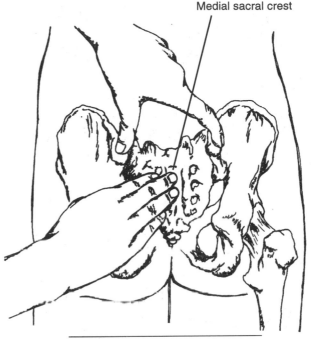

Medial sacral crest

(6.25) Using the PSIS as guides to exploring the medial sacral crest

(6.26) Palpating the edge of the sacrum

Reptiles and most birds have two sacral vertebrae while mammals have between three and five. Humans have more because, as upright creatures, the entire weight of the upper body is transferred through the sacrum to the pelvis and legs. All that remains of the spinous processes of the sacrum's vertebrae are the timeworn tips of the medial sacral crest.

sacrum   **sa**-krum   L. sacred, from its use in Roman animal sacrifice

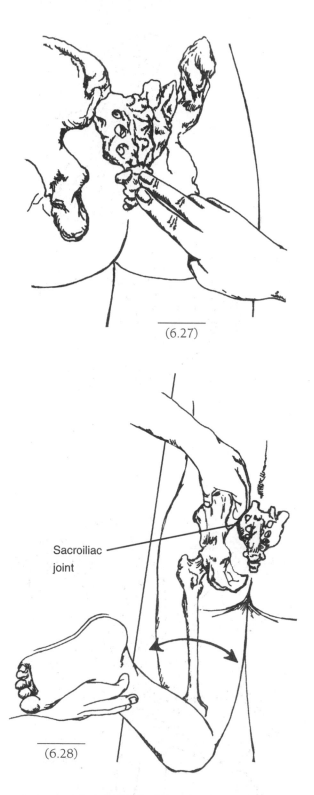

(6.27)

(6.28)

Sacroiliac joint

Coccyx

The coccyx attaches to the end of the sacrum and is located at the top of the gluteal cleft. Composed of three or four fused bones, it has a segmented, bumpy contour and can be an inch or more in length. Its tip may curve in toward the body or veer slightly to the left or right.

Because of its proximity to the gluteal cleft, initial palpation of the coccyx can be challenging for both you and your partner, so palpate your own coccyx before palpating your partner.

👁1) Prone. Walk your fingers down the medial sacral crest toward the gluteal cleft. At the top of the cleft, the bumpy surface of the coccyx will be felt.
2) Explore the coccyx surface and sides, noting how the wide upper part narrows to a tip. (6.27) The tip of the coccyx may not be accessible since it curves into the body.

✔ *Are you palpating the most inferior aspect of bone in this region? Can you sculpt out the edges of the coccyx and its shape?*

Sacroiliac Joint

The sacroiliac joint is the junction between the sacrum and the ilium. It is located medial to the PSIS and is deep to the thoracolumbar aponeurosis and sacroiliac ligaments (p. 236). The ilium overhangs the sacroiliac joint, leaving only the edge of the joint accessible.

👁1) Prone. Locate the PSIS. Move slightly inferior and medial to locate the sacroiliac joint.
2) Create a small widening at the joint by keeping one hand upon the joint while the other hand bends the knee to 90°. Then passively externally rotate the hip and feel for a small opening at the joint space. (6.28) Try internally rotating the hip as well.

✔ *Are you just medial and distal to the PSIS? Can you sculpt out the edge of the ilium as it overlaps the sacrum?*

The Greek philosopher Herophilus named the last segments of the vertebral column the "kokkyx" since it resembled a cuckoo's beak. However, during the Renaissance, the French anatomist Jean Riolan thought the term referred to the release of gas from the anus that sounded like the cry of a cuckoo. The coccyx is also referred to as the "tailbone" - an appropriate term in the human embryo. During early development a small, distinct tail extends off the sacrum, but by the eighth week it disappears, leaving just the coccyx.

coccyx   **kok**-siks   Gr. cuckoo

Trail 4 - "Hip Hike"
Greater Trochanter

(Refer to p. 200 for more information.)

👁 1) Prone. Locate the middle of the iliac crest.
2) Slide your fingerpads distally four or five inches along the side of the thigh. There you will feel the superficial hump of the greater trochanter.
3) Explore and sculpt around its two inch wide surface and all of its sides.

✔ *Flex the knee to 90°, holding the ankle. As your proximal hand palpates the trochanter, use the other hand to internally and externally rotate the hip. (6.29) Do you feel the trochanter swivel back and forth under your fingers?*

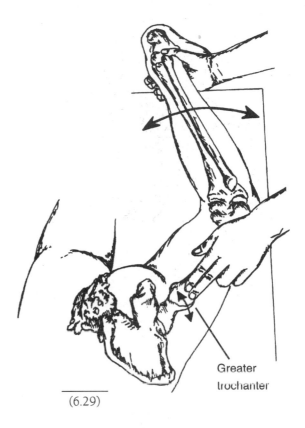

Greater
trochanter

(6.29)

Gluteal Tuberosity

The gluteal tuberosity is located distal to the posterior surface of the greater trochanter. It is a small ridge serving as an attachment site for the gluteus maximus muscle. Although it is surrounded by the gluteus maximus tendon and the upper fibers of the vastus lateralis muscles, the gluteal tuberosity is relatively superficial and accessible.

👁 1) Prone. Locate the posterior surface of the greater trochanter.
2) Slide one or two inches distally along the posterior shaft of the femur until you feel the solid surface of the tuberosity. (6.30) It may not feel like a ridge, but more like a flat, superficial portion of bone.

✔ *Can you press into the area you are palpating and feel the superficial surface of the femur? Are you directly lateral to the ischial tuberosity (p. 208)?*

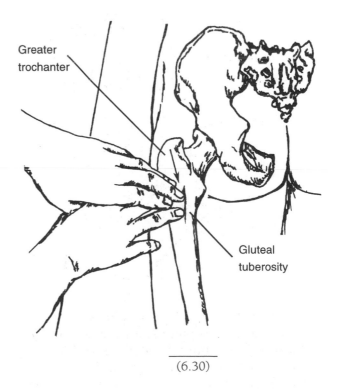

Greater
trochanter

Gluteal
tuberosity

(6.30)

Here are a few suggestions to make sure this route is comfortable for yourself and your partner:

a) Explain to your partner what it is you will be doing and obtain permission to proceed.

b) If your partner would be more comfortable, use his or her hand to palpate with your hand guiding on top.

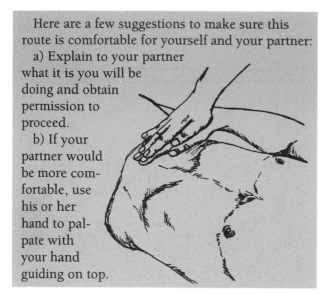

Trail 5 - "The Underpass"
Umbilicus

The umbilicus (or navel) will, of course, be visible when the abdomen is undraped. If not exposed, it can be felt at the center line of the body, superior to the level of the anterior superior iliac spines (ASIS).

Pubic Crest and Tubercles

Refer to p. 199 for more information about the pubic crest. The pubic tubercles are located on the superior aspect of the pubic crest. Each tubercle is shaped like a small horn, serving as an attachment site for the adductor longus muscle and the inguinal ligament. The tubercles may be one to two inches apart and are not always easily palpable.

1) Supine. Locate your partner's umbilicus.
2) Using firm, yet gentle pressure, walk your fingers inferiorly along the midline of the body.
3) Move five to seven inches until you feel the superficial pubic crest. (6.31) Explore the horizontal ridge of the crest. Remember that the pubic crest is the only horizontal stretch of bone in this vicinity.
4) Move laterally and explore for the tips of the pubic tubercles. Palpate both tubercles noting the distance between them.

Begin at the ASIS and follow the inguinal ligament (p. 235) inferiorly and medially at 45° to the pubic tubercle.

Do you feel a firm, bony prominence inferior and medial to the level of the ASIS? For the pubic tubercles, are the bony prominences you feel on the superior part of pubic crest? Are the tubercles on the same level as the greater trochanters?

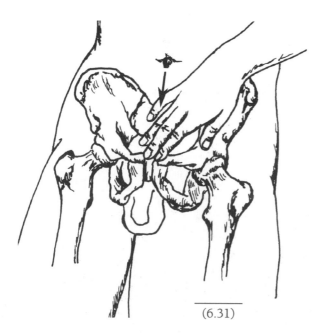

(6.31)

---

umbilicus   um-**bil**-i-kus   L. a pit
symphasis   **sim**-fi-sis   Gr. growing together

Superior Ramus of the Pubis

The superior ramus of the pubis spans 45° from the pubic tubercle toward the AIIS. It forms a ridge that serves as an attachment site for the pectineus muscle (p. 223). Since it is deep to the inguinal ligament and a neurovascular bundle, the superior ramus can be challenging to palpate.

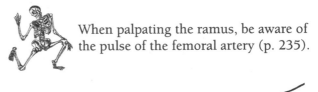

When palpating the ramus, be aware of the pulse of the femoral artery (p. 235).

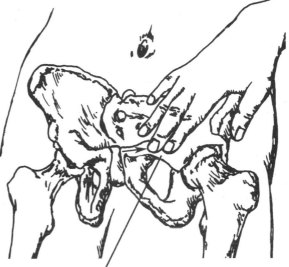

1) Supine. Place your flexed knee under your partner's knee. This position will flex and externally rotate the hip and allow for easier palpation.
2) Locate the pubic crest. (6.32) Slide laterally off the pubic crest toward the AIIS. Sink into the tissue, feeling for the buried ridge of the superior ramus.

✔ *Are you lateral and slightly superior to the pubic tubercle? If you cannot feel the edge of the ramus, can you sense its density beneath the superficial tissue?*

Superior ramus   (6.32)

Inferior Ramus of the Pubis and Ramus of the Ischium

The two rami are located along the inferior aspect of the pelvis and together form a bridge between the pubic crest and the ischial tuberosity. The ramus of the pubis, the anterior half of the bridge, serves as an attachment for the gracilis and adductor brevis muscles; both rami are attachment sites for the adductor magnus muscle. (6.33) When palpating the rami, use your fingertips and keep them close to the medial thigh. The angle of the rami will be wider on females than males.

(6.33) Muscle attachment sites along the rami

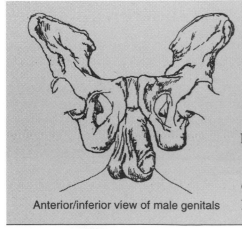

Anterior/inferior view of male genitals

"How do you access structures which are situated close to the genitals?" Actually, all of the bony landmarks, tendons and blood vessels in this region can be easily palpated without contacting the genitals. If you follow the given instructions, the comfort of your partner and yourself will be maintained.

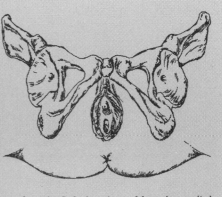

Anterior/inferior view of female genitals

ramus   **ray**-mus   L. branch

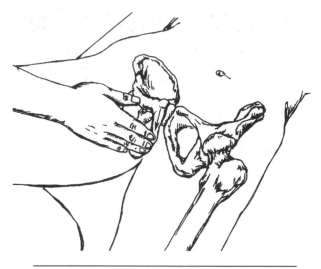

(6.34) Palpating the rami to the ischial tuberosity

✋👁 **1)** Supine. Place your flexed knee under your partner's knee. This position will flex and externally rotate the hip and allow for easier palpation.
2) Locate the pubic crest.
3) Move to the lateral edge of the pubic crest and slide posteriorly around the medial thigh. (6.34) Using slow, but firm pressure, palpate for the hard ridge of the rami. This "bridge of bone" is the only bony mass in the area, so if you are pressing on a solid line of bone, you have found it.
4) Continue around the leg until you reach the large ischial tuberosity.

✔ *As you follow the rami around the medial thigh do they lead you posteriorly around the inside of the thigh? As you move around the thigh do you feel the rami widen laterally? Can you feel any of the adductor tendons (p. 222) attach to the rami?*

Ischial Tuberosity

(Refer to p. 200 for more information.)

✋👁 **1)** Prone. Locate the gluteal fold, the horizontal line between the buttock and thigh. Place your fingers at the center of the gluteal fold and press superiorly and medially until your fingertips bump into the large surface of the ischial tuberosity. (6.35)
2) Explore all sides of its large mass and note its relationship to the greater trochanter.

◆ Sidelying. Flex the top hip. Place your hand on the medial thigh. Slide proximally to the gluteal fold and ischial tuberosity. (6.36)

✔ *Are you palpating between the inferior buttock and proximal thigh? Can you feel the large hamstring tendons attach to the ischial tuberosity?*

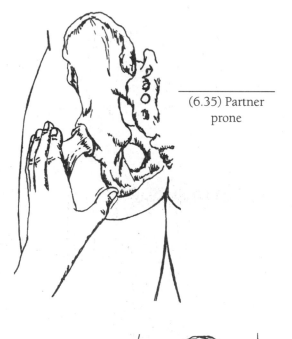

(6.35) Partner prone

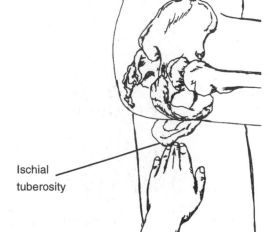

Ischial tuberosity

(6.36) Partner sidelying

Muscles of the Pelvis and Thigh

The muscles of the pelvis and thigh primarily create movement at the coxal (hip) and tibio-femoral (knee) joints. The majority of the hip and thigh muscles can be divided into five groups.

Two groups are located in the buttock region: Three *gluteal* muscles give shape to the buttock and lateral hip. Six small *lateral rotators* are deep to the gluteals.

Three groups form the mass of the thigh: Four *quadriceps* are located on the thigh's anterior and lateral surfaces. Three long *hamstrings* lie along the posterior thigh. Five *adductors* are tucked between the quadriceps and hamstrings along the medial thigh. Additional muscles include the iliopsoas, sartorius and tensor fascia latae.

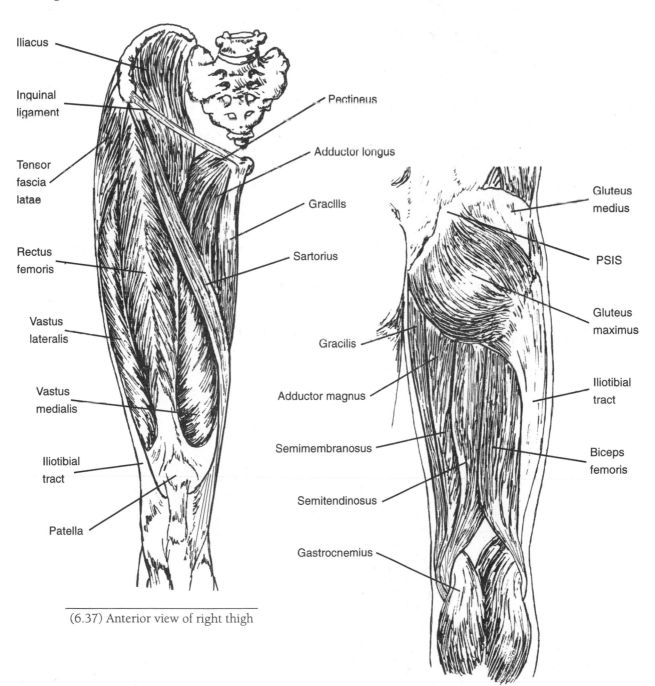

Iliacus

Inguinal ligament

Tensor fascia latae

Rectus femoris

Vastus lateralis

Vastus medialis

Iliotibial tract

Patella

Pectineus

Adductor longus

Gracilis

Sartorius

(6.37) Anterior view of right thigh

Gluteus medius

PSIS

Gluteus maximus

Iliotibial tract

Biceps femoris

Gracilis

Adductor magnus

Semimembranosus

Semitendinosus

Gastrocnemius

(6.38) Posterior view of right buttock and thigh

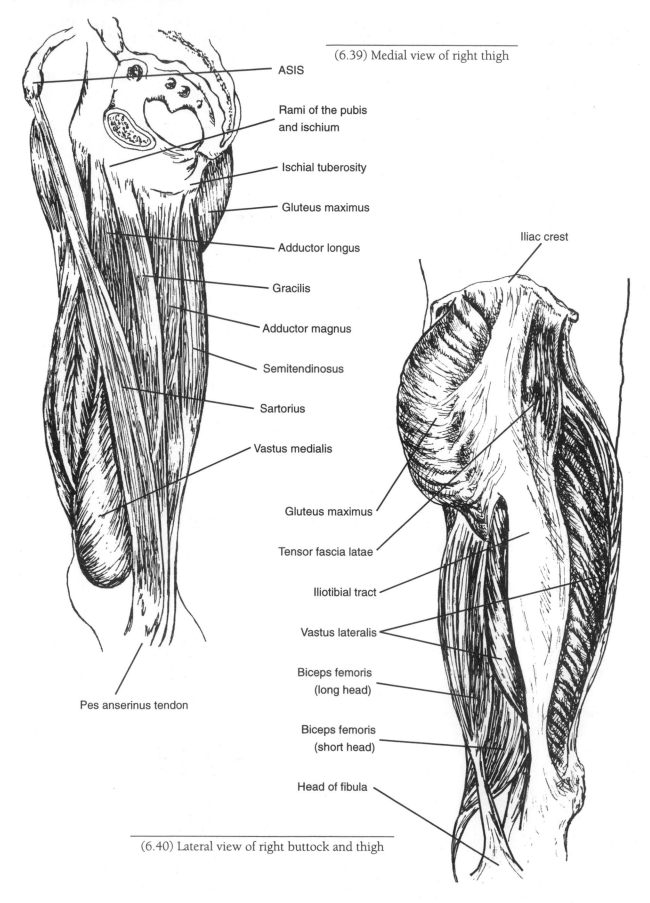

ASIS

Rami of the pubis
and ischium

Ischial tuberosity

Gluteus maximus

Adductor longus

Gracilis

Adductor magnus

Semitendinosus

Sartorius

Vastus medialis

Pes anserinus tendon

(6.39) Medial view of right thigh

Iliac crest

Gluteus maximus

Tensor fascia latae

Iliotibial tract

Vastus lateralis

Biceps femoris
(long head)

Biceps femoris
(short head)

Head of fibula

(6.40) Lateral view of right buttock and thigh

Quadriceps Femoris Group

Rectus Femoris
Vastus Medialis
Vastus Lateralis
Vastus Intermedius

The four, large quadriceps primarily extend the knee. The cylindrical, superficial *rectus femoris* is located on the anterior thigh and is the only quadricep that crosses two joints - the hip and knee. (6.37) *Vastus intermedius* is deep to the rectus femoris and only its edges are sometimes accessible when the rectus femoris is shifted to the side. (6.41)

The *vastus lateralis* single-handedly forms the musculature of the lateral thigh. Its posterior edge lies next to the biceps femoris, one of the hamstrings. Vastus lateralis is deep to the iliotibial tract (p. 224), but its fibers are still easily accessible. (6.40) The palpable aspect of *vastus medialis* forms a "tear drop" shape at the distal, medial thigh. (6.37)

The four quadricep muscles converge to a single tendon above the knee. The tendon connects to the top and sides of the patella before attaching to the tibial tuberosity.

A -
All:
     Extend the knee
Rectus Femoris:
     Flex the hip joint

O -
Rectus Femoris:
     Anterior inferior iliac spine
Vastus Lateralis:
     Lateral lip of linea aspera, gluteal tuberosity
Vastus Medialis:
     Medial lip of linea aspera
Vastus Intermedius:
     Anterior and lateral shaft of the femur

I - Tibial tuberosity

N - Femoral

The distal tendon of the quadriceps and the patellar ligament are the same structure. Because the tendon attaches a bone to another bone (the patella to the tibia), it is considered a ligament.

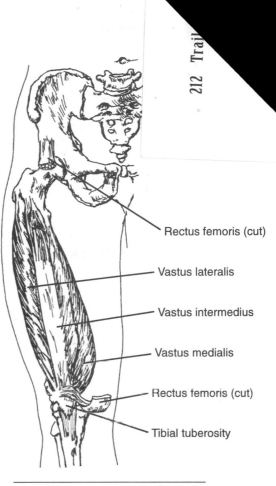

Rectus femoris (cut)
Vastus lateralis
Vastus intermedius
Vastus medialis
Rectus femoris (cut)
Tibial tuberosity

(6.41) Anterior view of right thigh

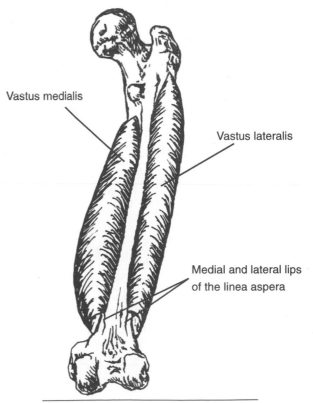

Vastus medialis
Vastus lateralis
Medial and lateral lips of the linea aspera

(6.42) Posterior view of right femur

quadriceps   **kwod**-ri-seps   L. four + head
rectus   **rek**-tus   L. straight

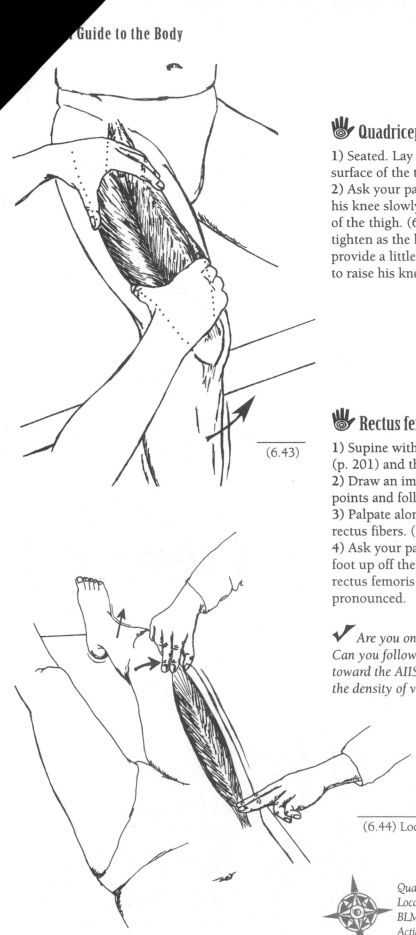

👁 Quadriceps as a group

1) Seated. Lay the flat of your hands on the anterior surface of the thigh.
2) Ask your partner to alternately extend and relax his knee slowly. Explore the lateral and medial sides of the thigh. (6.43) Do you feel the quadriceps tighten as the leg extends? For greater contraction, provide a little resistance below the knee as he tries to raise his knee.

(6.43)

👁 Rectus femoris

1) Supine with knee bolstered. Locate the AIIS (p. 201) and the patella (p. 246).
2) Draw an imaginary line between these two points and follow the path of the rectus femoris.
3) Palpate along this line and strum across the rectus fibers. (It will be two to three fingers wide.)
4) Ask your partner to flex his thigh and hold his foot up off the table. (6.44) This position contracts rectus femoris making its definition more pronounced.

✔ *Are you on the anterior surface of the thigh? Can you follow the muscle belly to the patella and toward the AIIS? Can you shift it to the side and feel the density of vastus intermedius beneath it?*

(6.44) Locating the rectus femoris

Quadriceps
Location - Superficial, anterior and lateral thigh
BLMs - Patella, tibial tuberosity
Action - "Straighten your knee" or "Flex your hip"

👁✋ Vastus medialis

1) Supine with knee bolstered. Ask your partner
to contract his quadriceps by extending his knee
completely. Palpate just medial and proximal to
the patella for the bulbous shape of the medialis.
2) Locate the rectus femoris and sartorius (p. 225),
noting how these muscles border the medialis,
forming its long "tear drop" shape (6.45).

✔ *Are you medial to the rectus femoris?*
Can you make out the round shape of the vastus
medialis and follow its fibers to the patella?

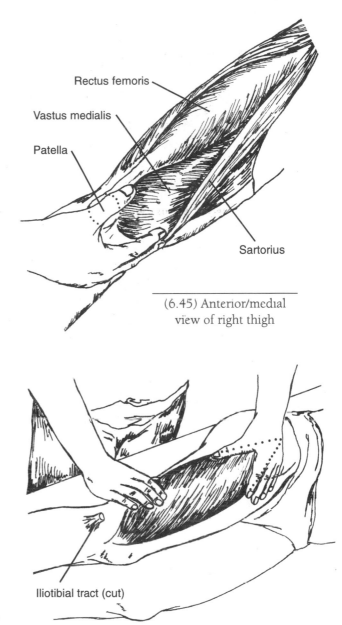

(6.45) Anterior/medial
view of right thigh

👁✋ Vastus lateralis

1) Sidelying. Place the flat of your hand on
the lateral side of the thigh while your partner
slowly extends and relaxes his knee. (6.46)
Note the vastus lateralis contracting and relaxing.
2) Palpate its entire belly - posteriorly to the
biceps femoris (p. 214) and proximally to the
greater trochanter. With the leg relaxed,
identify the direction and depth of the lateralis
fibers and the superficial iliotibial tract (p. 224).

✔ *Can you follow its fibers to the patella? Can you*
differentiate between the vertical fibers of the iliotibial
tract and the deeper, oblique fibers of the lateralis?

(6.46) Palpating the vastus
lateralis with partner sidelying

Ask your partner to extend his knee
fully by contracting his quadriceps.
Observe and palpate the distal ends of
the vastus medialis and vastus lateralis.
Do you notice how the vastus medialis
extends further distal than the vastus
lateralis?

The reason for this variance concerns the tracking (or movement) of
the patella. The angle of the femur,
combined with the pull of the quadriceps, creates a natural tendency for
the patella to track laterally. This is

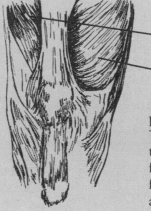

prevented, however, in two ways.
The edge of the lateral condyle of
the femur (p. 248) is elevated to
form a lateral wall and the distal
fibers of vastus medialis are set at an
angle, pulling the patella medially.

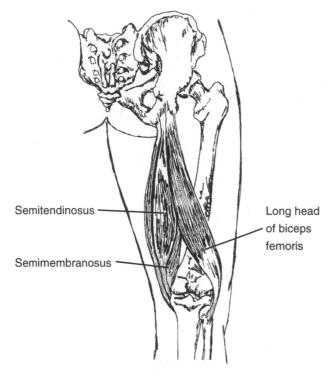

(6.47) Posterior view of right thigh
showing superficial hamstrings

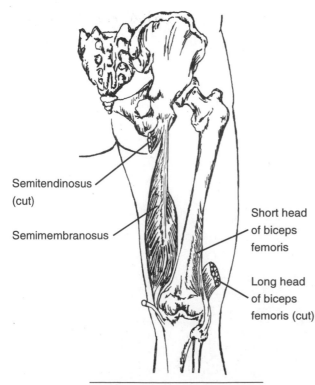

(6.48) Deep layer of hamstrings

Hamstrings

Semitendinosus
Semimembranosus
Biceps Femoris

The hamstrings are located along the posterior thigh between the vastus lateralis and adductor magnus. (6.47) Comparatively, the hamstrings are not as massive as the quadriceps femoris group, yet they are strong hip extensors and knee flexors. All three hamstrings have a common origin at the ischial tuberosity. Their tubular bellies extend superficially down the thigh before becoming long, thin tendons which stretch across the posterior knee. As a group, the hamstrings and their distal tendons are easily palpable.

Biceps femoris is the lateral hamstring. It has two heads - a superficial long head and a deep, inaccessible short head. (6.48) The two "semi" muscles are the medial hamstrings: the *semitendinosus* lies superficial to the wide *semimembranosus*.

A -
All:
> Flex the knee
> Extend the hip
> Tilt the pelvis posteriorly

Biceps Femoris:
> Laterally rotate the hip

Semis:
> Medially rotate the hip

O -
Semis and long head of Biceps Femoris:
> Ischial tuberosity

Short head of Biceps Femoris:
> Lateral lip of linea aspera

I -
Biceps Femoris:
> Head of the fibula

Semitendinosus:
> Proximal, medial shaft of the tibia
> at pes anserinus tendon

Semimembranosus:
> Posterior aspect of medial condyle of tibia

N -
Biceps Femoris: Peroneal and sciatic
Both Semis: Sciatic

semimembranosus   **sem**-eye-**mem**-bra-**no**-sus
semitendinosus   **sem**-eye-**ten**-di-**no**-sus

The term "hamstring" originated in eighteenth century England. Butchers would display pig carcasses in their shop windows by hanging them from the long tendons at the back of the knee.

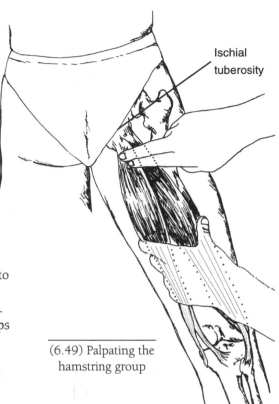

Ischial
tuberosity

(6.49) Palpating the
hamstring group

☝ Hamstrings as a group

1) Prone. Place a hand on the posterior thigh between the buttocks and knee. Ask your partner to flex his knee, holding his foot off the table. As the hamstrings contract, explore their mass and width.
2) Locate the ischial tuberosity. Slide your fingertips distally one inch and strum across the large, solid tendon of the hamstrings. (6.49)
3) Follow the tendon distally as it spreads out into the separate bellies of the hamstrings.

✔ *Follow the bellies proximally. Do they attach to the ischial tuberosity? Follow the bellies distally. Do you feel their skinny tendons along the posterior knee?*

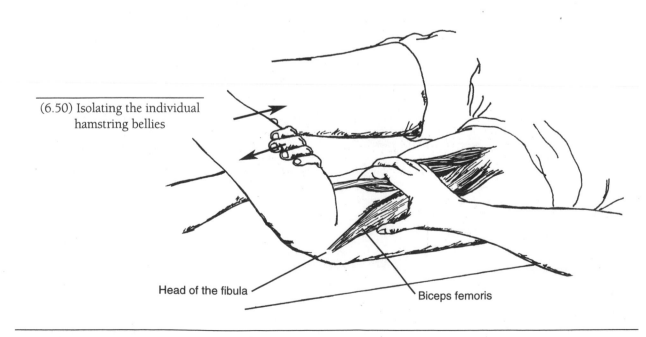

(6.50) Isolating the individual
hamstring bellies

Head of the fibula

Biceps femoris

---

biceps femoris   **bye**-ceps fe-**mor**-is
pes anserinus   pes **an**-ser-**i**-nus   L. *pedes*, foot, L. *anserine*, like a goose

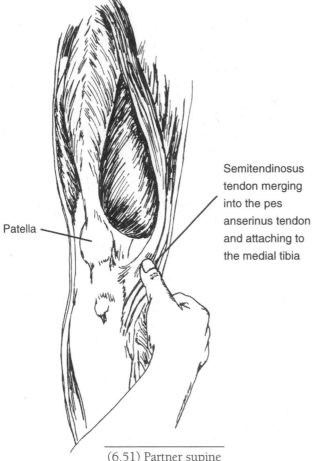

Patella

Semitendinosus tendon merging into the pes anserinus tendon and attaching to the medial tibia

(6.51) Partner supine

✋👁 Individual bellies and distal tendons

1) Ask your partner to hold his knee in a flexed position. Again, explore the belly of the hamstrings.
2) The lateral half of the hamstring belly is the biceps femoris. Its belly will lead toward the head of the fibula. Palpate on the lateral side of the knee for the long, prominent tendon of the biceps femoris and follow it toward the fibula. (6.50)
3) The medial half of the hamstrings is the layered bellies of the semitendinosus and semimembranosus. Move to the medial side of the knee and palpate for the tendons of these muscles.
4) The most superficial tendon will be the semitendinosus. Turn your partner supine and follow it distally as it merges with the pes anserinus tendon. (6.51) The semimembranosus is tucked deep to the semitendinosus and often difficult to isolate.

✔ *Are the tendons along the back of the knee slender and superficial? Does the biceps femoris tendon lead to the head of the fibula? Can you follow the semis as they seem to disappear into the medial knee?*

Hamstrings
Location - Superficial, posterior thigh
BLMs - Ischial tuberosity, tendons of posterior knee
Action - "Bend your knee" or "Extend your thigh"

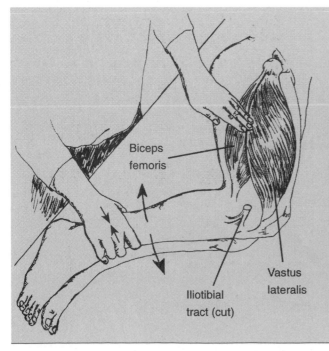

Biceps femoris

Iliotibial tract (cut)

Vastus lateralis

How do you differentiate the vastus lateralis from the biceps femoris on the posterior, lateral thigh? Have them do what comes naturally - be antagonists.

✋👁 **1)** Sidelying. Bend the top knee to 90° and clasp the ankle. Lay your other hand on the lateral thigh.
2) Ask your partner to alternate between flexing and extending his knee ever-so-slightly against your resistance. Sense how the vastus lateralis contracts while extending and how the biceps femoris remains limp. The opposite will happen when flexing the knee.
3) Often there will be a palpable dividing line or depression between the edges of these muscles.

Gluteals

Gluteus Maximus
Gluteus Medius
Gluteus Minimus

The three gluteal muscles are located in the buttock region, deep to the surrounding adipose tissue. The large, superficial *gluteus maximus* is the most posterior of the group and has parallel fibers running diagonally across the buttock. (6.52)

The *gluteus medius* is located on the side of the hip and is also superficial, except for the posterior portion which is deep to the maximus. (6.53) Both the gluteus maximus and medius are strong extensors and abductors of the hip. The gluteus medius could be thought of as the "deltoid muscle of the coxal joint" with its convergent fibers that pull the femur in multiple directions.

The *gluteus minimus* lies deep to the gluteus medius and is inaccessible. (6.54) However, the density of the gluteus minimus fibers can be felt deep to the gluteus medius. Because it attaches to the anterior surface of the greater trochanter, the gluteus minimus performs the opposite movement of the other gluteals by flexing and internally rotating the thigh.

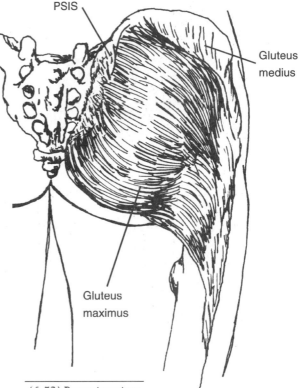

PSIS

Gluteus
medius

Gluteus
maximus

(6.52) Posterior view

Gluteus Maximus

A - Extend
   Laterally rotate
   Abduct hip
   Lower fibers adduct the hip

O - Coccyx, posterior sacrum, posterior iliac crest, sacrotuberous and sacroiliac ligaments

I - Gluteal tuberosity and iliotibial tract

N - Gluteal

Gluteus Medius

A - Abduct
   Flex
   Extend the hip
   May medially rotate and laterally rotate hip

O - External surface of the ilium between iliac crest and posterior and anterior gluteal line

I - Greater trochanter

N - Gluteal

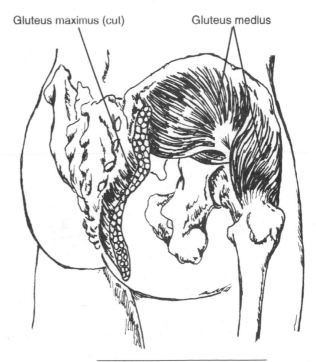

Gluteus maximus (cut)   Gluteus medius

(6.53) Posterior/lateral view

gluteus   **gloo**-te-us   Gr. *gloutos*, buttocks, which in turn is Anglo-Saxon for *buttuc*, meaning end

Gluteus medius (cut)　　　Gluteus minimus

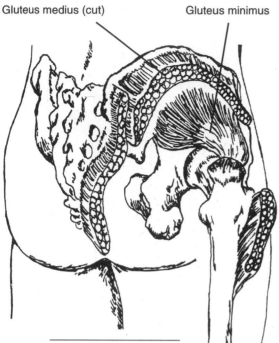

(6.54) Gluteus minimus

Gluteus Minimus

A - Abduct
　　Medially rotate
　　Flex the hip

O - External surface of the ilium between anterior and inferior gluteal lines

I - Anterior border of greater trochanter

N - Gluteal

👁 Gluteus maximus

1) Prone. Locate the coccyx, the edge of the sacrum, the PSIS and the posterior two inches of the iliac crest to isolate the landmarks which form the origin of the maximus.
2) Locate the insertion of the maximus at the gluteal tuberosity. (6.55)
3) Connect its origin to its insertion by drawing the fiber direction on your partner. Then palpate its thick, superficial fibers. Also notice the textural and depth differences between the adipose tissue of the buttock and the muscle fibers of the maximus. The adipose is superficial to the maximus and often has a soft, gel-like consistency.

✔ *Ask your partner to extend his hip. (6.56) Palpate the bulging fibers which lead to the gluteal tuberosity. If extending the leg is difficult while prone, try palpating with the knee flexed or with your partner standing.*

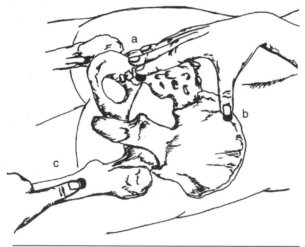

(6.55) Isolating the borders of the gluteus maximus: a) coccyx, b) posterior iliac crest, c) gluteal tuberosity

(6.56) Partner contracts gluteus maximus by extending the hip

👁 Gluteus medius and minimus

1) Sidelying. Isolate the shape of the gluteus medius by laying the webbing of one hand along the iliac crest (from PSIS to nearly the ASIS) while the other hand locates the greater trochanter.
2) Your hands will form the pie-shaped outline of the gluteus medius. (6.57)
3) Palpate in this area from just below the iliac crest to the greater trochanter for the dense fibers of the gluteus medius.
4) Sink your fingers deep to the gluteus medius and explore for the density and mass of the gluteus minimus.

✔ *Ask your partner to slightly abduct his thigh. (6.58) Do you feel the medius contract?*

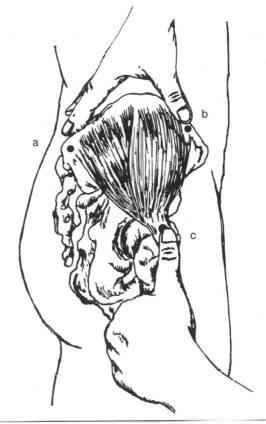

(6.57) Isolating the borders of the gluteus medius:
a) PSIS, b) iliac crest, c) greater trochanter

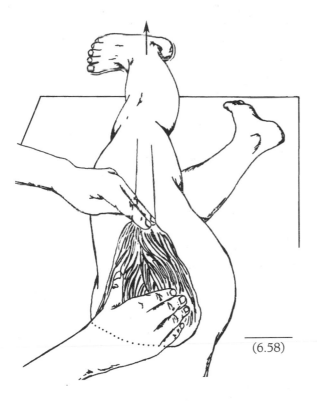

(6.58)

Gluteus Maximus
Location - Superficial, posterior hip
BLMs - Edge of sacrum, gluteal tuberosity
Action - "Extend your hip"

Gluteus Medius
Location - Partially superficial, lateral hip
BLMs - Iliac crest, greater trochanter
Action - "Abduct your hip"

Gluteus Minimus
Location - Deep to medius
BLMs - Greater trochanter, gluteus medius
Action - "Abduct your hip"

Humans are not only unique among animals in respect to their extra large brains, but also because of their buttocks. No other mammal has such deposits of adipose tissue in the gluteal region and no one seems to know *why* we have them. It was thought that the buttocks gave us something to sit upon, but we really sit on our ischial tuberosities. And for good reason; the gluteus maximus and gluteal fascia would be compressed beneath us if we did not.

Women have larger buttocks than men, so it was conjectured that the buttocks were food storage organs during pregnancy. Not so.

One thing is known: The gluteal fold - the crease between the buttock and thigh - helps localize the subcutaneous adipose at the top of the thigh. Biomechanically, it is easier to swing the thigh back and forth during walking with the tissue located proximally rather than dispersed down the thigh.

Adductor Group

Adductor Magnus
Adductor Longus
Adductor Brevis
Pectineus
Gracilis

Situated between the hamstrings and quadriceps femoris muscles, the five adductors are located along the medial thigh. (6.59) Their proximal tendons attach at specific locations along the base of the pelvis. These tendons together form a drape of connective tissue extending from the superior ramus of the pubis to the ischial tuberosity. (6.60)

When viewing the thigh anteriorly, the muscle bellies of the adductors lie in three layers. The *pectineus* and *adductor longus* are most anterior. Behind them is the *adductor brevis*, and furthest posterior is the *adductor magnus*. (6.61) Known as the "floor of the adductors," the broad span of adductor magnus lies anterior to the hamstrings. (6.62)

These four muscles tuck behind the quadriceps femoris group and insert on the posterior femur. *Gracilis*, the fifth adductor, lays superficially on the medial thigh and is the only adductor that crosses the knee.

As a group, the adductors are easily located, although their specific bellies can be challenging to isolate. When palpating the adductor tendons near the pubic bone, there will be one prominent tendon that extends off, or near, the pubic tubercle. The source of this superficial tendon is either the gracilis or adductor longus. In some cases it is a merging of both muscles.

In either case, the tendon can serve as an important guidepost not only for locating gracilis and adductor longus, but also pectineus and adductor magnus. The pectineus will be located on the anterior side of this tendon, while the adductor magnus will be located posterior to it.

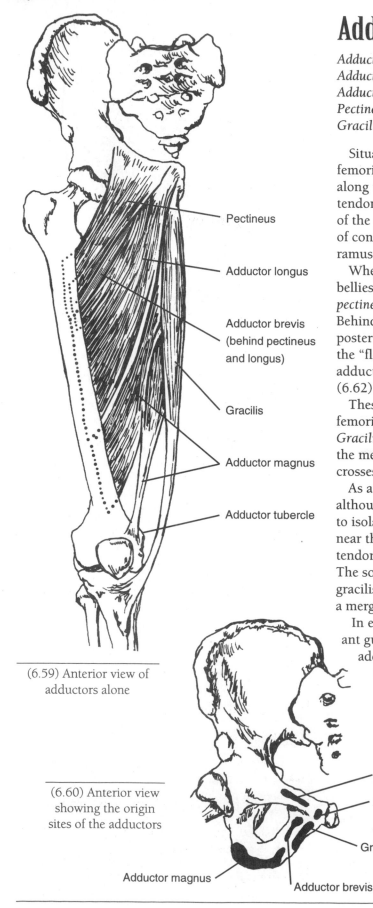

Pectineus

Adductor longus

Adductor brevis
(behind pectineus
and longus)

Gracilis

Adductor magnus

Adductor tubercle

(6.59) Anterior view of
adductors alone

(6.60) Anterior view
showing the origin
sites of the adductors

Pectineus

Adductor longus

Gracilis

Adductor magnus

Adductor brevis

gracilis	gra-**cil**-is	L. slender, graceful	brevis	L. short
pectineus	pek-**tin**-e-us	L. comb		

A -
All:
    Adduct
    Medially rotate the hip
All, except Gracilis:
    Assist to flex the hip
Gracilis:
    Flex and medially rotate the knee
Posterior fibers of Adductor Magnus:
    Extend the hip

Adductor Magnus

O - Inferior ramus of the pubis,
ramus of ischium and ischial tuberosity

I - Linea aspera and adductor tubercle

N - Obturator

Adductor Longus

O - Pubic tubercle

I - Medial lip of linea aspera

N - Obturator

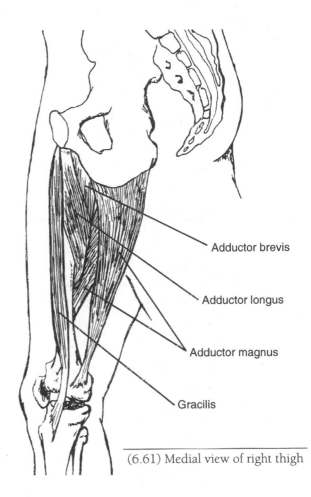

(6.61) Medial view of right thigh

- Adductor brevis
- Adductor longus
- Adductor magnus
- Gracilis

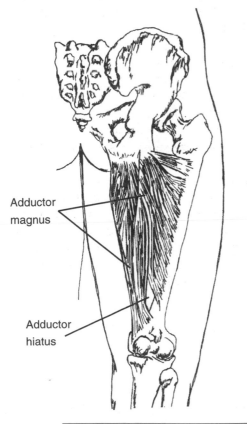

Adductor
magnus

Adductor
hiatus

(6.62) Posterior view of right thigh

Adductor Brevis

O - Inferior ramus of pubis

I - Pectineal line and medial lip of linea aspera

N - Obturator

Pectineus

O - Superior ramus of pubis

I - Pectineal line of femur

N - Obturator

Gracilis

O - Inferior ramus of pubis, ramus of ischium

I - Proximal, medial shaft of tibia
at pes anserinus tendon

N - Obturator

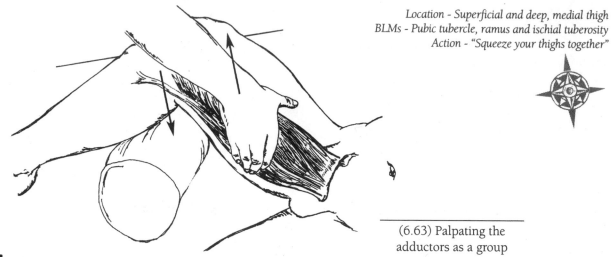

Location - Superficial and deep, medial thigh
BLMs - Pubic tubercle, ramus and ischial tuberosity
Action - "Squeeze your thighs together"

(6.63) Palpating the
adductors as a group

👁 Adductors as a group

1) Supine. Hip slightly flexed and laterally rotated. Place your hand along the medial thigh and ask your partner to adduct his hip against your resistance. (6.63) Do you feel the adductors tighten?
2) Ask your partner to alternately adduct and relax, as you palpate proximally to the adductor tendons

and anteriorly and posteriorly to the edges of the adductor bellies.

✔ *Are you on the medial side of the leg? Explore either side of the group of muscles you are palpating and determine if you are between the hamstrings and quadriceps femoris group.*

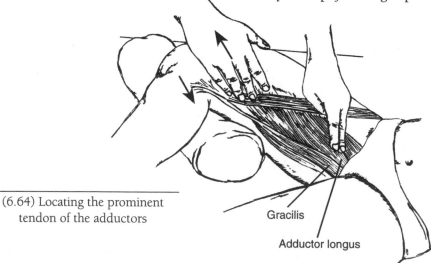

(6.64) Locating the prominent
tendon of the adductors

Gracilis

Adductor longus

👁 Gracilis and adductor longus

1) Supine. Hip slightly flexed and laterally rotated. Place the flat of your hand at the middle of the medial thigh. Ask your partner to adduct his hip slightly.
2) While your partner contracts, slide your fingers proximally to the pubic bone and locate the taut, prominent tendon(s) of the gracilis and adductor longus extending off or near to the pubic tubercle.
3) Strum your fingertip across this tendon and follow it distally as it develops into muscle tissue.

(6.64) If the muscle belly slowly angles into the medial thigh, you are palpating adductor longus. If the belly is slender and continues down the medial leg toward the knee, you are accessing gracilis.

✔ *The sartorius (p. 225) has a similar shape and location as gracilis. Distinguish the two by simply following the muscle you are palpating proximally. If it leads toward the ASIS it is sartorius; toward the pubis the muscle is gracilis.*

👁 Pectineus

1) Partner supine with the hip slightly flexed and laterally rotated. Place the flat of your hand at the middle of the medial thigh and ask your partner to adduct his hip slightly.

2) Locate the prominent tendon of the adductor longus or gracilis. Slide off the tendon laterally toward the ASIS. Slowly sink into the belly of pectineus. (6.65) You should be inferior to the superior ramus of the pubis (p. 207).

3) Ask your partner to alternately adduct and relax his hip and feel the fibers of pectineus contract.

✔ *Are you just anterior to the prominent adductor tendon? Do the fibers you are palpating contract upon adduction?*

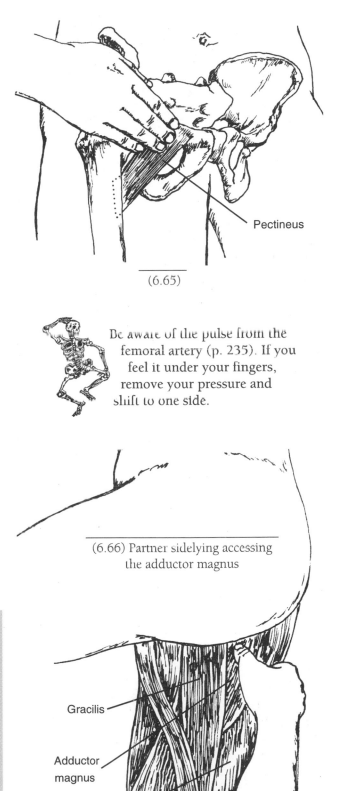

Pectineus

(6.65)

👁 Adductor magnus

1) Partner sidelying with top hip flexed. Begin by locating the ischial tuberosity.

2) Ask your partner to adduct his hip slightly. Locate the prominent tendon of adductor longus or gracilis. Slide off the tendon posteriorly. Palpate the wide tendon of adductor magnus as it stretches from this point to the ischial tuberosity. (6.66)

3) Follow the fibers of adductor magnus distally by strumming your fingers across its belly. Differentiating magnus fibers from biceps femoris fibers is difficult, yet the thin, distal tendon of magnus can be accessed as it attaches to the adductor tubercle (p. 249).

Be aware of the pulse from the femoral artery (p. 235). If you feel it under your fingers, remove your pressure and shift to one side.

(6.66) Partner sidelying accessing the adductor magnus

Gracilis

Adductor magnus

Hamstrings

The adductor's attachment on the posterior femur would make it seem only natural that they would rotate the coxal joint laterally, rather than medially. However, in anatomical position, some of the adductors medially rotate. If the femur is already medially rotated, all of the adductors will help to further this movement. And if the femur is laterally rotated, some of the adductors will laterally rotate.

👁 Supine. Lay your hand on the adductors. Ask your partner to alternately rotate his thigh medially and laterally. Grasping the ankle to create a little resistance may clarify the movement. Do you feel the adductors contract on medial rotation? What do they do during lateral rotation?

Tensor Fasciae Latae and Iliotibial Tract

The *tensor fasciae latae* (TFL) is a small, superficial muscle located on the lateral side of the upper thigh. (6.67) Approximately three fingers wide, the TFL is easily assessable between the upper fibers of the rectus femoris and the gluteus medius. Along with gluteus maximus, the TFL attaches to the iliotibial tract.

The *iliotibial tract* is a superficial sheet of fascia located along the lateral side of the thigh. It emerges from the gluteal fascia, is wide and dense over the vastus lateralis muscle (p. 213), funnels into a strong cable along the side of the knee, and inserts at the tibial tubercle (p. 248). The fibers of tensor fascia latae and some fibers of gluteus maximus (p. 217) attach to the proximal aspect of the iliotibial tract. The iliotibial tract has a thick, matted texture (similar to packing tape) which makes it a strong stabilizing component of the hip and knee.

For palpation purposes, the iliotibial tract has vertical fibers and is entirely palpable. Its distal, cable portion is anterior to the biceps femoris tendon and is the most easily isolated part of the iliotibial tract.

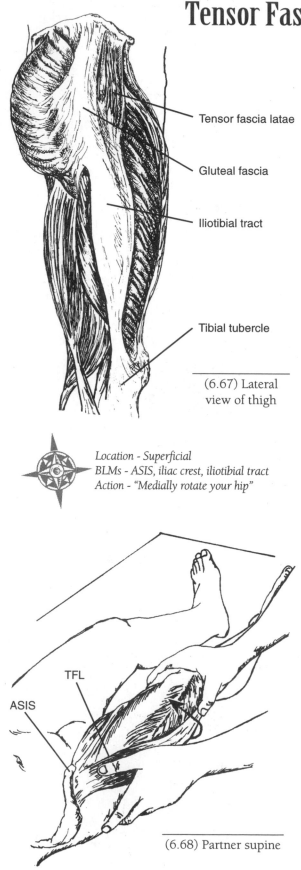

Tensor fascia latae

Gluteal fascia

Iliotibial tract

Tibial tubercle

(6.67) Lateral view of thigh

Location - Superficial
BLMs - ASIS, iliac crest, iliotibial tract
Action - "Medially rotate your hip"

TFL

ASIS

(6.68) Partner supine

A -
Flex
Medially rotate
Abduct the hip

O - Iliac crest, posterior to the ASIS

I - Iliotibial tract

N - Gluteal

✋ Tensor fascia latae

1) Supine. Locate the ASIS.
2) Place the flat of your hand posterior and distal to the ASIS and iliac crest.
3) Ask your partner to alternate between medial rotation and relaxing the hip. Upon medial rotation the TFL will contract into a solid, oval mound beneath your hand. (6.68)
4) Palpate its vertical fibers, outline its width and follow it distally until it blends into the iliotibial tract.

✔ *Are you posterior and distal to the anterior iliac crest? If you ask your partner to laterally rotate the hip, does the TFL contract? It should not.*

| tensor | **ten**-sor | L. a stretcher | latae | **la**-ta | L. broad |
| tract | | L. extent | | | |

👁 Distal end of the iliotibial tract

1) Sidelying. Locate the biceps femoris tendon (p. 215) just proximal to the back of the knee.
2) Slide anteriorly from the biceps femoris tendon to the lateral side of the thigh. Roll your fingers horizontally across the fibers of the iliotibial tract and explore for its tough, superficial quality. Its most distal aspect may feel similar in size and shape to the biceps tendon.
3) Follow it distally as it disappears toward the tibial tubercle. Explore proximally and note how it becomes broader and thinner as it progresses up the thigh. Feel the tension of the iliotibial tract change by asking your partner to alternately abduct and relax his hip. (6.69)

✔ *Are the fibers you are feeling superficial and stringy compared to the deeper, fleshier vastus lateralis fibers? Do the fibers run vertically down the thigh and converge to a thin, tendon-like cable at the tibial tubercle?*

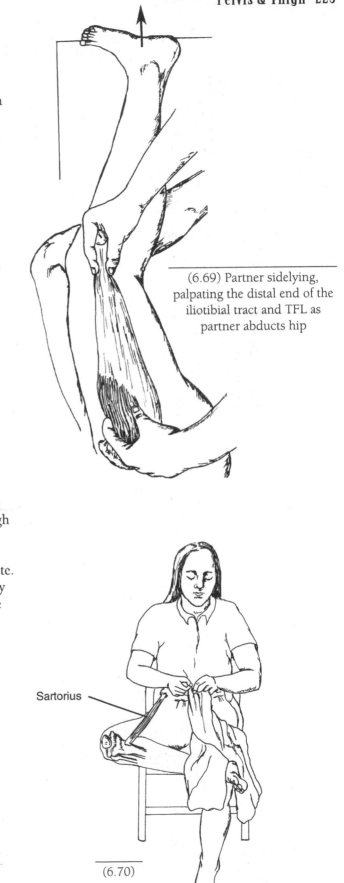

(6.69) Partner sidelying, palpating the distal end of the iliotibial tract and TFL as partner abducts hip

Sartorius

The sartorius is the longest muscle in the body, stretching from the anterior ilium, across the thigh to the medial knee. (6.71) Though it is entirely superficial, the slender belly of the sartorius, roughly two fingers wide, can be difficult to isolate. Its proximal fibers are lateral to the femoral artery (p. 235). Its name (Gr. *sartus,* tailor) refers to the sartorius' ability to bring the thigh and leg into a "cross-legged" position as a tailor might sit while sewing. (6.70)

A -
Flex
Laterally rotate
Abduct the hip

Flex
Medially rotate the flexed knee

O - Anterior superior iliac spine (ASIS)

I - Proximal, medial shaft of the tibia (pes anserinus tendon)

N - Femoral

Sartorius

(6.70)

sartorius   sar-**tor**-ee-us

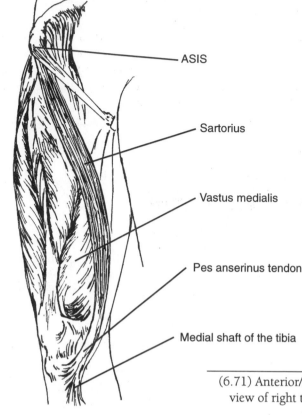

ASIS

Sartorius

Vastus medialis

Pes anserinus tendon

Medial shaft of the tibia

(6.71) Anterior/medial
view of right thigh

Location - Superficial, slender
BLMs - ASIS, pes anserinus region
Action - "Bring your knee toward your chest"

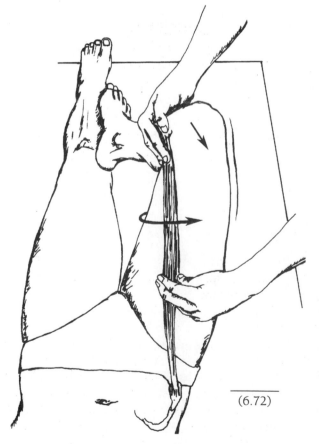

(6.72)

1) Supine or seated. Ask your partner to place his foot so it is resting on his opposite knee. The hip will be flexed and laterally rotated.
2) Lay your hand along the middle of the medial thigh. Ask your partner to raise his knee toward the ceiling (contracting the sartorius).
3) Strum your fingers across the slender sartorius, following it proximally to the ASIS and distally to the medial tibia. (6.72)
4) Maintain your hand placement and ask your partner to relax his hip. Continue to palpate, noticing how the sartorius curves from the ASIS to the medial side of the thigh.

✔ *Is the muscle belly you feel roughly two fingers wide and superficial? Distal to the ASIS, can you strum across its tendon? Are you medial to the vastus medialis belly?*
The sartorius and gracilis are slender, superficial muscles along the medial thigh. Differentiate between them by following their respective bellies proximally; the sartorius will lead toward the ASIS, the gracilis to the pubic tubercle.

Tendons of the Posterior Knee

There are five distinct tendons located on the posterior aspect of the knee. (6.73) Biceps femoris and the iliotibial tract are located on the lateral/posterior knee. Semitendinosus, gracilis and sartorius are bundled together on the medial/posterior knee.

Where is the semimembranosus tendon? Its distal tendon is very short and deep to the semitendinosus and gracilis. The distal aspect of semimembranosus can be accessed by palpating between the tendons of semitendinosus and gracilis.

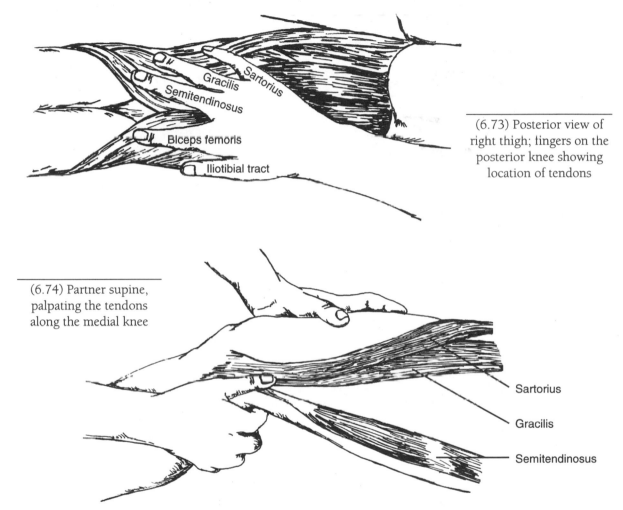

(6.73) Posterior view of right thigh; fingers on the posterior knee showing location of tendons

(6.74) Partner supine, palpating the tendons along the medial knee

👁 Lateral tendons

1) Prone. Ask your partner to flex and hold his knee at 45°. The tendons will become taut in this position. For greater clarity, place your hand on the ankle and give your partner some resistance.
2) The most prominent and often visible tendons will be biceps femoris and semitendinosus. The biceps femoris tendon will be slender and extend down to the head of the fibula.
3) Move laterally approximately one inch from the biceps tendon and palpate the iliotibial tract. Unlike the biceps femoris, the iliotibial tract is broader and located on the lateral side of the leg.

👁 Medial tendons

1) Prone or supine. Move to the medial side of the knee and palpate the thin, prominent tendon of semitendinosus.
2) Slide off semitendinosus anteriorly and palpate the equally slender tendon of gracilis. (6.74)
3) Situated anterior to gracilis will be sartorius. Unlike the long, skinny tendons of semitendinosus and gracilis, the sartorius has a short, wide tendon. For this reason, it can be challenging to isolate.
4) Follow the three tendons distally as they blend together to become the pes anserinus tendon and attach on the proximal medial shaft of the tibia.

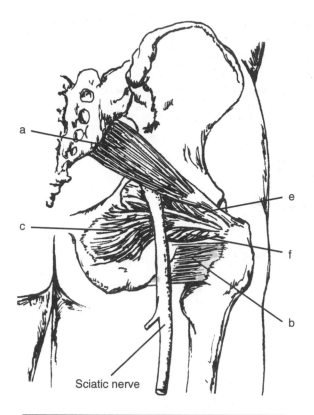

(6.75) Posterior view of buttock with gluteals removed, obturator externus (d) not seen

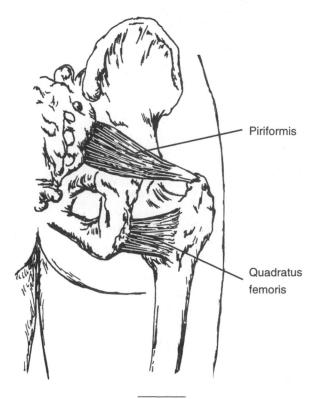

(6.76)

Lateral Rotators of the Hip

a) *Piriformis*
b) *Quadratus Femoris*
c) *Obturator Internus*
d) *Obturator Externus*
e) *Gemellus Superior*
f) *Gemellus Inferior*

Sometimes known as the "deep six", these small muscles are located deep to the gluteus maximus and create lateral rotation of the hip. They all attach to aspects of the greater trochanter and fan medially to attach to the sacrum and pelvis. (6.75, 6.76)

As a group they are accessible, with the *piriformis* and *quadratus femoris* the most distinct. Except for the piriformis, all of the lateral rotators are deep to the large sciatic nerve (p. 237). The piriformis lays superficial to the sciatic nerve and, when over-contracted, can create sciatic nerve compression.

Piriformis

A - Laterally rotate the hip
Abduct the thigh when hip flexed

O - Anterior surface of sacrum

I - Greater trochanter

N - Sacral

Quadratus Femoris

A - Laterally rotate the hip

O - Lateral border of ischial tuberosity

I - Posterior surface of femur, between greater and lesser trochanter

N - Sciatic

Obturator Internus

A - Laterally rotate the hip

O - Obturator membrane and pelvic surface

I - Medial surface of greater trochanter

N - Obturator

---

gemellus	jem-**el**-us	L. twins	piriformis	**pir**-i-form-is	L. pear-shaped
obturator	**ob**-tu-**rat**-or	L. occluded, stopped up	quadratus	**kwod**-rait-us	L. squared

Obturator Externus

A - Laterally rotate the hip

O - Superior and inferior rami of pubis

I - Trochanteric fossa of femur

N - Obturator

Gemellus Superior

A - Laterally rotate the hip

O - Spine of ischium

I - Upper border of greater trochanter

N - Sacral

Gemellus Inferior

A - Laterally rotate the hip

O - Ischial tuberosity

I - Upper border of greater trochanter

N - Sacral

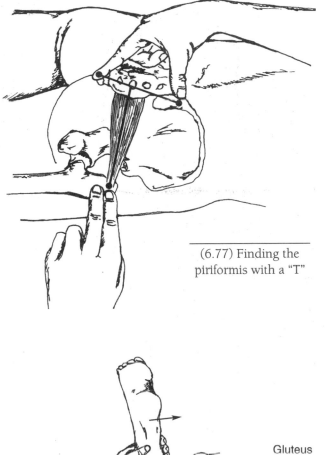

(6.77) Finding the piriformis with a "T"

Piriformis

1) Prone. Locate the greater trochanter, PSIS and coccyx. Together, these landmarks form a "T". The piriformis is located along the base of the "T". (6.77)

2) Lay your fingers along this line. Working through the thick gluteus maximus, roll your fingers across the belly of the slender piriformis.

3) Strum across its belly to clarify its location, staying aware of the deep sciatic nerve. (6.78)

✔ *Are you compressing through the thick gluteus maximus fibers? With your fingers on the piriformis, bend the knee to 90° and ask your partner to laterally rotate his hip against your gentle resistance. You may feel gluteus maximus contract, but can you feel piriformis contract beneath it?*

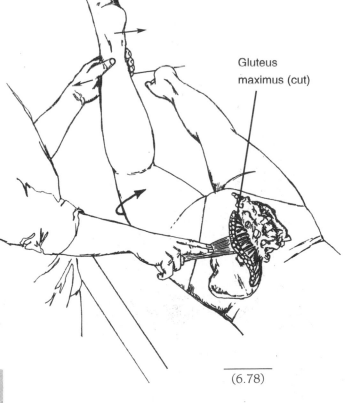

Gluteus maximus (cut)

(6.78)

In regards to its evolutionary history, the piriformis is a remnant of its former glory. It is a descendant of the great caudofemoral elevator muscles, still seen today extending from a reptile's femur to its tail. These large muscles provide these animals with tremendous thrusting force to extend the femur when running.

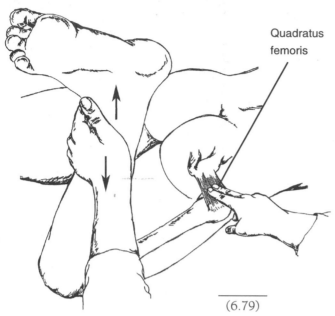

Quadratus femoris

(6.79)

👁 Quadratus femoris

1) Prone. Locate the distal, posterior aspect of the greater trochanter and the ischial tuberosity. Place your fingertips between these two landmarks.
2) Pressing firmly through the gluteus maximus fibers, strum vertically across the fibers of quadratus femoris.

✔ *Does the belly stretch between the ischial tuberosity and the distal trochanter? Rolling your fingers over its belly, can you feel its horizontal fibers? Flex the knee to 90˚ and passively rotate the hip medially and laterally. Can you sense how the tension of the muscle changes as it shortens and lengthens? (6.79)*

Piriformis
Location - Deep to gluteus maximus
BLMs - Coccyx, PSIS, greater trochanter
Action - "Laterally rotate your hip"

Quadratus Femoris
Location - Deep to gluteus maximus
BLMs - Greater trochanter, ischial tuberosity, sacrum
Action - "Laterally rotate your hip"

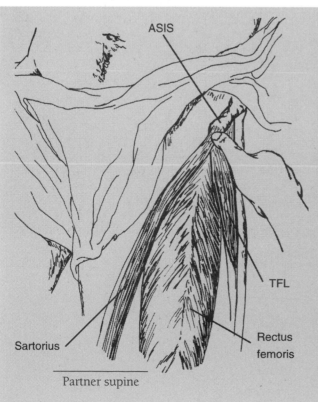

ASIS

Sartorius

Partner supine

TFL

Rectus femoris

The tendons of the sartorius, tensor fascia latae and rectus femoris attach very close to each other on the anterior pelvis. The sartorius extends off the ASIS, the TFL lies alongside on the iliac crest, just posterior to the ASIS. The rectus femoris originates at the AIIS, just inferior to the ASIS. How can you differentiate them?

👁 1) Supine, with the knee bolstered to flex the hip. Locate the thin tendon of sartorius by locating the ASIS. To feel the tendon tighten, ask your partner to laterally rotate his hip.
2) Access the wide attachment of the tensor fascia latae by accessing the iliac crest behind the ASIS and asking your partner to medially rotate his hip.
3) Ask your partner to medially and laterally rotate his hip back and forth. Can you feel the antagonistic contraction happening between these two muscles?
4) Isolate the tendon of rectus femoris by sliding distal from the ASIS to the AIIS and asking your partner to alternately flex and relax his hip.

Iliopsoas

Psoas Major
Iliacus

The iliacus and psoas major, together called the iliopsoas, are major hip flexors. The long, slender *psoas major* is located deep to the abdominal contents and stretches from the lumbar vertebrae, underneath the inguinal ligament to the lesser trochanter. (6.80) The stocky *iliacus* is located deep to the abdomen in the iliac fossa. (6.81) Due to their location, both muscles are only partially accessible and can be challenging to palpate.

Psoas Major

A -
Flex
Laterally rotate
Adduct the hip

O - Bodies and transverse processes of lumbar vertebrae

I - Lesser trochanter of the femur

N - Lumbar plexus

Iliacus

A -
Flex
Laterally rotate
Adduct the hip

O - Iliac fossa

I - Lesser trochanter of the femur

N - Femoral

Psoas Major
Location - Deep in abdomen, lateral to navel
BLMs - Bodies of lumbar vertebrae
Action - "Bring your knee toward your chest"

Iliacus
Location - Deep in abdomen
BLMs - Iliac fossa and crest
Action - "Bring your knee toward your chest"

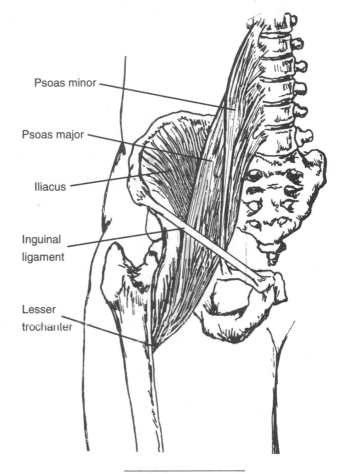

(6.80) Anterior view

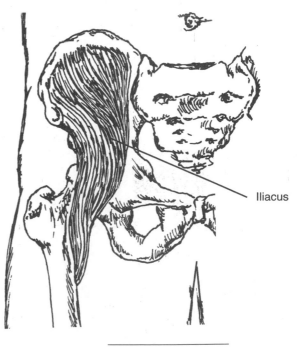

(6.81) Anterior view

psoas	**so**-as	Gr. muscle of the loin
iliacus	i-**li**-a-cus	

When accessing either the psoas or iliacus, palpate slowly and remain in close communication with your partner. If at any point he does not feel safe or comfortable, slowly remove your hands.

The psoas major lies just lateral to the abdominal aorta (p. 154). If you feel a strong pulse directly beneath your fingers when accessing the muscle, realign your fingers further laterally.

Psoas major

1) Supine, with the hip slightly flexed and laterally rotated. Locate the navel and ASIS, placing your fingerpads hand-on-hand between these points.
2) Slowly compress your fingerpads into the abdomen, moving only when your partner exhales. (Upon initial exhalations, compressing in small circles will assist in moving the abdominal contents to the side.) As you compress further, keep your fingerpads stationary and direct your fingers down toward the table. (6.82)
3) Check that you are palpating the psoas, not the surrounding tissues, by asking your partner to flex his hip ever-so-slightly. If your fingers are accessing the psoas, you will feel a definite, solid contraction. (6.83)

✔ *Are you between the ASIS and navel? Is the direction of your fingers at a slight angle toward the spine? Have you compressed slowly, allowing the overlying tissue to relax? If you did not feel the muscle contract, try again with the fingers repositioned slightly inferiorly.*

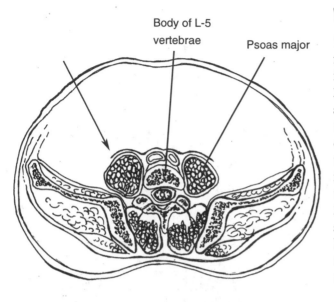

Body of L-5 vertebrae

Psoas major

(6.82) Cross-section of the trunk at the level of L-5, arrow showing direction of fingers when accessing the psoas major

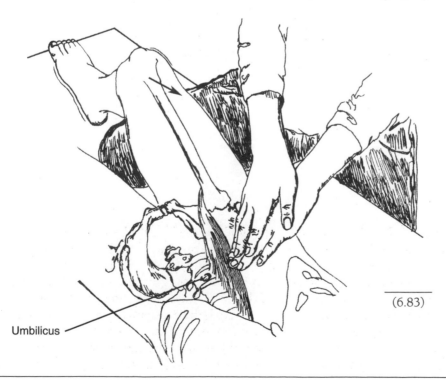

Umbilicus

(6.83)

👁 Iliacus

1) Supine, with the hip slightly flexed and laterally rotated. Locate the anterior portion of the iliac crest and place your fingerpads hand-on-hand along its ridge.

2) Slowly curl your fingers around the iliac crest into the iliac fossa, moving only when your partner exhales. (6.84) Your fingers may sink only a short distance into the tissue.

3) Ask your partner to flex his hip slightly, with your fingers in place. You will feel the strong iliacus contract.

◆ Follow the same procedure, but palpate across the body to the opposite iliacus. Try curling into the iliac fossa with your thumbs.

✔ *Are you in the iliac fossa? Have you compressed slowly, allowing the overlying tissue to relax?*

ASIS

(6.84) Partner supine

Psoas major primarily flexes the hip. But when the femur is stabilized, the psoas, in conjunction with iliacus, can increase the lordotic curvature in the lumbar spine and create downward rotation of the pelvis. It has also been proposed that only the superficial fibers of the psoas increase the lordotic curve, while the deeper fibers may decrease it.

Roughly 40% of the population has a psoas minor. It is a small muscle which extends from the lumbar vertebrae to the superior ramus of the pubis. If present, psoas minor assists in upward rotation of the pelvis - the opposite action of the psoas major. Interestingly, because of the pelvic/vertebral arrangement on a quadruped, the psoas minor is an important muscle of locomotion on a dog or cat. On a bipedal human, it is relatively insignificant, except when doing the horizontal rumba dance!

Ligaments, Nodes and Vessels of the Hip and Thigh

The femoral triangle is located on the anterior, medial surface of the thigh. (6.85) It is formed by the inguinal ligament, adductor longus and sartorius. Several important vessels, including the femoral artery, nerve and vein, pass superficially through the femoral triangle.

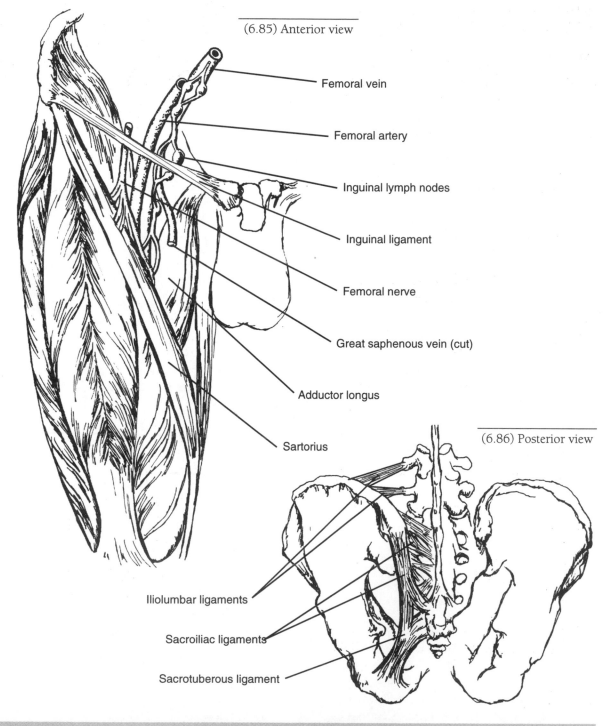

(6.85) Anterior view

Femoral vein

Femoral artery

Inguinal lymph nodes

Inguinal ligament

Femoral nerve

Great saphenous vein (cut)

Adductor longus

Sartorius

(6.86) Posterior view

Iliolumbar ligaments

Sacroiliac ligaments

Sacrotuberous ligament

The great saphenous vein is a superficial vessel which travels the length of the lower extremity. Often visible, it begins near the ankle, passes along the medial aspect of the tibia, and follows the sartorius up the thigh to empty into the femoral vein at the femoral triangle. Since it is long and accessible, portions of the saphenous vein are often used as grafts for coronary bypass surgery.

Inguinal Ligament

The inguinal ligament is a superficial band stretching between the ASIS and the pubic tubercle. It forms the superior border of the femoral triangle and the lower edge of the abdominal aponeurosis. It is an attachment site for the external oblique muscle.

1) Partner supine. Soften the surrounding tissue of the ligament by bolstering your partner's knee.
2) Locate the ASIS and slide diagonally in the direction of the pubic tubercle. (6.87)
3) Strum gently across the slender ligament, feeling its cord-like quality.

✔ *Can you feel a thin, superficial band just beneath the skin? Does the band stretch from the ASIS and extend to the pubic tubercle?*

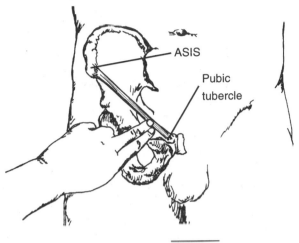

(6.87)

Femoral Artery, Nerve and Vein

The femoral artery, nerve and vein form a neurovascular bundle that courses through the femoral triangle. These vessels lay beneath the inguinal ligament and extend distally into the tissue of the thigh. The bundle is relatively superficial; the pulse of the femoral artery can be easily felt.

Pulse of the femoral artery

1) Partner supine. Slip your flexed knee behind your partner's knee. This position will flex and laterally rotate the hip, allowing for easier palpation.
2) Place the flat of your fingerpads halfway between the ASIS and the pubic tubercle just distal to the inguinal ligament. Palpate for the strong pulse of the artery. (6.88)

✔ *Are you distal to the inguinal ligament? Are you between the ASIS and the pubic tubercle?*

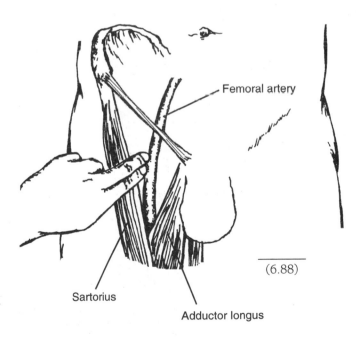

(6.88)

The penis contains no muscle tissue. During sexual arousal, the arteries of the penis dilate and a small muscle (ischiocavernosus) at the base of the penis helps to maintain an erection.

The testicles are enwrapped by the cremaster muscle. It protects the sperm inside by lowering the testes when the body becomes warm and pulls them up to the body when the testes become cold.

| inguinal | **ing**-gwi-nal | L. pertaining to the groin |
| penis | | L. tail |

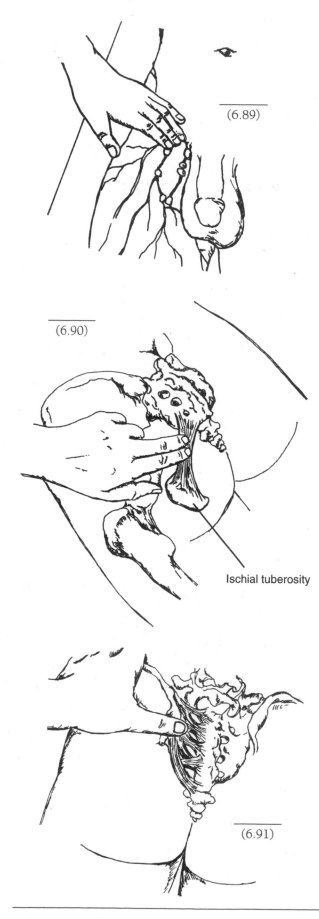

(6.89)

(6.90)

Ischial tuberosity

(6.91)

Inguinal Lymph Nodes

Located distal to the inguinal ligament, the superficial inguinal lymph nodes are easily palpable. They number between eight and ten and may vary in size from a small lentil to a raisin.

1) Supine, with the knee bolstered. This position will flex and laterally rotate the hip, allowing for easier palpation.
2) Locate the inguinal ligament. Slide inferiorly and explore for the superficial nodes. (6.89)

Sacrotuberous Ligament

This broad, solid ligament stretches between the ischial tuberosity and the edge of the sacrum. Deep to the gluteus maximus muscle, it is distinctly palpable and may feel like a span of bone.

1) Prone. Locate the ischial tuberosity.
2) Locate the edge of the sacrum (p. 203).
3) Slide your fingertips off the tuberosity toward the edge of the sacrum. Using firm pressure, palpate through the gluteus maximus belly and strum broadly across the ligament. (6.90)

Are you deep to the gluteus maximus fibers? Is the structure you are rolling over roughly an inch wide and inflexible? Does it stretch from the ischial tuberosity toward the sacrum?

Sacroiliac Ligament

Located superficial to the sacroiliac joint, the dense sacroiliac ligament supports the union of the posterior sacrum and the ilium. It has several segments which attach from the sacrum to the area around the PSIS. The ligament is deep to the thoracolumbar aponeurosis, and its oblique fibers can be challenging to distinguish.

1) Prone. Locate the surface of the sacrum.
2) Using firm pressure, strum your fingertips across the dense fibers of the sacroiliac ligament. (6.91)

Are you medial to the PSIS, on top of the sacroiliac joint space (p. 204)?

---

saphenous	**sa**-fe-nus	origin unclear, perhaps Arabic *saphin*, standing; or Greek *saphen*, clearly visible
sciatic	si-**at**-ik	Gr. *ischion*, hip joint

Iliolumbar Ligaments

The iliolumbar ligaments are located between the transverse processes of the fourth and fifth lumbar vertebrae and the posterior iliac crest. The strong, horizontal fibers of these ligaments are important in stabilizing L-5. Deep to the thoracolumbar aponeurosis, the thick multifidi (p. 141), and the quadratus lumborum (p. 146), these ligaments are difficult to palpate. However, their location and density can be determined.

👁 1) Prone. Locate the PSIS.
2) Slide your thumb straight superior from the PSIS to the level of L-4 and L-5. Your thumb should be between the iliac crest and the transverse processes of the lumbar vertebrae.
3) Using firm pressure, sink into the dense muscles of the low back and attempt to roll vertically across the ligament's taut fibers. (6.92)

✔ *Can you palpate its dense, horizontal fibers?*

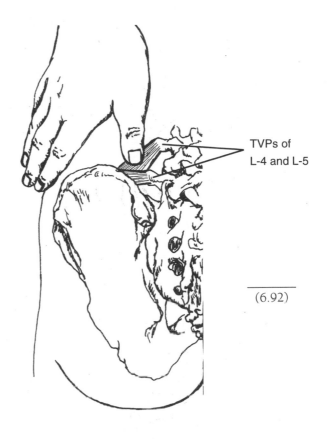

TVPs of L-4 and L-5

(6.92)

Sciatic Nerve

The sciatic nerve is the largest nerve in the body - sometimes measuring three-quarters of an inch in diameter. It is formed by the spinal nerves of L-4 through S-3. The nerve passes through the greater sciatic notch, between the ischial tuberosity and greater trochanter and extends down the posterior thigh. (6.75) Distally, it branches into the tibial and peroneal nerves.

The sciatic nerve runs deep to the piriformis muscle (p. 228), creating a potential for nerve entrapment. In the gluteal region, the sciatic nerve is difficult to isolate and best avoided.

👁 1) Prone. Outline the placement of the sciatic nerve by locating the edge of the sacrum.
2) Draw a line down the buttock between the ischial tuberosity and greater trochanter. Continue down the middle of the posterior thigh.
3) To access the sciatic nerve, turn your partner sidelying and flex the hip. Locate the ischial tuberosity and greater trochanter.
4) Palpate between these landmarks for the pathway of the sciatic nerve. (6.93) You can avoid pinching the nerve by palpating with the flat of your fingerpads.

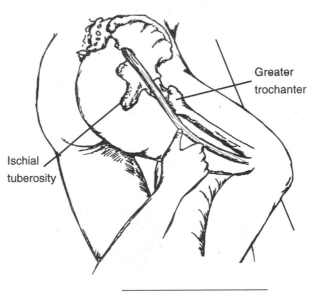

Greater trochanter

Ischial tuberosity

(6.93) Partner sidelying

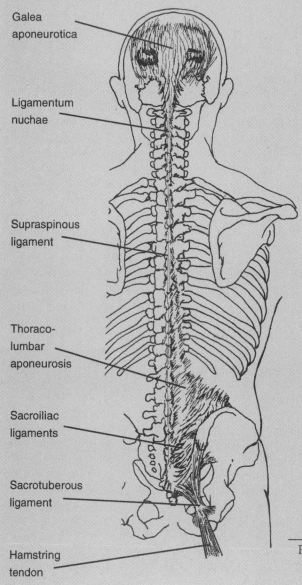

Galea
aponeurotica

Ligamentum
nuchae

Supraspinous
ligament

Thoraco-
lumbar
aponeurosis

Sacroiliac
ligaments

Sacrotuberous
ligament

Hamstring
tendon

Ligaments, tendons, fascia and retinaculum are all forms of connective tissue. They are constructed of virtually the same ingredients (collagen, elastin and ground substance) and differ only in the proportions of these materials. For anatomical purposes, these bands and sheets have been specifically categorized; yet they are not separate structures. Together they form an incredible, supportive matrix spreading throughout the entire body.

Now that you have explored the location of several connective tissue structures, here is an exercise to help you feel how a few of them are connected together.

1) Locate the proximal hamstring tendon (p. 215) as it attaches to the ischial tuberosity. 2) Follow the tendon superiorly as it melds into the sacrotuberous ligament (p. 236) and then to the sacroiliac ligament (p. 236) on the sacrum. 3) Continue superiorly as the sacroiliac ligaments blend into the thoracolumbar aponeurosis (p. 155) and supraspinous ligament (p. 154) between the spinous processes of the vertebrae. 4) Ultimately, follow the supraspinous ligament all the way up the spine to the ligamentum nuchae (p. 153) and finally to the galea aponeurotica (p. 182) which surrounds the cranium.

Posterior view

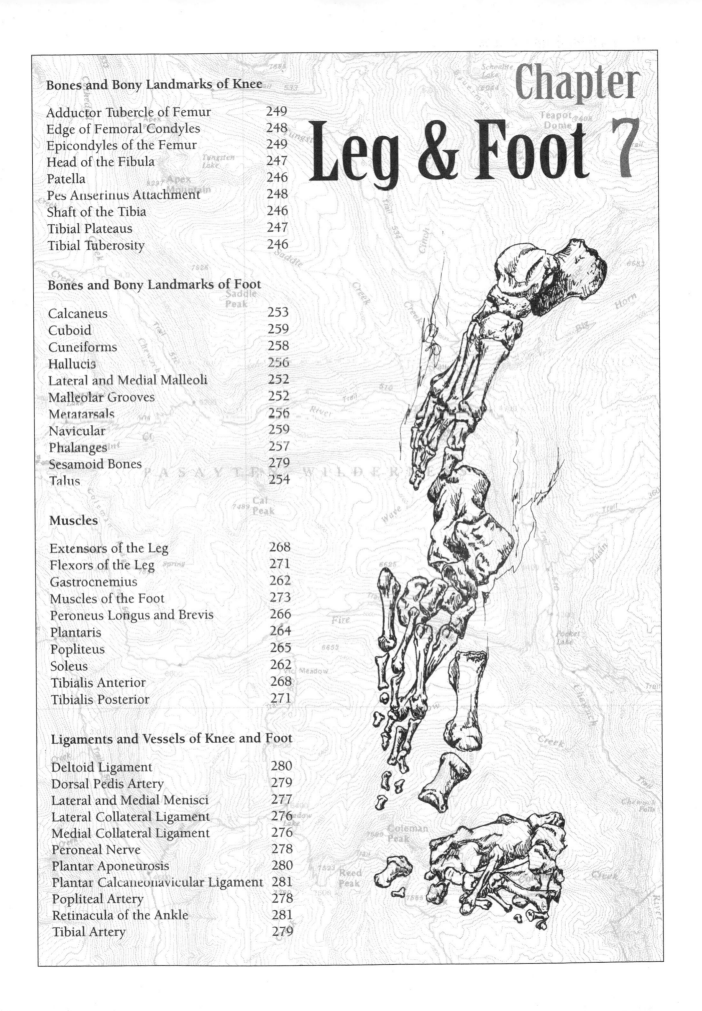

Chapter
Leg & Foot 7

Topographical Views

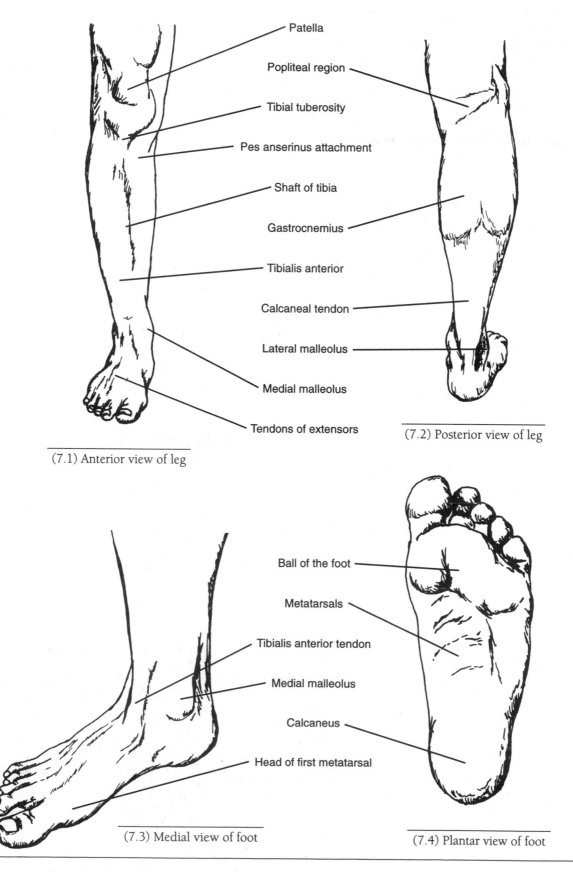

Patella

Popliteal region

Tibial tuberosity

Pes anserinus attachment

Shaft of tibia

Gastrocnemius

Tibialis anterior

Calcaneal tendon

Lateral malleolus

Medial malleolus

Tendons of extensors

(7.1) Anterior view of leg

(7.2) Posterior view of leg

Ball of the foot

Metatarsals

Tibialis anterior tendon

Medial malleolus

Calcaneus

Head of first metatarsal

(7.3) Medial view of foot

(7.4) Plantar view of foot

Bones of the Knee, Leg and Foot

The knee is formed by the articulation of the distal femur and proximal tibia. (7.5) Together they form the tibiofemoral (or knee) joint, the largest synovial joint in the body. The tibiofemoral is a modified hinge joint: it is capable of flexion and extension and also capable of medial and lateral rotation of the tibia when the knee is flexed. (7.8)

The region of the knee also includes the small *patella* ("knee cap") and the proximal fibula. The bony surfaces of the knee joint are superficial and easily accessible.

The leg consists of the tibia and the fibula. The *tibia* ("shin bone") is superficial from the knee to the ankle, similar to the depth of the ulna in the forearm. Like the radius' relationship to the ulna, the small *fibula* is lateral to the tibia and primarily deep to surrounding muscles. The fibula bears only 10% of the body weight and rightfully so: in proportion to its length, it is the thinnest bone in the body.

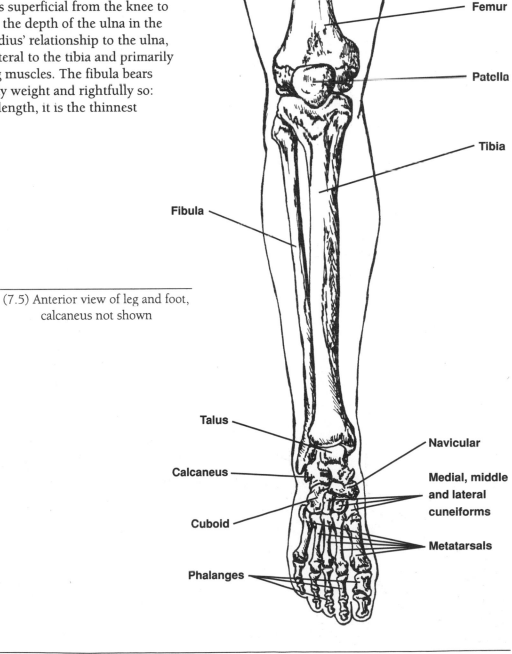

(7.5) Anterior view of leg and foot, calcaneus not shown

Femur

Patella

Tibia

Fibula

Talus

Calcaneus

Cuboid

Phalanges

Navicular

Medial, middle and lateral cuneiforms

Metatarsals

| fibula | **fib**-u-la | L. pin or buckle | tibia | **tib**-e-a | L. shinbone |
| patella | pa-**tel**-a | | | | |

Bony Landmark of the Knee

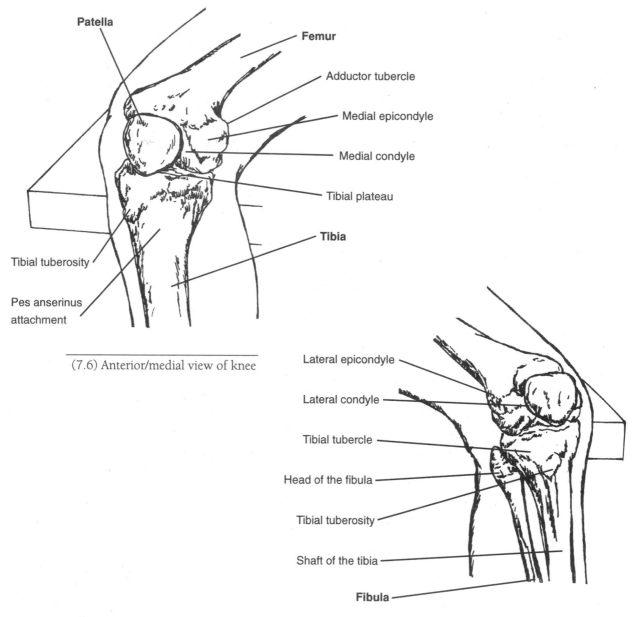

Patella

Femur

Adductor tubercle

Medial epicondyle

Medial condyle

Tibial plateau

Tibia

Tibial tuberosity

Pes anserinus attachment

(7.6) Anterior/medial view of knee

Lateral epicondyle

Lateral condyle

Tibial tubercle

Head of the fibula

Tibial tuberosity

Shaft of the tibia

Fibula

(7.7) Anterior/lateral view of knee

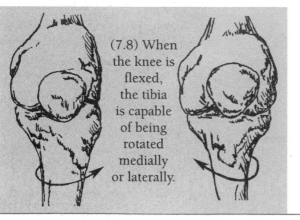

(7.8) When the knee is flexed, the tibia is capable of being rotated medially or laterally.

Bony Landmark Trails of the Knee

Trail 1 - "Landmark Trail" links together the most prominent landmarks of the knee. (7.9)

   a) Patella
   b) Tibial tuberosity
   c) Shaft of the tibia
   d) Head of the fibula

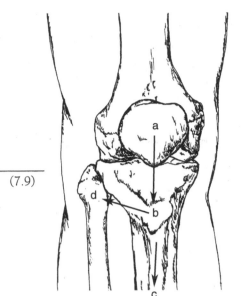

(7.9)

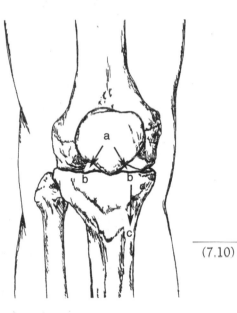

(7.10)

Trail 2 - "Waddle Walk" has two paths exploring the medial and lateral aspects of the proximal tibia. (7.10) It ends at the pes anserinus attachment ("goose foot" in Latin).

   a) Patella
   b) Medial and lateral tibial plateaus
   c) Pes anserinus attachment

Trail 3 - "Hills on Both Sides" explores the bumps of the distal end of the femur. (7.11)

   a) Edge of the medial and lateral femoral condyle
   b) Medial and lateral epicondyle of the femur
   c) Adductor tubercle

The thickest layering of cartilage in the body can be found on the posterior surface of the patella. Its eighth-of-an-inch thick coating protects the patella from the incredible pressure applied by the quadriceps during knee flexion. Simply walking up or down stairs can place as much as six-hundred pounds of pressure on the patella.

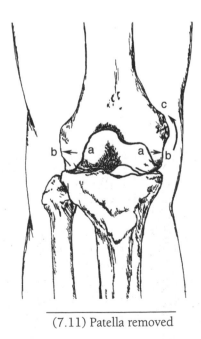

(7.11) Patella removed

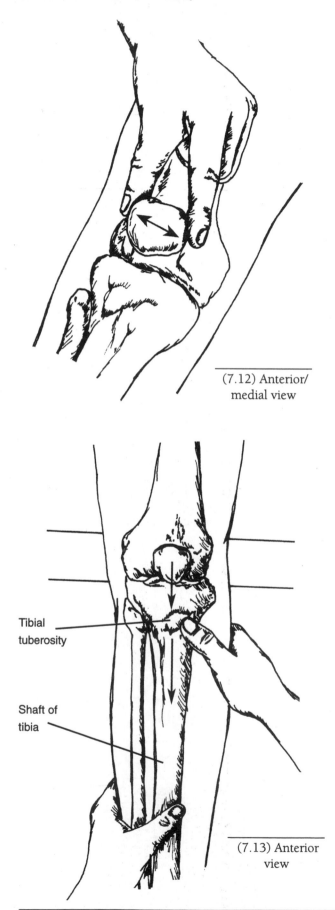

(7.12) Anterior/
medial view

Tibial
tuberosity

Shaft of
tibia

(7.13) Anterior
view

Trail 1 - "Landmark Trail"
Patella

The patella is located on the anterior surface of the knee. It is superficial and round with an apex pointing distally. The largest sesamoid bone in the body, the patella is an attachment site for the quadriceps femoris tendon (p. 211). When the knee is flexed, the patella seems to disappear as it sinks into the space between the proximal tibia and femoral condyles.

1) Partner supine with the knee extended. In this position the quadriceps tendon is shortened and the patella is more mobile and easier to access.
2) Locate the patella on the anterior knee and palpate its round surface and edges. (7.12) Note any bumps or crevices along its edges.
3) Have your partner sit with his legs hanging off the table. Passively flex and extend the knee as you explore the patella's movements and its relationship to the quadriceps tendon.

Tibial Tuberosity and Shaft of Tibia

The *tibial tuberosity* is a superficial knob located distal to the patella on the shaft of the tibia. It is roughly half an inch in diameter and serves as an attachment site for the patellar ligament. It is sometimes visibly protruding.

The *shaft of the tibia* lies superficially along the anterior leg. Its flat surface and edges are clearly palpable from the tibial tuberosity to the medial malleolus (p. 252).

1) Partner seated with knee flexed. Locate the patella. Slide your fingers three or four inches straight inferior from the patella and, using your fingerpads, explore for the tuberosity. (7.13)
2) Continue to palpate inferiorly along the shaft of the tibia. Determine the width of the shaft by palpating along its edges. Follow it down to the medial malleolus.

✔ *With your fingers at the tibial tuberosity, ask your partner to extend his knee slightly. With this action, the patellar ligament will tighten, and you will be able to feel it at its attachment to the tuberosity. When palpating the tibial shaft, does it have distinct edges and lead to the medial ankle?*

---

sesamoid   **ses**-a-moyd   L. resembling a sesame seed
tuberosity   tu-ber-**os**-i-te

Head of the Fibula

The head of the fibula is located on the lateral side of the leg and is sometimes visibly protruding. It is the attachment site for the biceps femoris muscle, a portion of the soleus muscle, and the lateral collateral ligament.

👁 **1)** Partner seated with knee flexed. Locate the tibial tuberosity.
2) Slide your fingers laterally three to four inches to the lateral side of the leg. Palpate for the head of the fibula. (7.14) Explore its tip and inch wide tip.

◆ With your partner prone, bend the knee at 90˚, and follow the biceps femoris tendon (p. 215) distally to where it inserts into the head of the fibula.

✔ *Is the knob you are palpating lateral to the tibial tuberosity? Can you sculpt a circle around it outlining its shape? Does the biceps femoris tendon lead to the head of the fibula?*

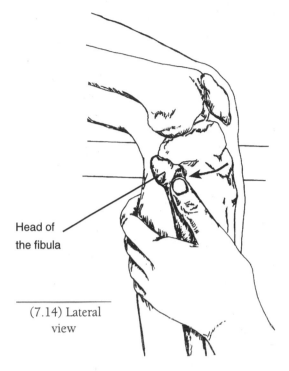

Head of
the fibula

(7.14) Lateral
view

Be aware of the common peroneal nerve (p. 278) lying along the posterior aspect of the head of the fibula.

Trail 2 - "Waddle Walk"
Tibial Plateaus

The medial and lateral plateaus are located on the proximal end of the tibia. Situated inside the knee joint, the plateaus cannot be palpated, but their edges can be easily accessed. These edges are superficial, and are located on either side of the patellar ligament.

👁 **1)** Partner seated with knee flexed. Place your thumbs on either side of the patella.
2) Slide inferiorly compressing into the tissue. You will feel a softening in the knee as your thumbs sink into the joint space between the femur and tibia.
3) Continue inferiorly until you feel the plateau edge. (7.15) Palpate both edges, following them in either direction.

✔ *Can you follow the edges of the plateaus horizontally to the sides of the knee? Can you feel the soft joint space superior to them? If you passively extend the knee with one hand, while the other palpates the edges, do you feel the edges move closer to the patella?*

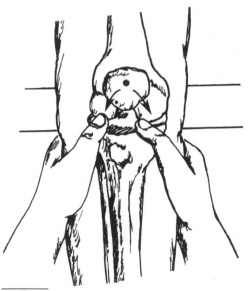

(7.15)

The patella, tibial tuberosity and head of the fibula together form a triangle. Can you create this shape by placing a finger on each of these three structures?

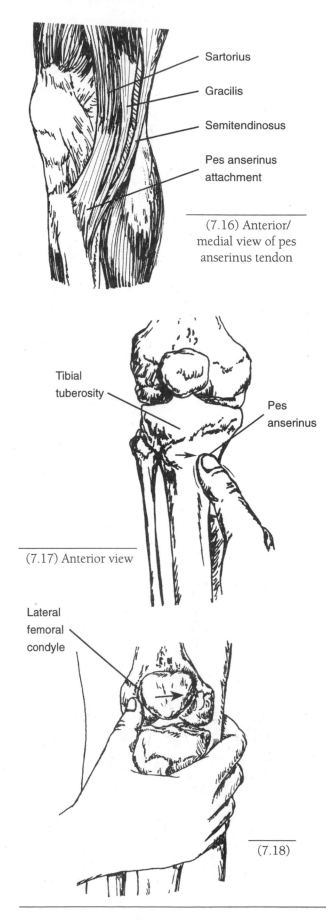

Sartorius

Gracilis

Semitendinosus

Pes anserinus
attachment

(7.16) Anterior/
medial view of pes
anserinus tendon

Tibial
tuberosity

Pes
anserinus

(7.17) Anterior view

Lateral
femoral
condyle

(7.18)

✋👁 Distal to the lateral plateau is a swelling of bone called the tibial tubercle. It is commonly the attachment site of the iliotibial tract (p. 224). Slide distally off the lateral plateau and explore the girth of the tubercle. When the knee is extended, the tubercle lays generally between the patella and head of the fibula.

Pes Anserinus Attachment

Three tendons of the thigh converge at the medial knee to form the pes anserinus tendon. The pes anserinus tendon attaches to the proximal, medial shaft of the tibia. (7.16) More specifically, it is the flat area medial to the tibial tuberosity.

✋👁 1) Partner seated with knee flexed. Locate the tibial tuberosity.
2) Slide medially one inch and explore its flat surface and any palpable tendons. (7.17)

✔ *Is the region you are isolating medial to the tibial tuberosity? Is it on the anterior shaft of the tibia?*

Trail 3 - "Hills on Both Sides"
Edge of Femoral Condyles

Most of the large, round femoral condyles are inaccessible. However, the edges of the condyles, located on either side of the patella, are accessible. The two ridges play an important role in the tracking of the patella during flexion and extension of the knee (p. 213).

✋👁 1) Partner supine with the knee fully extended. Locate the sides of the patella.
2) Shift the patella medially. Slide off the patella onto the condyle. Explore its distinct edge. (7.18)
3) Follow the edge distally as it continues toward the joint space.
4) Palpate the edge of the lateral condyle in the same manner. Compare the size and height of the medial and lateral edges and their relationship to the patella.

✔ *Are the edges slightly underneath the patella? Can you follow the edges distally toward the joint space of the knee?*

pes anserinus　　pes **an**-ser-i-nus　　L. *pedes*, foot; *anserine*, like a goose

Epicondyles of the Femur

The lateral epicondyle is a bald, knobby area located on the lateral side of the knee. It serves as an attachment site for the lateral collateral ligament. It is deep to the iliotibial tract and anterior to the biceps femoris tendon.

The medial epicondyle is deep to the tendon of the sartorius, distal to the vastus medialis muscle and serves as an attachment site for the medial collateral ligament.

👁✋ **1)** Supine with knee extended. Locate the patella.
2) Slide directly medial from the patella to the medial side of the knee. Note the epicondyle's superficial quality, round surface and location superior to the tibiofemoral joint. (7.19)
3) Return to the patella and slide to the lateral epicondyle on the lateral knee. Explore this region, noting its location proximal to the head of the fibula.

✔ *Can you palpate the vastus medialis (p. 213) proximal to the medial epicondyle? Is the head of the fibula distal to the lateral epicondyle?*

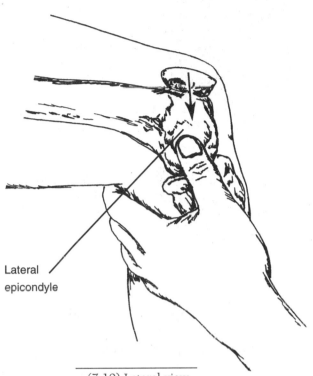

Lateral epicondyle

(7.19) Lateral view

Adductor Tubercle

The adductor tubercle is located proximal to the medial epicondyle, between the belly of the vastus medialis and the hamstring tendons. Its small tip sticks out from the top of the medial epicondyle and is an attachment site for the adductor magnus tendon (p. 220). It can often be tender to the touch.

👁✋ **1)** Partner supine with the knee extended. Locate the medial epicondyle of the femur.
2) Slide superiorly along the medial side of the femur. As the outline of the femur drops off into the soft tissue, explore for the small point of the tubercle. (7.20)
3) Strum across the adductor magnus tendon by rubbing your fingertip anteriorly and posteriorly.

✔ *Are you directly proximal to the medial epicondyle? With a finger on the proximal aspect of the tubercle (on the adductor magnus tendon), have your partner gently adduct his hip. Does the tendon of the magnus become taut, pressing into your finger?*

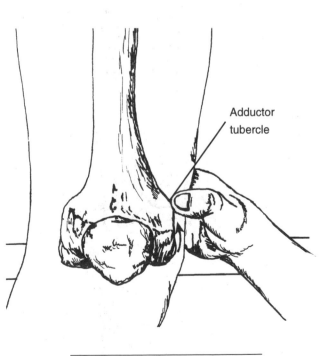

Adductor tubercle

(7.20) Anterior view of right knee

Bones and Bony Landmarks of the Foot

The foot contains twenty-six bones. (7.21-7.23)
The hindfoot is formed by the talus and calcaneus.
The *talus* articulates with the tibia and fibula to
form the talocrural or ankle joint. The large,
chunky *calcaneus* is the bone at the heel of the foot.

The midfoot is composed of five tarsals. Small
and each uniquely shaped, the *tarsals* are tightly
wedged together. They are accessible along the
dorsal surface of the foot.

The forefoot is formed by the long, superficial
metatarsals and phalanges. Similar to the metacarpals, each *metatarsal* has a proximal base, a
shaft and a distal head. The big toe is formed
by two sizable *phalanges*, and the remaining
toes each have three. All sides of the phalanges
are accessible.

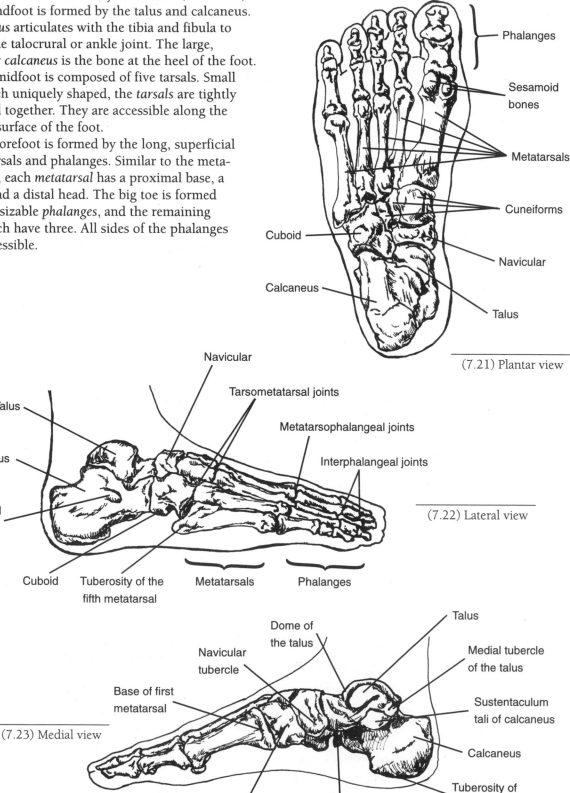

Phalanges

Sesamoid
bones

Metatarsals

Cuneiforms

Cuboid

Navicular

Calcaneus

Talus

(7.21) Plantar view

Navicular

Tarsometatarsal joints

Metatarsophalangeal joints

Interphalangeal joints

Talus

Calcaneus

Peroneal
tubercle

Cuboid   Tuberosity of the
         fifth metatarsal

Metatarsals   Phalanges

(7.22) Lateral view

Dome of
the talus

Navicular
tubercle

Base of first
metatarsal

Talus

Medial tubercle
of the talus

Sustentaculum
tali of calcaneus

Calcaneus

Tuberosity of
the calcaneus

(7.23) Medial view

Medial cuneiform   Head of the talus

Bony Landmark Trails of the Foot

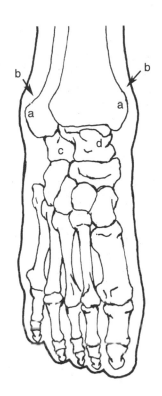

(7.24)

The trails of the foot present the hind and forefoot first, followed by the more challenging structures of the midfoot.

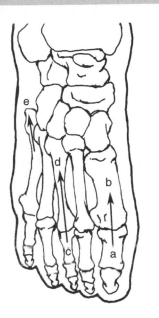

(7.25)

Trail 1 - "The Back Road" locates the bones and landmarks of the hindfoot and ankle. (7.24)

a) Lateral and medial malleoli
b) Malleolar grooves
c) Calcaneus
   Tuberosity of calcaneus
   Sustentaculum tali
   Peroneal tubercle
d) Talus
   Head
   Dome
   Medial tubercle

Trail 2 - "Little Piggies" palpates the bones and joints of the toes and forefoot. (7.25)

a) Hallucis
b) First metatarsal
c) Second through fifth phalanges
d) Second through fifth metatarsals
e) Tuberosity of fifth metatarsal

Trail 3 - "The Archway" explores the bones of the midfoot located at the arch of the foot. (7.26)

a) Navicular and navicular tuberosity
b) Medial, middle and lateral cuneiforms
c) Cuboid

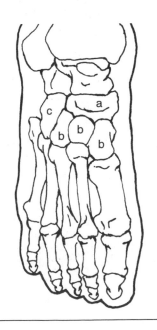

(7.26)

| metatarsal | **met**-a-**tar**-sal | Gr. after or beyond | tarsal | **tar**-sul | Gr. wickerbasket |
| phalanx | **fal**-anks | Gr. closely knit row | | | |

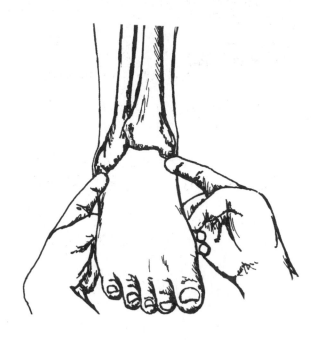

(7.27) Palpating the level of each malleolus

Trail 1 -"The Back Road"
Lateral and Medial Malleoli

The medial and lateral malleoli are the large visible knobs on either side of the ankle. The broader medial malleolus is located at the distal end of the tibia. The more slender lateral malleolus protrudes from the distal fibula.

👁 Seated or supine. Explore and compare the shapes and sizes of the malleoli. Palpate all sides of their surfaces.

✔ *Can you follow proximally from the medial malleolus along the shaft of the tibia to the tibial tuberosity? (7.27) Is the medial malleolus more proximal than the lateral malleolus?*

Malleolar Grooves

The medial and lateral malleoli each have a small vertical groove carved into their posterior surfaces. These grooves are designed to offer stability and leverage for tendons bending around the ankle. Because these tendons lie inside or beside the groove, it can be difficult to feel the groove's actual depressions.

👁 1) Supine or seated. Locate the medial malleolus.
2) Slide roughly half an inch posteriorly and palpate the posterior aspect of the malleolus for a slender, vertical groove. (7.28)
3) Shorten the surrounding tissue by passively inverting the foot and explore the length of the groove and the superficial tendons.
4) Try this same method along the lateral malleolus. Passively evert the foot to shorten the surrounding tissue.

✔ *Since each groove runs vertically, can you roll your finger horizontally across it to determine its location and shape?*

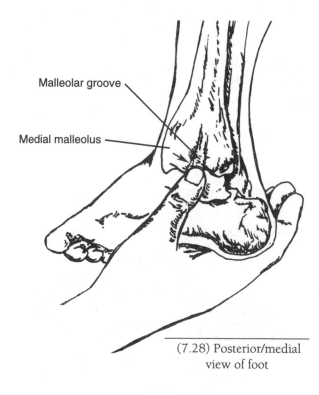

Malleolar groove

Medial malleolus

(7.28) Posterior/medial
view of foot

calcaneus	kal-**kay**-nee-us	L. heel
malleolus	mal-**e**-o-lus	L. little hammer

Calcaneus
Tuberosity, Sustentaculum Tali and Peroneal Tubercle

The large, solid *calcaneus* forms the heel of the foot. It is situated beneath the talus and projects two inches posteriorly from the malleoli. The medial and lateral sides of the calcaneus are deep to tendons, yet clearly palpable. The *tuberosity of the calcaneus* is a flat region located along the posterior surface of the calcaneus. The calcaneal tendon attaches to the superior aspect of the tuberosity.

The *sustentaculum tali* is located roughly one inch distal to the medial malleolus on the medial side of the calcaneus. (7.29) Shaped like a plank, the tali supports the talus on the calcaneus. The tali is an attachment site for the deltoid ligament (p. 280) and is deep to the flexor tendons. Only its small tip is accessible.

The *peroneal tubercle* is located on the lateral side of the foot. (7.30) Roughly an inch distal to the lateral malleolus, the tubercle is a small, superficial prominence that expands from the calcaneal surface to help stabilize the peroneal muscles (p. 266).

Calcaneus

1) Supine or seated. Walk your fingers distally from the malleoli down to the heel. Palpate and explore the shape and girth of the posterior calcaneus.
2) Move to the plantar surface to isolate the tuberosity at the base of the heel. (7.31) The tuberosity will not feel like a distinct bump, but more like a flat region.

✔ *Place one hand at the malleoli and the other at the tuberosity. Take note of how far the calcaneus extends posteriorly.*

Sustentaculum tali

1) Supine or seated. Place the ankle in a neutral position and locate the medial malleolus.
2) Slide approximately one inch distal to the small tip of the tali. (7.32) Passively inverting the foot will soften the surrounding tissues.
3) Sculpt around its sides noting the soft tissues located just distal to it.

✔ *Are you distal to the medial malleolus? If you slide distally off the sustentaculum tali do you feel the thick tissues at the sole of the foot?*

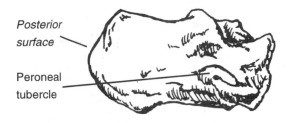
(7.29) Medial view of calcaneus

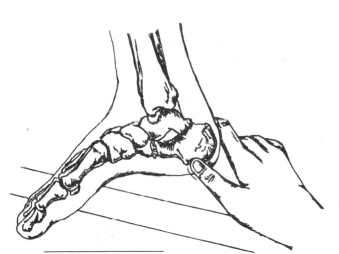
(7.30) Lateral view of calcaneus

(7.31) Medial view

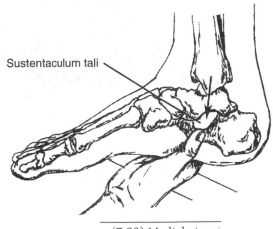
(7.32) Medial view

sustentaculum   **sus**-ten-**tak**-u-lum   L. support

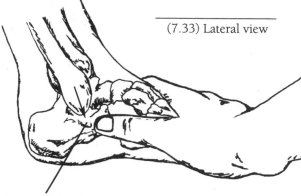

_____
(7.33) Lateral view

Peroneal tubercle

✋👁 Peroneal tubercle

1) Supine or seated. With the ankle in a neutral position, locate the lateral malleolus.
2) Slide roughly an inch inferiorly and explore for the small, superficial tubercle. It may feel like a short ridge on the surface of the calcaneus. (7.33) Passively everting the foot will soften the surrounding tissues.
3) Sculpt around its edges, noting the soft tissues located just distal.

✔ *Are you distal to the lateral malleolus? If you slide off the tubercle distally, do you feel the thick tissues of the foot? Ask your partner to alternately evert and relax her foot. Do the peroneal tendons pass along either side of the tubercle?*

> Of the two-hundred primates in the world, humans are the only ones with a non-grasping big toe. Since we are no longer tree climbers, the foot evolved from its hand-like design to serve as a platform for the human body.
>
> This does not mean, however, that the toes were designed to be stationary. A baby's foot has twenty times the toe-grasping capacity of a shoe-wearing adult. And in shoeless cultures, the foot's prehensile abilities remain throughout adulthood and are used for sewing and even threading needles.

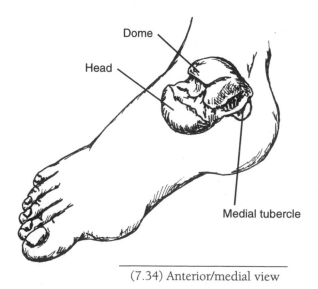

Dome

Head

Medial tubercle

_____
(7.34) Anterior/medial view

Talus

Head, Dome and Medial Tubercle

The *talus* consists of a head, neck and body. (7.34) The *head* is the round, anterior portion of the talus that articulates with the navicular. The body of the talus has a large superior prominence called the *"dome of the talus,"* wedged between the distal ends of the fibula and tibia.

There are three bony landmarks of the talus which can be accessed: The medial aspect of the head is accessible posterior to the navicular tubercle (p. 259). The anterior part of the dome of the talus is located between the malleoli. The *medial tubercle* of the talus is small and is located posterior to the medial malleolus. The tubercle is an attachment site for the deltoid ligament (p. 280).

_____

talus   **ta**-lus   L. ankle

👁✋ Head of the talus

1) Supine or seated, with the ankle in a neutral position. Locate the navicular tubercle (p. 259).
2) Slide proximally off the tubercle to the head of the talus. Compared to the tubercle, the head may feel like a depression.
3) Passively invert and evert the foot to obtain a clear distinction between these landmarks. When the foot is everted, the talar head will be more pronounced. When the foot is inverted, the navicular tubercle becomes more prominent.

✔ *If you draw a line between the medial malleolus and navicular tubercle, the head of the talus will be located along that line.* (7.35)

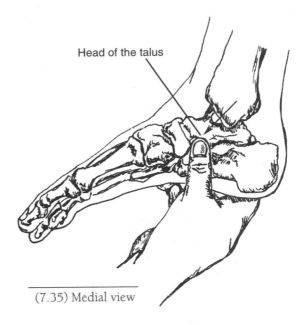

Head of the talus

(7.35) Medial view

👁✋ Dome of the talus

1) Passively invert and plantar flex the foot.
2) Draw a horizontal line connecting the malleoli, drop inferiorly off the center of the line, looking for a bony prominence. The dome will be deep to the overlying tendons, more prominent near the lateral malleolus. (7.36)

✔ *Is the tissue you are feeling hard and immovable like bone or firm and mobile like tendon? If you passively move the foot back to neutral, does the bony mound you are palpating seem to disappear into the ankle?*

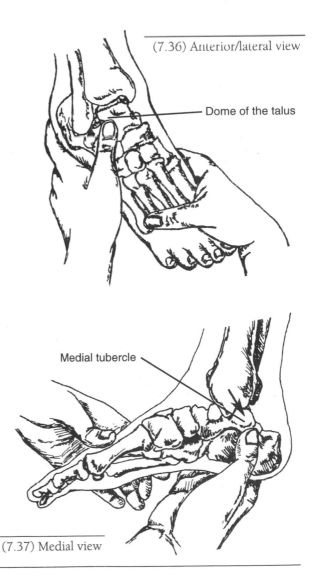

(7.36) Anterior/lateral view

Dome of the talus

👁✋ Medial tubercle

1) Locate the medial malleolus.
2) As opposed to sliding straight distally for the sustentaculum tali (p. 253), slide just off the malleolus posteriorly at a 45° angle to locate the medial tubercle. (7.37)
3) Passively dorsiflex and plantar flex the ankle, noting how the tubercle slides around the malleolus.

Medial tubercle

(7.37) Medial view

The calcaneus, talus and cuboid are all roughly cube-shaped. Soldiers of ancient Rome used these bones (probably from horses) to hack out playing dice. For this reason, the talus is sometimes called the astragalus, which in Latin means *die*, the root of the plural *dice*.

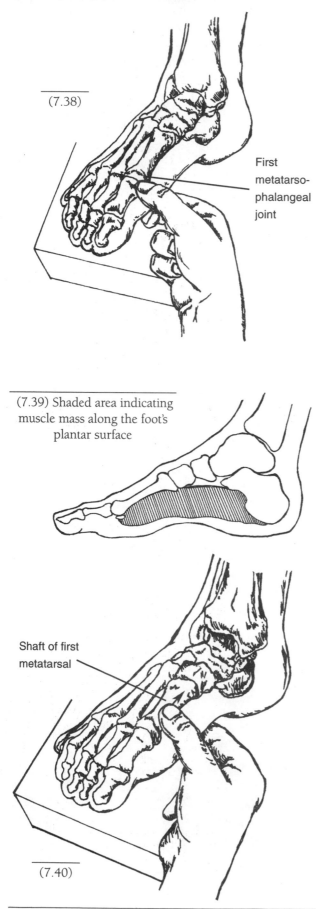

(7.38)

First metatarso-phalangeal joint

(7.39) Shaded area indicating muscle mass along the foot's plantar surface

Shaft of first metatarsal

(7.40)

Trail 2 - "Little Piggies"
Hallucis

The hallucis is composed of two phalanges. The joint between the phalanges, the inter-phalangeal, is a hinge joint wrapped in supportive ligaments. The first metatarsophalangeal joint is located at the ball of the foot. It is an ellipsoid joint with a large, bulbous shape.

👁 **1)** Seated or supine. Palpate the entire surface of the big toe noting the thickness and textural differences between its dorsal and plantar surfaces. **2)** Explore the surface of each joint by passively moving it through its range of motion. (7.38)

✔ *Is the proximal phalange nearly twice as long as the distal phalange?*

First Metatarsal

Unlike the long, slender metatarsals of toes two through five, the first metatarsal is short and stocky. Its dorsal and medial sides are superficial and easily accessible; its plantar surface is deep to several thick muscles. (7.39) The proximal end of the first metatarsal flares to articulate with the medial cuneiform. This articulation often forms a visible crest on the top of the foot and can be irritated when wearing tight shoes.

👁 Seated or supine. Locate the metatarsal shaft along the medial side of the foot. Explore the shaft's size and length by sliding across its entire surface. Palpate the junction and crest at the metatarsal head and medial cuneiform. (7.40)

✔ *Are the head and base broader than the shaft of the metatarsal? Can you feel the cylindrical shape of the shaft?*

Strapping on shoes has certainly protected our feet and reduced the number of sprained ankles, but it has also wreaked havoc on the arches of the feet. With such external support, the arches need not adapt to uneven terrain, so the supportive musculature weakens. Eventually, the arch diminishes on the medial side of the foot, a condition commonly known as "flat foot."

hallux   **hal**-uks   L. big toe

Phalanges of Toes

Unlike the hallucis, the second through fifth toes each contain three phalanges. There are two articulations in each toe, the proximal interphalangeal (or "pip" joint) and the distal interphalangeal (or "dip" joint). The metatarsophalangeal joints are proximal to the webbing between the toes and are much smaller than the first metatarsophalangeal joint.

Seated or supine. Palpate along all surfaces of the toes, noting the thin tissue along their inner sides. Explore one toe at a time, slowly moving it through each range of motion. (7.41)

Second through Fifth Metatarsals

The long, slender second through fifth metatarsals each have an enlarged base and head. The bases are set closely together next to the tarsals. The space between the metatarsals is filled with the small intrinsic muscles of the toes, and is accessible along the dorsal surface of the foot.

The tuberosity of the fifth metatarsal is a superficial point that extends laterally from the base of the fifth metatarsal. It is the attachment site for the peroneus brevis (p. 266).

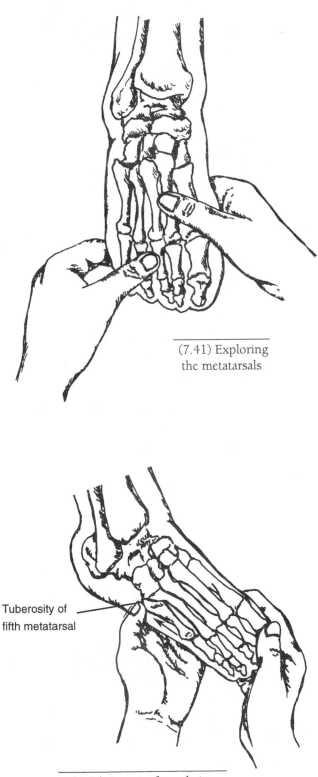

(7.41) Exploring the metatarsals

Metatarsals

1) Seated or supine. Grasp the foot with both hands and palpate the heads of each metatarsal on the dorsal side of the foot.
2) Use both thumb tips to explore the length of each bone and its surrounding spaces. Follow the shaft of each metatarsal proximally. Note how it widens to form the base of the metatarsal.

Tuberosity of fifth metatarsal

1) Seated or supine. Locate the shaft of the fifth metatarsal.
2) Follow the shaft proximally to where the base bulges laterally. (7.42) Explore the superficial shape of the tuberosity and surrounding landmarks as it projects from the side of the foot.

✔ *When the ankle is dorsiflexed, are you roughly two inches anterior to the lateral malleolus? Is the tip you are palpating connected to the fifth metatarsal?*

Tuberosity of fifth metatarsal

(7.42) Anterior/lateral view

Mammals such as cats and dogs are called digitigrades - they walk on their digits. When standing, the metatarsals and tarsals of digitigrades are off the ground and form what appears to be the "leg." For this reason, the ankle of a dog or cat is often mistaken for the knee - giving the "knee" the appearance of hyperextending.

Digitigrades have raised themselves up on their toes to bring their bodies higher. This additional height allows for greater sensory perception and increases the length of their stride.

Hoofed animals (unguligrades)

go one step further than digitigrades by lifting all of their phalanges but the distal one. With a wide, four-point stance, these animals can literally walk on the tips of their toes.

Walking "tippy-toe" will quickly tell you neither of these designs work for humans. We are plantigrades - "one who walks on the soles." As bipeds, we must spread our feet out and press all of the bones of the feet firmly on the ground to keep our balance.

Knee
Ankle
Metatarsals

Hind leg of a dog

Trail 3 - "The Archway"
Medial, Middle and Lateral Cuneiforms

The three cuneiforms lie in a row between the navicular, talus and metatarsals. The medial cuneiform serves as an attachment for the tibialis anterior and tibialis posterior tendons. The medial cuneiform can be isolated along its dorsal and medial surfaces. The middle and lateral cuneiform, sandwiched between the medial cuneiform and the cuboid, are accessible on their dorsal surfaces.

1) Seated or supine. Locate the base of the first metatarsal.
2) Glide proximally to the skinny ditch of the first tarsometatarsal joint.
3) Continue proximally onto the surface of the medial cuneiform. (7.43)
4) Slide laterally from the medial cuneiform along the dorsal surface of the foot and explore the surfaces of the middle and lateral cuneiforms.

The tibialis anterior tendon (p. 268) runs superficially down the front of the ankle and leads directly to the medial cuneiform. Have your partner dorsiflex his foot and follow the tendon distally as it blends into the medial side of the medial cuneiform.

✔ *Are you proximal to the base of the first metatarsal and can you isolate the joint between these two bones? If you follow the tibialis anterior tendon, does it lead to the same location where you were palpating the medial cuneiform?*

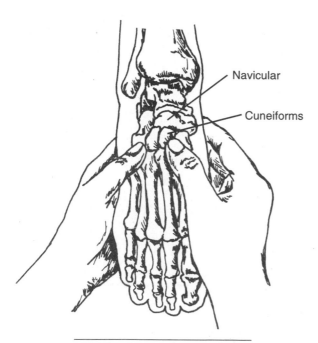

Navicular

Cuneiforms

(7.43) Palpating the cuneiforms

---

cuneiform   ku-**ne**-i-form   L. wedge-shaped

Navicular

The bean-shaped navicular is sandwiched between the medial and middle cuneiforms and the calcaneus. Its dorsal and medial surfaces are superficial and palpable. The superficial tuberosity bulges out the medial side of the foot and is an attachment for the tibialis posterior (p. 271) and the spring ligament (p. 281).

1) Seated or supine. Locate the base of the first metatarsal.
2) Sliding along the foot's medial side, move proximally across the surface of the medial cuneiform and the slender joint between the medial cuneiform and the navicular.
3) As you move onto the surface of the navicular, explore the shape and size of the navicular tuberosity. (7.44) The tuberosity will lie approximately one to two inches distal to the medial malleolus.

✔ *Does the bone you are palpating project more medially than the surfaces of the other bones on the medial foot? If you place a finger at the tuberosity of the fifth metatarsal and the navicular tuberosity simultaneously, does the fifth metatarsal lie slightly distal to the navicular tuberosity?*

Cuboid

As its translation suggests, the cuboid is cube-shaped. It is surrounded on three of its four sides by the fourth and fifth metatarsals, the lateral cuneiform and the calcaneus. The cuboid's dorsal surface is partially covered by the belly of the extensor digitorum brevis (p. 273). Due to its crowded location and covering, the cuboid is only partially accessible.

1) Seated or supine. Draw an imaginary line from the tuberosity of the fifth metatarsal to the lateral malleolus.
2) Following this line, at roughly half an inch from the tuberosity, is the cuboid. (7.45)

✔ *Are you proximal to the tuberosity of the fifth metatarsal? With the foot dorsiflexed, are you roughly an inch distal to the lateral malleolus?*

(7.44) Medial view

Navicular tubercle

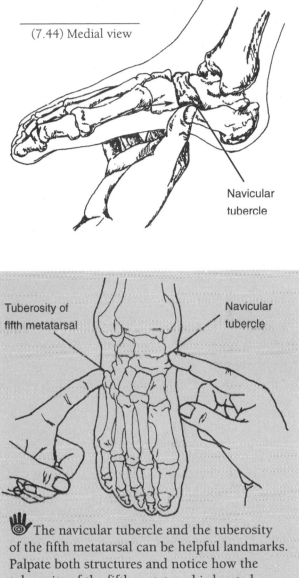

Tuberosity of fifth metatarsal

Navicular tubercle

The navicular tubercle and the tuberosity of the fifth metatarsal can be helpful landmarks. Palpate both structures and notice how the tuberosity of the fifth metatarsal is located further distal than the navicular.

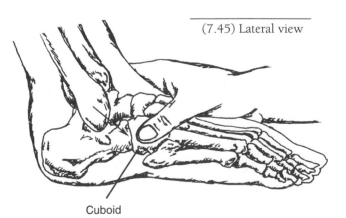

(7.45) Lateral view

Cuboid

navicular	na-**vik**-u-lar	L. boat-shaped
cuboid	**ku**-boyd	Gr. cube-shaped

Muscles of the Leg and Foot

Similar to the forearm and hand, the leg and foot contain numerous muscles. Most are directly or partially accessible, and their names reveal a great deal about their actions.

The majority of muscles of the leg can be divided into four groups:

- The large *gastrocnemius* and *soleus* form the "calf muscles" of the posterior leg.

- The *peroneus longus and brevis* are slender muscles located along the lateral side of the leg.

- The *"extensors"* of the foot and ankle (tibialis anterior, extensor digitorum longus and extensor hallucis longus) are layered together on the anterior leg and dorsum of the foot.

- The small *"flexors"* of the foot and ankle include tibialis posterior, flexor digitorum longus and flexor hallucis longus. They are deep to the gastrocnemius and soleus on the posterior leg.

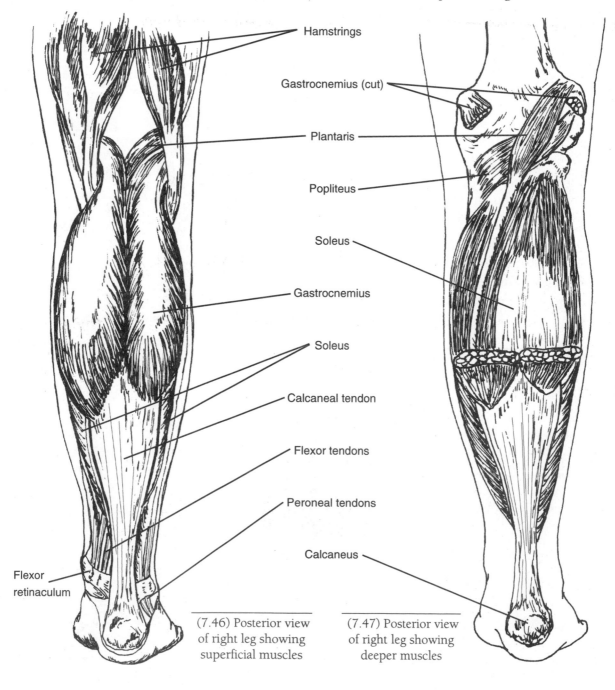

Hamstrings

Gastrocnemius (cut)

Plantaris

Popliteus

Soleus

Gastrocnemius

Soleus

Calcaneal tendon

Flexor tendons

Peroneal tendons

Calcaneus

Flexor retinaculum

(7.46) Posterior view of right leg showing superficial muscles

(7.47) Posterior view of right leg showing deeper muscles

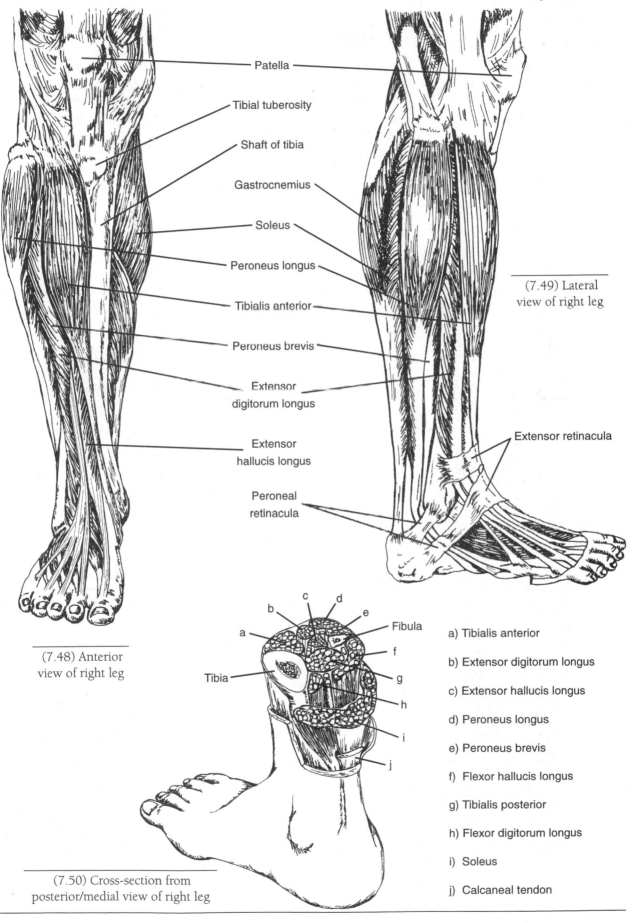

Patella

Tibial tuberosity

Shaft of tibia

Gastrocnemius

Soleus

Peroneus longus

Tibialis anterior

Peroneus brevis

Extensor digitorum longus

Extensor hallucis longus

Peroneal retinacula

(7.49) Lateral view of right leg

Extensor retinacula

(7.48) Anterior view of right leg

Tibia

Fibula

c
b
d
a
e
f
g
h
i
j

a) Tibialis anterior

b) Extensor digitorum longus

c) Extensor hallucis longus

d) Peroneus longus

e) Peroneus brevis

f) Flexor hallucis longus

g) Tibialis posterior

h) Flexor digitorum longus

i) Soleus

j) Calcaneal tendon

(7.50) Cross-section from posterior/medial view of right leg

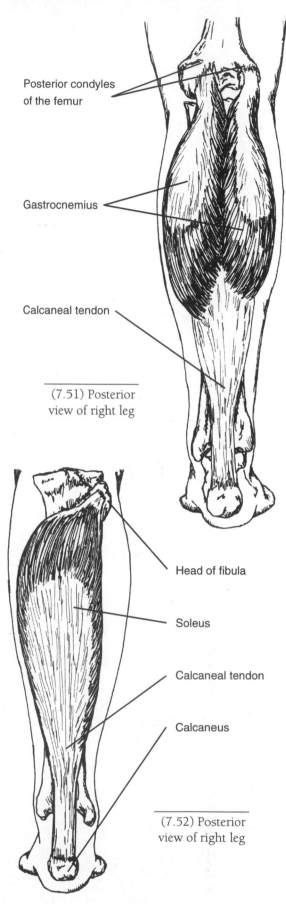

Posterior condyles
of the femur

Gastrocnemius

Calcaneal tendon

(7.51) Posterior
view of right leg

Head of fibula

Soleus

Calcaneal tendon

Calcaneus

(7.52) Posterior
view of right leg

Gastrocnemius and Soleus

The large muscle mass of the posterior leg is composed of the two-headed gastrocnemius and the soleus muscles. Together they form what is known as the triceps surae, which attaches to the strong calcaneal (Achilles) tendon. Both the gastrocnemius and soleus are easily accessible.

The superficial *gastrocnemius* has two heads and crosses two joints - the knee and ankle. (7.51) Emerging from between the hamstring tendons, the short gastrocnemius heads extend halfway down the leg before blending into the calcaneal tendon. Although its name (Latin for "belly muscle of the knee") suggests the gastrocnemius heads are rotund, they are actually thin compared to the thick soleus belly.

The *soleus* is deep to the gastrocnemius, yet its superficial, distal fibers bulge out the side of the leg and extend more distally than the gastrocnemius heads. (7.52) The soleus is sometimes called the "second heart" because of the important role its strong contractions play in returning blood from the leg to the heart.

Gastrocnemius

A -
Flex the knee
Plantar flex the ankle

O - Posterior condyles of the femur

I - Calcaneus via calcaneal tendon

N - Tibial

Soleus

A -
Plantar flex the ankle

O - Soleal line and posterior surface of tibia, proximal, posterior fibula

I - Calcaneus via calcaneal tendon

N - Tibial

---

gastrocnemius **gas**-trok-**ne**-me-us Gr. *gaster*, stomach + *kneme*, leg
soleus so-**lay**-us L. *solea*, sole of the foot

👁Gastrocnemius and soleus

1) Standing. Ask your partner, supported by a chair, to stand on her toes.
2) Palpate the posterior leg, sculpting out the gastrocnemius' oval heads. Follow both heads proximally to the back of the knee. Then follow them distally, noting how the medial head extends further distal than the lateral head. (7.53)
3) Move distal to the gastrocnemius and palpate the distal portion of soleus. Also explore the medial and lateral sides of soleus which bulge out from the gastrocnemius.
4) Follow both muscles distally as they blend into the calcaneal tendon.

✔ *Can you follow the gastrocnemius heads proximally between the hamstring tendons? Is the medial gastrocnemius head slightly longer than the lateral? Can you feel the textural differences between the fleshy muscle bellies and the tough, dense calcaneal tendon?*

👁Soleus

1) Prone. Bend the knee to 90° and investigate the soft, massive bellies of the gastrocnemius and soleus and the thick calcaneal tendon. (7.54)
2) When the knee is flexed, the gastrocnemius muscle is shortened and ineffective as a plantar flexor. Isolate the soleus by asking your partner to gently plantar flex against your resistance. Notice how the thick soleus contracts while the thin, superficial bellies of the gastrocnemius remain limp.

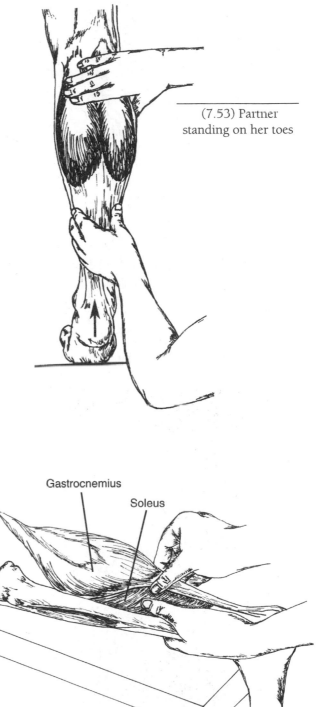

(7.53) Partner standing on her toes

Gastrocnemius

Soleus

(7.54) Medial view with partner prone

Gastrocnemius
Location - Superficial, two heads
BLMs - Calcaneal tendon
Action - "Bend your knee" or
"step down on your foot"

Soleus
Location - Deep to gastrocnemius
BLMs - Calcaneal tendon
Action - "Step down on your foot"

Why was the calcaneal tendon originally called the "Achilles" tendon? In Greek mythology, Achilles was dipped into the River Styx by his mother to make him invulnerable. All was immersed except the ankle by which she held him. After fighting successfully in the Trojan Wars, Achilles was mortally wounded by an arrow which penetrated his heel. Hence, "Achilles heel" refers to a seemingly small, but actual mortal weakness.

triceps surae   **sir**-eye   L. calf of the leg

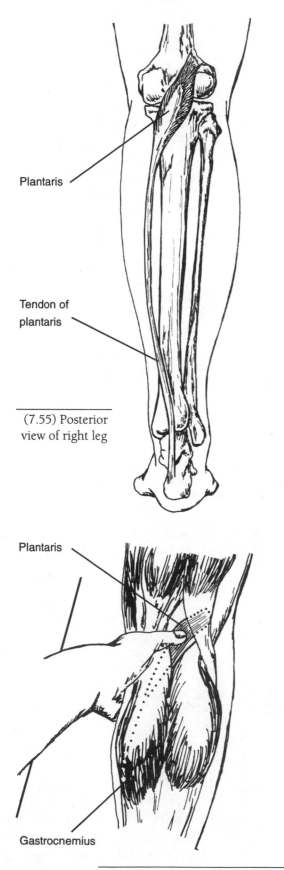

Plantaris

Tendon of
plantaris

(7.55) Posterior
view of right leg

Plantaris

Gastrocnemius

(7.56) Posterior view of right leg

Plantaris

The plantaris has a short muscle belly and a long, skinny tendon. Its belly lies at an oblique angle along the posterior knee between the gastrocnemius heads; its tendon extends the length of the leg and attaches to the calcaneus. (7.55) Although the plantaris belly is situated in a small, cramped area, its belly can be accessed quite readily.

In terms of its evolution, the plantaris is thought to be all that remains of a large plantar flexor of the foot. Its importance is still evident in reptiles where it serves as an important muscle of propulsion.

A -
Weak plantar flexion of ankle

O - Lateral epicondyle of femur

I - Calcaneus via calcaneal tendon

N - Tibial

1) Prone, with the knee flexed. Locate the head of the fibula.
2) Move your thumb medial into the popliteal space, between the gastrocnemius heads. (Sliding your thumb a little proximal in the popliteal space will position it off the gastrocnemius' heads.)
3) With your thumb between the gastrocnemius heads, slowly sink into the tissue of the posterior knee. (7.56) Explore for a one-inch wide belly that runs laterally to medially (at an oblique angle). When you believe you have located the plantaris, outline its shape by strumming your thumb across its belly.

✔ *Are you medial from the head of the fibula rather than being further distal? Are you accessing between the gastrocnemius heads? Is the belly you are palpating one to two fingers wide, with oblique fibers?*

Location - Deep to gastrocnemius, posterior knee
BLMs - Between heads of the gastrocnemius
Action - "Bend your knee"

Perhaps not so coincidentally, the plantaris is similar to the palmaris longus (p. 100) in the forearm. Both muscles have weak abilities, short bellies followed by long tendons and are absent in nearly 10% of the population.

Popliteus

As its name suggests, the popliteus is located in the popliteal region. It has a small, short muscle belly with diagonal fibers. (7.57) It is the deepest muscle of the posterior knee, laying beneath the upper fibers of the gastrocnemius and plantaris. Because of its depth, popliteus is inaccessible. However, its tendonous insertion on the posterior tibia can be palpated.

Popliteus is a weak flexor of the knee, but is vital in unlocking the knee from an extended position. Hence, its nickname "the key which unlocks the knee."

A -
Internally rotate
Flex the knee

O - Lateral condyle of the femur

I - Proximal, posterior aspect of tibia

N - Tibial

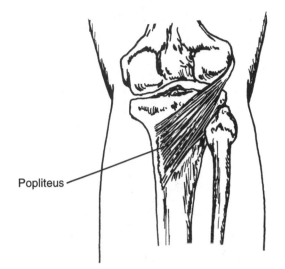

(7.57) Posterior view of right leg

 1) Prone, with the knee flexed. Access a portion of the popliteus by locating the tibial tuberosity and sliding medially around the tibia to the posterior surface of the tibial shaft.
2) Explore the posterior surface of the tibia by pushing the overlying edge of the soleus and gastrocnemius muscles to the side. (7.58)
3) The popliteus will not present itself as a palpable structure; however, if you are accessing the posterior region of the tibial shaft, you are on its tendonous attachment site.

Location - Deep to gastrocnemius on posterior knee
BLMs - Proximal, posterior shaft of tibia
Action - "Bend your knee"

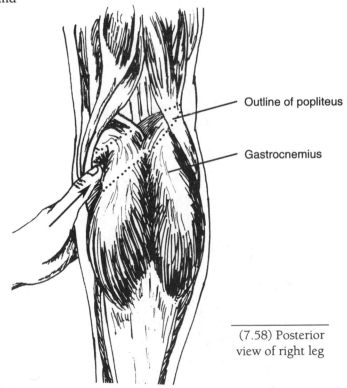

Outline of popliteus

Gastrocnemius

(7.58) Posterior view of right leg

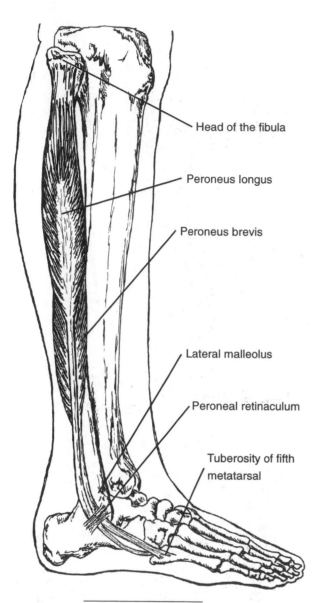

Head of the fibula

Peroneus longus

Peroneus brevis

Lateral malleolus

Peroneal retinaculum

Tuberosity of fifth
metatarsal

(7.59) Lateral view
of right leg

Location - Superficial, lateral side of leg
BLMs - Head of fibula, lateral malleolus
Action - "Evert your foot"

Peroneus Longus and Brevis

The slender peroneal muscles are located on the lateral side of the fibula. Specifically, they lie between the extensor digitorum longus and soleus. The peroneus longus lies on top of a portion of the peroneus brevis, yet both are accessible. (7.59) Their distal tendons are superficial and palpable behind the lateral malleolus and along the side of the heel. (7.60)

Peroneus Longus

A -
Evert the foot
Assist in plantar flexion of ankle

O - Proximal two-thirds of lateral fibula

I - Base of the first metatarsal, medial cuneiform

N - Peroneal

Peroneus Brevis

A -
Evert the foot
Assist in plantar flexion of ankle

O - Distal two-thirds of lateral fibula

I - Tuberosity of fifth metatarsal

N - Peroneal

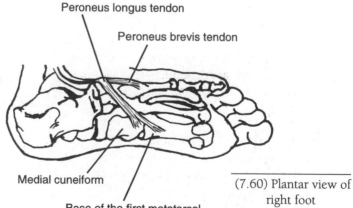

Peroneus longus tendon

Peroneus brevis tendon

Medial cuneiform

Base of the first metatarsal

(7.60) Plantar view of
right foot

| peroneus | per-o-**ne**-us | Gr. to pin |
| tertius | **ter**-she-us | L. third |

1) Supine, prone or sidelying. Place a finger at the head of the fibula and the lateral malleolus. The peroneal bellies are located between these landmarks. (7.61)

2) Lay your fingers between these landmarks and ask your partner to alternately evert and relax her foot. During eversion feel the peroneals tighten. This action sometimes creates a visible dimple or depression along the side of the leg.

3) As your partner continues to evert and relax, follow the peroneus longus proximally toward the head of the fibula. Follow both muscles distally to where their tendons wrap around the back of the lateral malleolus.

4) Follow the peroneus brevis tendon to the base of the fifth metatarsal. (7.62)

✔ *Are you on the lateral side of the leg between the head of the fibula and the lateral malleolus? Can you differentiate the slender peroneals from the lateral edge of the large gastrocnemius and soleus? Can you feel the tendon of the brevis attach to the base of the fifth metatarsal?*

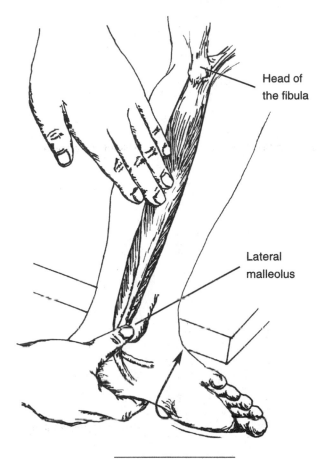

(7.61) Lateral view
of right leg

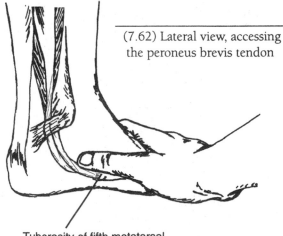

(7.62) Lateral view, accessing
the peroneus brevis tendon

It is rare, but not uncommon for a third peroneal to be present. The peroneus tertius, if present, will be found anterior to the lateral malleolus on the front of the ankle. Oddly, it is actually a short branch of the extensor digitorum longus. It attaches, along with peroneus brevis, at the tuberosity of the fifth metatarsal.

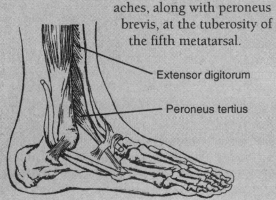

Extensor digitorum

Peroneus tertius

Tuberosity of fifth metatarsal

Extensors of the Leg

Tibialis Anterior
Extensor Digitorum Longus
Extensor Hallucis Longus

These extrinsic muscles are located on the anterior aspect of the leg between the shaft of the tibia and the peroneal muscles. The tendons of all three muscles cross beneath the extensor retinaculum at the ankle (p. 281).

The *tibialis anterior* is large and superficial, and is the most clearly isolated of the group. It lies directly lateral to the tibial shaft. (7.63)

Extensor digitorum longus is partially superficial, squeezed between the tibialis anterior and the peroneal muscles. The digitorum has four tendons which are clearly palpable on the dorsal surface of the foot. (7.64)

The muscle belly of the *extensor hallucis longus* lies deep to the other two muscles and can only be accessed indirectly. Like the extensor digitorum, however, its distal tendon is easily found on the dorsal surface of the foot as it leads toward the big toe.

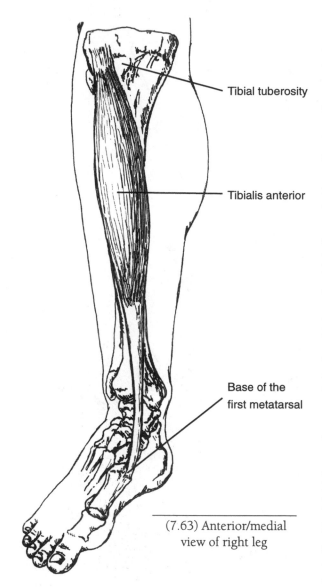

Tibial tuberosity

Tibialis anterior

Base of the
first metatarsal

(7.63) Anterior/medial
view of right leg

Tibialis Anterior

A -
Invert the foot
Dorsiflex the ankle

O - Lateral condyle of tibia, proximal lateral surface of tibia, interosseous membrane

I - Medial cuneiform, base of the first metatarsal

N - Peroneal

Extensor Hallucis Longus

A -
Extend the big toe
Dorsiflex the ankle
Invert the foot

O - Middle anterior surface of fibula, interosseous membrane

I - Distal phalanx of big toe

N - Peroneal

Extensor Digitorum Longus

A -
Extend toes two through five
Dorsiflex the ankle
Evert the foot

O - Lateral condyle of tibia, proximal anterior shaft of fibula, interosseous membrane

I - Middle and distal phalanges of lateral four toes

N - Peroneal

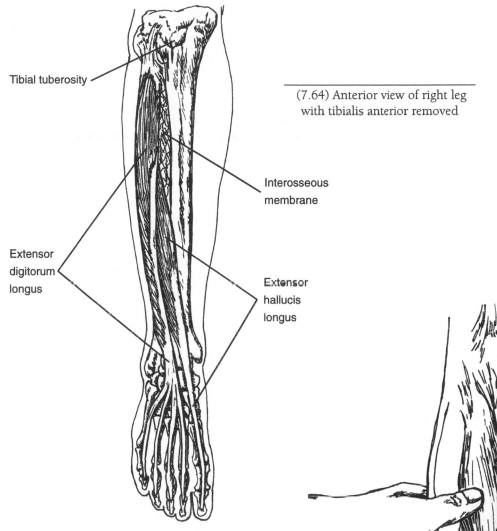

Tibial tuberosity

(7.64) Anterior view of right leg
with tibialis anterior removed

Interosseous
membrane

Extensor
digitorum
longus

Extensor
hallucis
longus

Shaft of
the tibia

Tibialis
anterior

(7.65) Anterior view
of right leg

👁 Tibialis anterior

1) Supine. Locate the shaft of the tibia. Slide off the shaft laterally onto the tibialis anterior.
2) Ask your partner to dorsiflex his foot and palpate its long, inch-wide belly. (7.65)
3) With the foot dorsiflexed, palpate the muscle distally as it becomes a thick, tendonous cord. Follow it to the medial side of the foot as it disappears at the medial cuneiform.

✔ *If your partner alternately dorsiflexes and relaxes his ankle, can you feel and see the tendon cross over the top of the ankle? Ask your partner to invert his foot and note if the tibialis anterior is involved. Can you palpate where the tendon passes under the extensor retinaculum (p. 281)?*

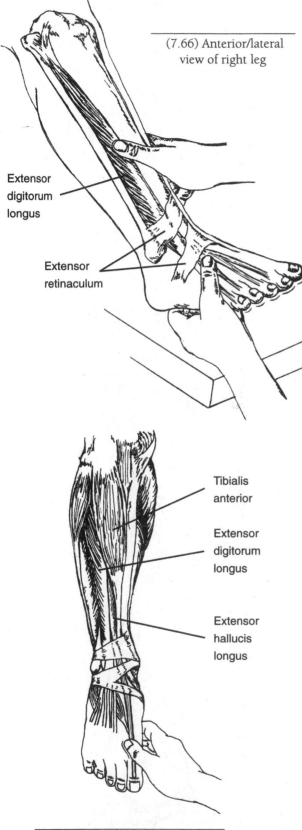

(7.66) Anterior/lateral
view of right leg

Extensor
digitorum
longus

Extensor
retinaculum

Tibialis
anterior

Extensor
digitorum
longus

Extensor
hallucis
longus

(7.67) Anterior view of right leg
with foot plantar flexed

☙ Extensor digitorum longus

1) Supine. It is easiest to begin by palpating the distal tendons of the digitorum. Ask your partner to extend his toes. Palpate and visibly identify the four tendons of the digitorum on the top of the foot.
2) With the toes still extended, follow the tendons toward the ankle. Note how they merge into a single tendonous bundle that loops underneath the extensor retinaculum. (7.66)
3) Follow this tendon proximally as it merges into its muscle belly. Explore the slender belly of the digitorum as it squeezes between the tibialis anterior and the peroneal muscles.

✔ *Locate the digitorum and tibialis anterior tendons on the top of the ankle. With the ankle dorsiflexed, ask your partner to slowly invert and evert her foot. During inversion, do you feel the tibialis tighten; during eversion, the digitorum?*

☙ Extensor hallucis longus

1) Supine. Ask your partner to extend his big toe. Palpate and visibly identify the solid tendon running along the dorsal surface of the foot to the big toe. (7.67)
2) With the toe still extended, follow the tendon toward the ankle. Note how it snuggles between and underneath the extensor digitorum and tibialis anterior tendons.

✔ *Can you follow the tendon from the big toe to the front of the ankle? Along the front of the ankle, can you distinguish the three separate tendons of these muscles?*

Extensors of the Leg
Location - Superficial between tibia and peroneals
BLMs - Lateral shaft of tibia, dorsum of foot
Action - "Bring your foot/toes toward your knee"

3

Flexors of the Leg

Tibialis Posterior
Flexor Digitorum Longus
Flexor Hallucis Longus

Buried deep to the gastrocnemius and soleus on the posterior leg are the *tibialis posterior*, *flexor digitorum longus*, and *flexor hallucis longus*. (7.68) All three muscles are virtually inaccessible. Only their most distal fibers and their tendons which cross behind the medial malleolus are directly palpable. Their tendons lie deep to the flexor retinaculum (p. 281). At the medial ankle, the tibial artery and tibial nerve are situated between the tendons.

Tibialis Posterior

A -

Invert the foot
Plantar flex the ankle

O - Proximal posterior shaft of tibia, proximal fibula, interosseous membrane

I - Navicular, cuneiform, cuboid, bases of second through fourth metatarsals

N - Tibial

Flexor Digitorum Longus

A -

Flex toes two through five
Weak plantar flexion of ankle
Invert the foot

O - Middle posterior surface of tibia

I - Distal phalanges of lateral four toes

N - Tibial

Flexor Hallucis Longus

A -

Flex the big toe
Weak plantar flex of ankle
Invert the foot

O - Middle half of posterior fibula

I - Distal phalanx of big toe

N - Tibial

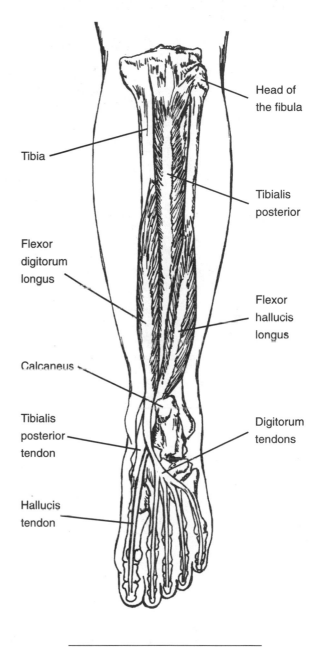

Head of the fibula

Tibia

Tibialis posterior

Flexor digitorum longus

Flexor hallucis longus

Calcaneus

Tibialis posterior tendon

Digitorum tendons

Hallucis tendon

(7.68) Posterior view of right leg with foot plantar flexed

Flexors of the Leg
Location - Deep to gastrocnemius and soleus
BLMs - Medial malleolus, posterior shaft of tibia
Action - "Plant and invert your foot"

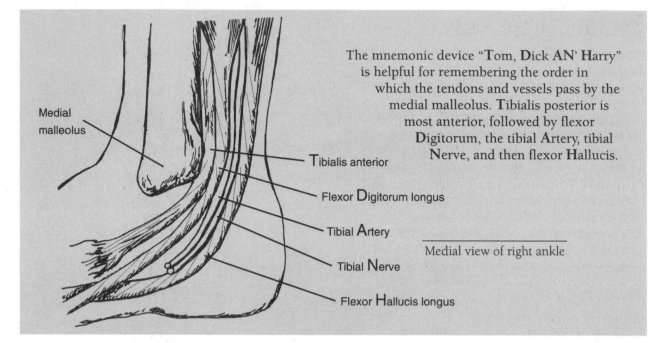

The mnemonic device "Tom, Dick AN' Harry" is helpful for remembering the order in which the tendons and vessels pass by the medial malleolus. Tibialis posterior is most anterior, followed by flexor Digitorum, the tibial Artery, tibial Nerve, and then flexor Hallucis.

Medial malleolus

Tibialis anterior

Flexor Digitorum longus

Tibial Artery

Medial view of right ankle

Tibial Nerve

Flexor Hallucis longus

1) Prone, supine or sidelying. Locate the medial malleolus. Slide off the malleolus posteriorly and proximally into the space between the posterior shaft of the tibia and the calcaneal tendon. 2) Explore this region for the distal bellies and tendons of these muscles. (7.69) Follow the tendons distally around the back of the medial malleolus. 3) It is difficult to isolate a specific tendon, but the tendon furthest anterior will be tibialis posterior. Have your partner invert his foot and follow this tendon around the ankle to the underside of the foot.

Medial malleolus

Shaft of tibia

Flexor retinaculum

Flexors of the leg

Calcaneal tendon

(7.69) Medial view of right leg

✔ *Place your fingers on the distal bellies and ask your partner to slowly wiggle all his toes. Can you feel the muscles or tendons shift? Can you locate the malleolar groove (p. 252) and feel the tendons in and posterior to it? Can you locate the pulse of the tibial artery?*

Muscles of the Foot

Extensor Digitorum Brevis
Flexor Digitorum Brevis
Abductor Hallucis
Abductor Digiti Minimi

Unlike the layers of muscle on the plantar surface of the foot, the dorsal surface is home to only one muscle. The small belly of the *extensor digitorum brevis* lies deep to the extensor digitorum longus tendons and is palpable. (7.70)

The thick tissue of the plantar surface of the foot is made up of several layers of muscle. The first layer of muscle is located deep to the plantar aponeurosis (p. 280) and is formed by three muscles which lie side by side. (7.71) The center muscle is the *flexor digitorum brevis*. It extends down the center of the foot from the calcaneus to the phalanges.

Medial to the flexor digitorum brevis is the thick, superficial *abductor hallucis*. Lateral to flexor digitorum brevis is the superficial *abductor digiti minimi*. Both abductors are easily palpable and are often visible along the sides of the foot.

These three muscles are relatively superficial with respect to the thick skin along the sole of the foot and the plantar aponeurosis, and are therefore palpable.

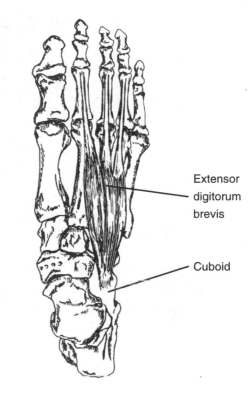

(7.70) Superior view of right foot

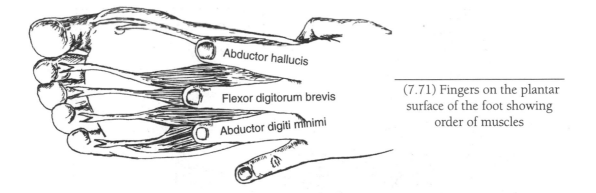

(7.71) Fingers on the plantar surface of the foot showing order of muscles

Extensor Digitorum Brevis

A - Extend toes two through four

O - Calcaneus (dorsal surface)

I - Toes two through four via the extensor digitorum longus tendons

N - Peroneal

Flexor Digitorum Brevis

A - Flex middle phalanges of toes two through four

O - Calcaneus (plantar surface)

I - Middle phalanges of toes two through four

N - Plantar

---

minimi   **min**-i-me   L. smaller

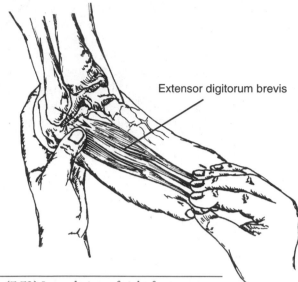

(7.72) Lateral view of right foot, partner extends toes against your resistance

Abductor Hallucis

A - Abduct
Assist in flexion of big toe

O - Calcaneus (plantar surface)

I - Proximal phalange on big toe (medial side)

N - Medial plantar

Abductor Digiti Minimi

A - Flex
Assist in abduction of fifth toe

O - Calcaneus (plantar surface)

I - Proximal phalange of little toe (lateral side)

N - Lateral plantar

👁 Extensor digitorum brevis

1) Supine, with the feet off the end of the table. Locate the lateral malleolus. Slide two inches off the malleolus toward the pinkie toe. Palpate beneath the extensor digitorum longus tendons to locate the small belly of extensor digitorum brevis.
2) Ask your partner to extend her toes against your resistance to feel the muscles contract. (7.72) Upon contraction, note how the belly forms a dense mound over the cuboid and lateral cuneiform.

👁 Flexor digitorum brevis

1) Supine, with the feet off the end of the table. Locate the plantar surface of the heel and the second through fifth toes.
2) Visualize this muscle's location by drawing imaginary lines between these points.
3) Palpating along the arch of the foot, sink your thumbs along these lines and roll across its fibers. (7.73) Ask your partner to alternately flex and relax her phalanges. It may be challenging to isolate its belly, but have faith that you are in the correct location of the flexor digitorum brevis.

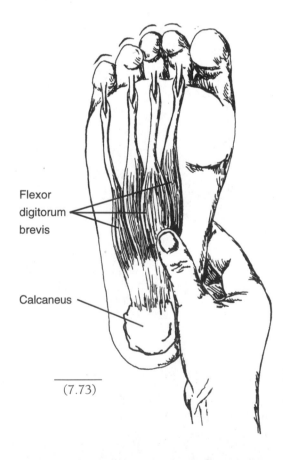

Flexor digitorum brevis

Calcaneus

(7.73)

If the discomfort of wearing high-heel shoes were not enough, try this on for size: The point of a spike-heel worn by an average-size woman is subjected to nearly 2,000 pounds of pressure per square inch. With every step this force is shot into the heel and reverberates up the entire body.

Actually, when air travel was in its infancy, women wearing high-heels were prohibited from boarding an airplane, given the potential danger of piercing the thin metal floors.

Bearing the weight of the body during standing, walking and running, the feet are sometimes known as the "little soldiers." Compared to standing, walking increases the pressure upon the feet roughly twofold. Running increases the stress fourfold.

Such demands require the foot to do more than lie flat and idle on the ground. Instead, the bones and ligaments of the foot are arranged to form three arches (right).

The three points of contact for the arches are the calcaneus and the heads of the first and fifth metatarsals.

Together the arches raise the center of the foot and create an ideal shape to absorb and distribute the weight of the body throughout the entire foot. The arches also assist the foot's plantar surface to adapt to uneven surfaces during hiking or climbing.

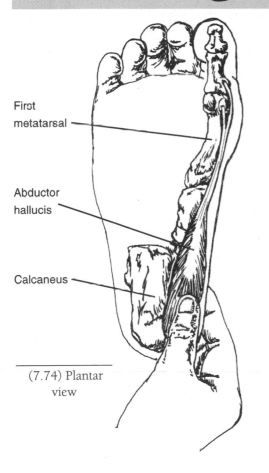

First metatarsal

Abductor hallucis

Calcaneus

(7.74) Plantar view

(7.75) Plantar view

Abductor digiti minimi

Calcaneus

👁 Abductor hallucis

1) Supine, with the feet off the end of the table. Locate the medial surface of the heel and the medial side of the big toe.
2) Palpate between these points and note the thick, superficial tissue running alongside the medial/plantar surface of the foot. (7.74)
3) Ask your partner to flex his big toe against your resistance and note the strength and density of the abductor hallucis fibers.

👁 Abductor digiti minimi

1) Supine, with the feet off the end of the table. Locate the plantar surface of the heel and the lateral surface of the pinkie toe.
2) Palpate between these points for the thick, superficial tissue running alongside the lateral/plantar surface of the foot. (7.75)
3) Ask your partner to abduct or flex his pinkie toe against your resistance to feel its fibers contract.

Ligaments, Nodes and Vessels of the Knee and Foot

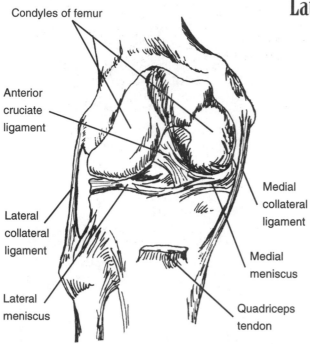

Condyles of femur

Anterior cruciate ligament

Lateral collateral ligament

Lateral meniscus

Medial collateral ligament

Medial meniscus

Quadriceps tendon

(7.76) Anterior view of knee with patella removed

Lateral and Medial Collateral Ligaments

The *lateral collateral ligament* is a strong, thin strap that crosses the knee joint from the lateral epicondyle of the femur to the head of the fibula. (7.76) It is superficial and located between the biceps femoris tendon and the iliotibial tract.

The broad *medial collateral ligament* lies superficial to the joint capsule of the knee and may not be as easy to isolate as the lateral collateral ligament. It stretches nearly two inches distal to the knee joint and is deep to the pes anserinus tendon. (7.76)

Both collateral ligaments resist against external rotation of the tibia. The lateral collateral also stabilizes against genu varum stresses upon the knee (often seen in bowed-legged cowboys). The medial collateral protects against genu valgum (knock-knee) stresses, like a blow from a football helmet to the lateral side of the knee.

👁 Lateral collateral ligament

1) Partner seated with knee flexed. Locate the head of the fibula and the lateral epicondyle. (7.77)
2) Slide your finger between these points and gently strum horizontally across this superficial ligament.

Ask your partner to cross his leg so the ankle is resting on top of the opposite knee. This position allows the lateral collateral ligament to be easily accessed. Roll your finger between the epicondyle and the head of the fibula and palpate the ligament. (7.78)

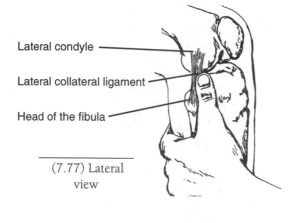

Lateral condyle

Lateral collateral ligament

Head of the fibula

(7.77) Lateral view

✔ *Is the band of tissue you feel the width of a pencil? Does it run from the epicondyle to the fibular head? Is it anterior to the biceps femoris tendon?*

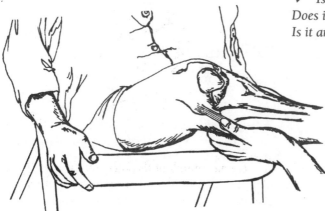

(7.78) Palpating the lateral collateral ligament with partner seated

---

genu valgum　　**je**-noo **val**-gum
genu varum　　**je**-noo **va**-rum

collateral　　ko-**lat**-er-al　　L. together, pertaining to the side

👁 Medial collateral ligament

1) Seated with knee flexed. Locate the medial epicondyle of the femur. Slide distally to the joint space, the thin crevice between the tibia and femur.
2) Strum your fingertip horizontally across this space, exploring for the broad fibers of the ligament. (7.79)

✔ *Are you on the medial side of the knee, just distal to the medial epicondyle of the femur?*

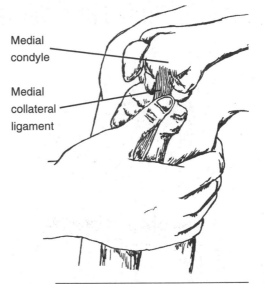

(7.79) Medial view of right knee

Anterior surface

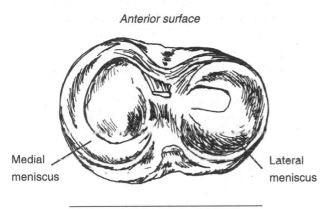

(7.80) Superior view of right tibia

Lateral and Medial Menisci

The menisci are fibrocartilaginous disks attached to the tibial condyles. (7.80) They play an important role in weight distribution and friction reduction, as well as assisting the round femoral condyles to sit comfortably upon the flat tibial plateaus. The edge of the medial meniscus can be palpated just above the edge of the medial tibial plateau. The smaller, more mobile lateral meniscus is difficult to access.

👁 Medial meniscus

1) Partner seated with knee flexed. Place your thumb superior to the medial tibial plateau in the joint space between the femur and tibia.
2) Grasp the foot or ankle with your other hand and slowly rotate the knee internally. (7.81)
3) As the medial side of the tibia rotates posteriorly, the edge of the medial meniscus will be pushed anteriorly into your thumb. The sensation may be quite subtle - a gentle pressure against your thumbpad.

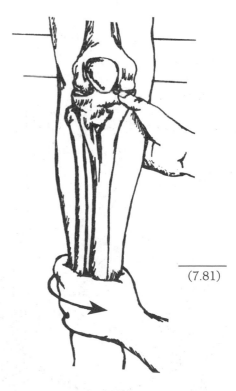

(7.81)

✔ *Is your thumb in the knee joint space? If you slowly alternate from internal to external rotation of the knee, do you feel a difference at your thumb?*

---

meniscus   men-**is**-kus   Gr. crescent-shaped

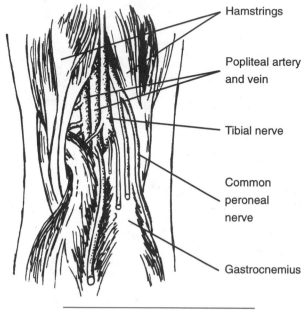

Hamstrings

Popliteal artery
and vein

Tibial nerve

Common
peroneal
nerve

Gastrocnemius

(7.82) Posterior view of right knee

Common Peroneal Nerve

Branching off the sciatic nerve, the common peroneal nerve runs along the lateral side of the popliteal space (or posterior knee). It passes between the biceps femoris tendon and the lateral head of the gastrocnemius, and posterior to the head of the fibula. (7.82) The nerve is most vulnerable behind the head of the fibula and is usually palpable.

👁✋ 1) Partner seated or prone with the knee flexed. Locate the head of the fibula.
2) Slide your finger around the posterior aspect of the head and, using gentle pressure, palpate for the slender, moveable peroneal nerve. If your partner complains of tingling or numbness along the lateral side of the foot, you are impinging the nerve.

✔ *Locate the biceps femoris tendon by asking your partner to flex his knee against your resistance. Follow the tendon to the head of the fibula, noting the nerve pathway alongside it.*

Popliteal Artery

The popliteal artery branches from the femoral artery to pass through the popliteal fossa at the back of the knee. It is situated deep in the fossa and, for this reason, its pulse can be difficult to feel.

👁✋ 1) Supine. Flex your partner's knee to soften the overlying tissues.
2) Hold the knee so the fingertips of both hands are at the midline of the posterior knee.
3) Sink your fingerpads deep into the popliteal fossa and explore for the subtle pulse. (7.83)

◈ If the pulse is undetectable, follow the same instructions with your partner prone.

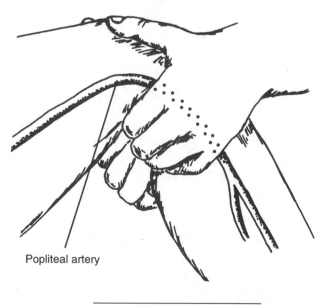

Popliteal artery

(7.83) Medial view of knee

Posterior Tibial Artery

The posterior tibial artery extends from the popliteal artery. It is superficial and its pulse can be felt just inferior and posterior to the medial malleolus.

1) Supine. Locate the medial malleolus.
2) Using two fingerpads, slide behind the malleolus and feel for its pulse. (7.84)

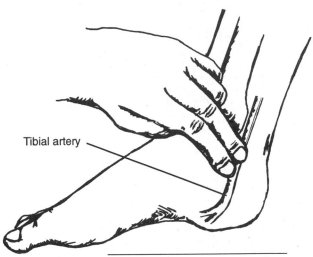

Tibial artery

(7.84) Medial view of right foot

Dorsal Pedis Artery

Located between the first and second metatarsal bones, the dorsal pedis artery lies superficially along the dorsal side of the foot.

1) Supine. Locate the first and second metatarsals. Place two fingerpads between the two bones and, using gentle pressure, explore for the pulse of the dorsal pedis artery. (7.85)

✔ *Are you lateral to the extensor hallucis longus tendon? If the pulse is undetectable, move slightly lateral.*

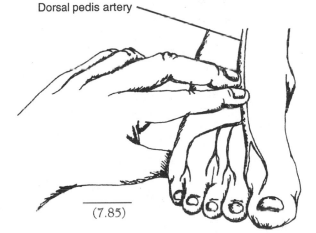

Dorsal pedis artery

(7.85)

Sesamoid Bones of First Metatarsal

The sesamoid bones of the first metatarsal are located along the plantar surface of the first metatarsal head. There are usually two, but sometimes more are present. The sesamoids are spherical and invested in the tendon of the flexor hallucis brevis. Often when palpating the sesamoids only their location and density can be determined, not their specific shapes.

1) Seated or supine. Locate the head of the first metatarsal.
2) Slide around to its plantar surface at the ball of the foot.
3) Using two thumbpads, explore this surface for the small sesamoid bones. Passively flex and extend the big toe to soften the surrounding tissues. (7.86)

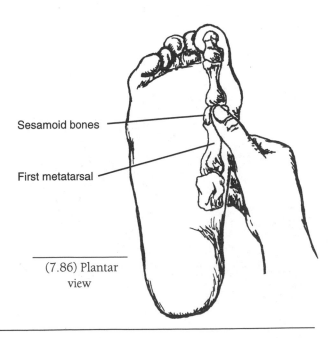

Sesamoid bones

First metatarsal

(7.86) Plantar view

| dorsal pedis | **dor**-sal **peh**-dis | L. *dorsum*, back; *pes*, foot |
| sesamoid | **ses**-a-moyd | L. resembling a sesame seed |

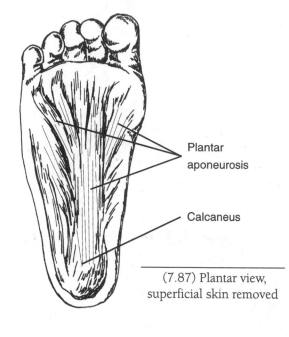

Plantar aponeurosis

Calcaneus

(7.87) Plantar view, superficial skin removed

Plantar Aponeurosis

The plantar aponeurosis is a thick, superficial band of fascia stretching from the heel to the ball of the foot. (7.87) Originating at the tuberosity of the calcaneus and expanding to the metatarsal heads, it plays an important role in supporting the longitudinal arch of the foot. Because the aponeurosis is located between the skin and muscles of the foot, it can be challenging to isolate from the surrounding tissues.

👁 **1)** Seated or supine. Draw an imaginary triangle from the heel to the ball of the foot.
2) Within this triangle explore the superficial layers of tissue along the sole of the foot. Passively flex and extend the toes and note how this movement affects the tension of the aponeurosis.

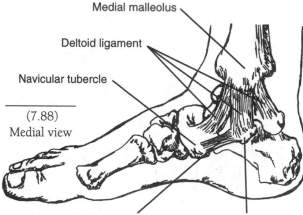

Medial malleolus

Deltoid ligament

Navicular tubercle

(7.88) Medial view

Spring ligament   Sustentaculum tali

Deltoid Ligament

The deltoid ligament is composed of several ligaments which originate at the medial malleolus and fan distally to attach at the talus, sustentaculum tali and navicular. (7.88) The ligament is designed to protect against medial stress of the talocrural joint. The deltoid ligament is deep to the flexor retinaculum and posterior flexor tendons (p. 271), yet palpable.

👁 **1)** Supine or seated. Locate the medial malleolus and sustentaculum tali (p. 253).
2) Place your finger between these points and strum your finger horizontally to isolate the ligament fibers.
3) Slide distally from the medial malleolus at a 45° angle and palpate its angled fibers to define the anterior and posterior aspects of the deltoid ligament. (7.89)

✔ *Are you palpating the space between the medial malleolus and sustentaculum tali? Do the fibers you are palpating fan out from the medial malleolus and have a firm, dense texture?*

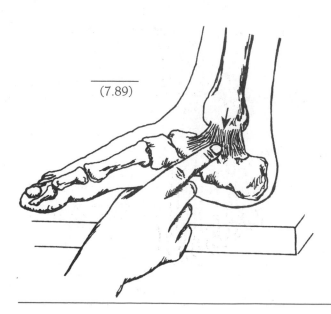

(7.89)

---

aponeurosis   **ap**-o-nu-**ro**-sis   Gr. *apo*, from + *neuron*, nerve or tendon

Plantar Calcaneonavicular (Spring) Ligament

The spring ligament is a small, tough band of tissue that plays an important role in stabilizing the medial longitudinal arch of the foot. (7.88) Located along the medial side of the foot, the ligament stretches from the sustentaculum tali to the navicular tubercle and may be positioned deep to the tibialis posterior tendon. The spring ligament may be extremely tender and should be accessed slowly and in close communication with your partner.

1) Supine or seated. Passively invert the foot to soften any surrounding tissue and locate the sustentaculum tali and navicular tubercle. (7.90)
2) Palpating between these bony landmarks, use a fingertip to slowly explore the taut surface of the spring ligament.

✔ *Are you between the sustentaculum tali and navicular tubercle? Can you roll your fingertip slowly across the surface of the ligament?*

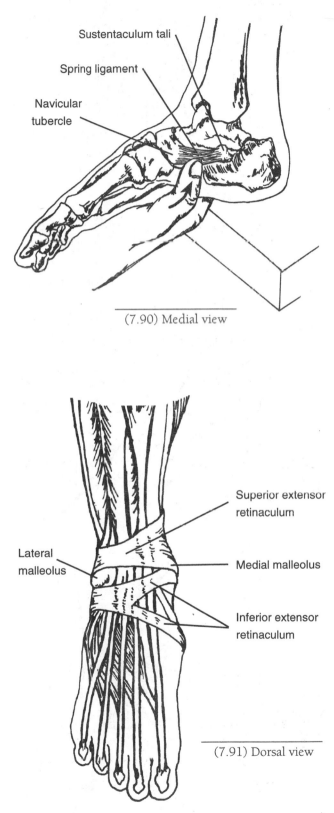

(7.90) Medial view

Retinacula of the Ankle

The tendons of the extensor muscles (p. 268) are supported by both the *superior and inferior extensor retinacula*. The superior retinaculum is broad and crosses the front of the ankle just proximal to the malleoli. The inferior retinaculum is "Y" shaped and begins distal to the lateral malleolus on the calcaneus. It spans the ankle and then divides, with one fork attaching at the medial malleolus and the other connecting to the navicular. (7.91)

The peroneal muscles are stabilized by the *superior and inferior peroneal retinacula*. The superior peroneal retinaculum stretches from the lateral malleolus to the calcaneus, and the inferior peroneal retinaculum holds the peroneal tendons to the peroneal tubercle.

The *flexor retinaculum* is a broad strap extending from the medial calcaneus to the medial malleolus. It is designed to hold the tendons of the flexor muscles and the tibial artery and nerve in place.

(7.91) Dorsal view

---

retinaculum   **ret-i-nak**-u-lum   L. halter

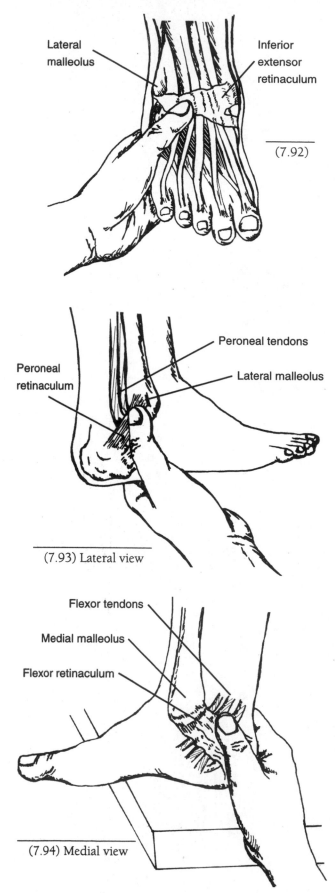

Lateral
malleolus

Inferior
extensor
retinaculum

(7.92)

Peroneal tendons

Peroneal
retinaculum

Lateral malleolus

(7.93) Lateral view

Flexor tendons

Medial malleolus

Flexor retinaculum

(7.94) Medial view

✋👁 Extensor retinacula

1) Partner supine. Ask your partner to dorsiflex her ankle and extend her toes. The pressure from the bulging tendons will make the retinacula more pronounced.
2) Palpate an inch proximal to the malleolus for the broad fibers of the superior extensor retinaculum.
3) Locate the inferior extensor retinaculum by moving distally to the level of the malleolus. (7.92) Explore either side of the large tibialis anterior tendon, for easy access of this retinaculum.

✔ *Are the fibers you are accessing superficial and perpendicular to the extensor tendons? Ask your partner to relax her ankle and notice how the retinacula softens.*

✋👁 Peroneal retinaculum

1) Ask your partner to evert his foot. The tension from the peroneal tendons will make the retinaculum more pronounced.
2) Locate the peroneal tendons between the lateral malleolus and lateral calcaneus. (7.93)
3) Roll your fingers along either side of the peroneal tendons for the small, short fibers of the retinaculum.

✔ *For the superior retinaculum, does the tissue you feel strap across the peroneal tendons from the lateral calcaneus to the lateral malleolus?*
For the inferior retinaculum, do you feel a short band crossing over the peroneal tubercle? Ask your partner to relax her ankle and see if you can still isolate the retinaculum.

✋👁 Flexor retinaculum

1) Ask your partner to dorsiflex and invert his foot. The strain from the flexor tendons will elevate the retinaculum closer to the surface.
2) Locate the medial malleolus and the medial side of the calcaneus.
3) Palpate between these landmarks, strumming across the broad, superficial fibers of the flexor retinaculum. (7.94)

✔ *Are you between the medial calcaneus and medial malleolus? Continue to explore the retinaculum with the foot relaxed.*

Bibliography

Alexander, R. McNeill, *The Human Machine*, Columbia University Press, New York, 1992

Anson, Barry, *An Atlas of Human Anatomy*, W.B. Saunders, Philadelphia, 1963

Asimov, Isaac, *The Human Body*, Houghton Mifflin Co., Boston, 1963

Backhouse, Kenneth and Hutchings, Ralph, *Color Atlas of Surface Anatomy*, Williams & Wilkins, Baltimore, 1986

Bates, Barbara, *A Guide to Physical Examination and History Taking*, 4th ed., J. B. Lippincott, Philadelphia, 1987

Bergman, Ronald; Thompson, Sue Ann and Adel K., Afifi, *Catalog of Human Variation*, Urban and Schwarzenberg, Baltimore, 1984

Bodanis, David, *The Body Book*, Little, Brown and Company, Boston, 1984

Calais-Germain, Blandine, *Anatomy of Movement*, Eastland Press, Seattle, 1993

Cartmill, Hylander and Shafland, *Human Structure*, Harvard University Press, Cambridge, 1987

Chaitow, Leon, *Palpatory Literacy*, Thorsons, London, 1991

Chaitow, Leon, *Palpatory Skills*, Churchill Livingstone, New York, 1997

Clemente, Carmine, *Anatomy: A Regional Atlas of the Human Body*, 3rd edition, Urban & Schwarzenberg, Baltimore, 1987

Clemente, Carmine, *Gray's Anatomy*, 30th edition, Lea & Febiger, Philadelphia, 1985

Craig, Marjorie, *Miss Craig's Face Saving Exercises*, Random House, New York, 1970

Cyriax, JH and Cyriax PJ, *Cyriax's Illustrated Manual of Orthopaedic Medicine*, 2nd ed., Butterworth/Heinemann Ltd., Oxford, 1992

Dorcas MacClintock, *A Natural History of Giraffes*, Charles Scribner's Sons, New York, 1973

Dorland's Illustrated Medical Dictionary, 24th edition, W.B. Saunders, Philadelphia, 1965

Eaton, Theodore Jr., *Comparative Anatomy of the Vertebrates*, 2nd edition, Harper and Brothers, 1971

Feher, Gyorgy and Szunyoghy, Andras, *Cyclopedia Anatomicae*, Black Dog & Levental Publishers, New York, 1996

Field, E. J., *Anatomical Terms: Their Origin and Derivation*, W. Heffer & Sons, Cambridge, UK, 1947

Gebo D, Plantigrady and foot adaptation in African apes: implications for hominid origins. *American Journal of Physical Anthropology* 89: 29-58, 1992

Gehin, Alain, *Atlas of Manipulative Techniques for the Cranium and Face*, Eastland Press, Seattle, 1985

Greene, Lauriann, *Save Your Hands! Injury Prevention for Massage Therapists*, Infinity Press, Seattle, 1995

Guillen, Michael, *Five Equations That Changed the World*, Hyperion, New York, 1995

Hamrick, MW and Inouye, SE, "Thumbs, tools, and early humans," *Science*, p. 586-7, April 1994

Handy, Chester, A History of Cranial Osteopathy, *Journal of American Osteopathic Association*, vol. 47, pp. 269-272, January 1948

Hertling, Darlene and Kessler, Randolph M., *Management of Common Musculoskeletal Disorders*, 3rd ed., JB Lippincott, Philadephia, 1996

Hildebrand, Milton, *Analysis of Vertebrate Structure*, 4th ed., John Wiley & Sons, New York, 1995

Hole, John, *Essentials of Human Anatomy and Physiology*, 4th edition, Wm. C. Brown, Dubuque, 1992

Hoppenfeld, Stanley, *Physical Examination of the Spine and Extremities*, Appleton & Lange, Norwalk, 1976

Jamieson, E. B., *Illustrations of Regional Anatomy, Sections I - VII*, E.S. Livingstone, Edinburgh, 1946

Jenkins, David, *Hollinshead's Functional Anatomy of the Limbs and Back*, 6th edition, W.B. Saunders, Philadelphia, 1991

Juhan, Deane, *Job's Body: A Handbook for Bodywork*, Station Hill, Barrytown, New York, 1987

Kapandji, I. A., *The Physiology of the Joints, Volumes 1, 2 & 3*, 5th ed., Churchill Livingstone, New York, 1982

Kapit, Wynn and Elson, Lawrence, *The Anatomy Coloring Book*, 2nd edition, HarperCollins College Publishers, 1993

Kendall, FP, McCreary EK, Provance PG, *Muscles: Testing and Function*, 4th edition, Williams & Wilkins, Baltimore, 1993

Kent, George, *Comparative Anatomy of the Vertebrates*, 6th edition, Mosby, St. Louis, 1987

Koch, Tankred, *Anatomy of the Chicken and Domestic Birds*, Iowa State University Press, Ames, Iowa, 1973

Lumley, John, *Surface Anatomy*, Churchill Livingstone, Edinburgh, 1990

Luttgens, Kathryn and Wells, Katharine, *Kinesiology: Scientific Basis of Human Motion*, Saunders College Publishing, Philadelphia, 1982

MacClintock, Dorcas, *A Natural History of Giraffes*, Charles Scribner's Sons, New York, 1973

Magee, David, *Orthopedic Physical Assessment*, 2nd edition, W.B. Saunders, Philadelphia, 1992

Marzke, MW, Evolutionary development of the human thumb, *Hand Clinics*, p. 1-9, Feb 1992

McAleer, Neil, *The Body Almanac*, 1st ed., Doubleday & Co., New York, 1985

McMinn, R.M.H., Hutchings, R.T., *Color Atlas of Human Anatomy*, Year Book Medical Publishers, Chicago, 1985

Montagna, William, *Comparative Anatomy*, John Wiley and Son, 1970

Napier, John, *Hands*, Princeton Science Library, Princeton, 1993

Netter, Frank, *Atlas of Human Anatomy*, CIBA-GEIGY, Summit, New Jersey, 1989

Norkin, Cynthia and Levangie, Pamela, *Joint Structure and Function*, 2nd edition, F.A. Davis, Philadelphia, 1992

Olsen, Andrea, *Bodystories: A Guide to Experiential Anatomy*, Station Hill Press, Barrytown, New York, 1991

Olsen, Todd, *A.D.A.M: Student Atlas of Anatomy*, Williams and Wilkins, Baltimore, 1996

Parker, Steve, *Natural World*, Dorling Kindersley, London, 1994

Peck, Stephen Rogers, *Atlas of Human Anatomy*, Oxford University Press, Oxford, 1982

Platzer, Werner, *Color Atlas and Textbook of Human Anatomy, Volume 1: Locomotor System*, Thieme Inc. New York, 3rd edition, 1986

Rohen, Johannes and Yokochi, Chihiro, *Color Atlas of Anatomy*, 3rd edition, Igaku-Shoin Publishers, New York, 1993

Rolf, Ida, *Rolfing and Physical Reality*, Healing Arts Press, Rochester, Vermont, 1990

Rolf, Ida, *Rolfing: Integration of Human Structures*, Harper Row, New York, 1977

Rossi, William, Shoes and "Normal" Foot, *Podiatry Management*, February, 1997

Schider, Fritz, *An Atlas of Anatomy for Artists*, 3rd edition, Dover Publishing, New York, 1957

Searfoss, Glenn, *Skulls and Bones*, Stackpole Books, Mechanicsburg, Pennsylvania, 1995

Seig, Kay and Adams, Sandra, *Illustrated Essentials of Musculoskeletal Anatomy*, 2nd Edition, Megabooks, Gainesville, 1993

Stern, Jack, *Core Concepts in Anatomy*, Little, Brown and Company, Boston, 1997

Stern, Jack, *Essentials of Gross Anatomy*, F.A. Davis, Philadelphia, 1988

Stone, Robert and Stone, Judith, *Atlas of the Skeletal Muscles*, Wm. C. Brown, Dubuque, 1990

Sutcliffe, Jenny and Duin, Nancy, *A History of Medicine*, Barnes and Noble, New York, 1992

Taber's Cyclopedic Medical Dictionary, 17th edition, F.A. Davis, Philadelphia, 1993

Thompson, Clem, *Manual of Structual Kinesiology*, 11th edition, Times Mirror/Mosby College, St. Louis, 1989

Thompson, Diana, *Hands Heal: Documentation for Massage Therapy*, Diana Thompson, Seattle, 1993

Todd, Mabel Elsworth, *The Thinking Body*, Dance Horizons, Brooklyn, 1979

Tortora, Gerald, *Principles of Human Anatomy*, 5th edition, Harper & Row, New York, 1989

Traupman, John, *New College Latin and English Dictionary*, Bantam Books, New York, 1995

Travell, Janet and Simons, David, *Myofascial Pain and Disfunction: Trigger Point Manual, Volume 1*, Williams and Wilkins, Baltimore, 1983

Travell, Janet and Simons, David, *Myofascial Pain and Disfunction: Trigger Point Manual, Volume 2*, Williams and Wilkins, Baltimore, 1992

Upledger, John and Vredevoogd, Jon, *Craniosacral Therapy*, Eastland Press, Seattle, 1983

Walker, Judith, *NeuroMuscular Therapy I - IV*, International Academy of NMT, St. Petersburg, 1994

Walker, Warren, *A Study of the Cat in Reference to the Human*, 5th edition, Saunders College Publishers, Fort Worth, 1993

Walker, Warren, *Functional Anatomy of the Vertebrates: An evolutionary perspective*, Saunders College Publishers, Fort Worth, 1987

Way, Robert, *Dog Anatomy - Illustrated*, Dreenan Press, Ltd. New York, 1974

Zihlman, Adrienne, *Human Evolution Coloring Book*, Harper & Row, New York, 1982

Glossary of Terms

abdomen - The region between the diaphragm and the pelvis.

acetabulum - The rounded cavity on the external surface of the coxal bone. The head of the femur articulates with the acetabulum to form the coxal joint.

adhesions - Abnormal adherence of collagen fibers to surrounding structures during immobilization, following trauma, or as a complication of surgery, which restricts normal elasticity of the structures involved.

agonist - A contracting muscle whose action is opposed by another muscle (antagonist).

anatomical position - Erect posture with face forward, arms at sides, forearms supinated so that palms of the hands face forward, and fingers and thumbs in extension.

antagonist - A muscle that works in opposition to another muscle (agonist); opponent.

antecubital - The anterior side of the elbow.

anterior - Toward the front or ventral surface.

anterior tilt - Pelvic tilt in which the vertical plane through the anterior-superior spines is anterior to the vertical plane through the symphysis pubis.

appendage - A structure attached to the body such as the upper and lower extremities.

arm - The portion of the upper limb between the elbow and shoulder joints.

arthrology - The study of joints.

articulation - A joint or connection of bones.

bicipital - A muscle that has two heads.

bilateral - Pertaining to two sides.

bursa - A synovial-lined sac existing between tendons and bone, muscle and muscles and any other site in which movement of structure occurs.

caudal - Downward, away from the head; (toward the tail).

collagen - The protein of connective tissue fibers.

concentric contraction - A shortening of the muscle during a contraction, a type of isotonic exercise.

contraction - An increase in muscle tension, with or without change in overall length.

coronal axis - A horizontal line extending from side to side, around which the movements of flexion and extension take place.

cramp - A spasmodic contraction of one or many muscles.

cranial - Upward, toward the head.

cutaneous - Referring to the skin.

distal - Farther from the center or median line, or from the thorax.

eccentric muscle contraction - An overall lengthening of the muscle while it is contracting or resisting a work load.

facet - A small plane or concave surface.

fascia - a general term for a layer or layers of loose or dense fibrous connective tissue.

flexibility - The ability to readily adapt to changes in position or alignment; may be expressed as normal limited, or excessive.

forearm - The portion of the upper limb between the wrist and elbow joints.

genu valgum - Knock-knees, defined as a lateral displacement of the distal end of the distal bone in the joint.

genu varum - Bowlegs, defined as a medial displacement of the distal end of the distal bone in the joint.

impingement - An encroachment on the space occupied by soft tissue, such as nerve or muscle. In this text, impingement refers to nerve irritation (i.e., from pressure or friction) associated with muscles.

insertion - The more mobile attachment site of a muscle to a bone. The opposite end is the origin.

interstitial - The space within an organ or tissue.

interstitial fluid - The fluid that surrounds cells.

isometric - Increase in tension without change in muscle length.

isotonic - Increase in tension with change in muscle length (in the direction of shortening); concentric contraction.

isotonic contraction (dynamic) - A concentric or eccentric contraction or a muscle. A muscle contraction performed with movement.

kinesiology - The study of movement.

kyphosis - A condition characterized by an abnormally increased convexity in the curvature of the thoracic spine as viewed from the side.

lateral - Away from the midline.

lateral tilt - Pelvic tilt in which the crest of the ilium is higher on one side than on the other.

leg - The portion of the lower extremity between the ankle and knee joints.

ligament - A fibrous connective tissue that connects bone to bone.

longitudinal axis - A vertical line extending in a cranial/caudal direction about which movements of rotation take place.

lordosis - An abnormally increased concavity in the curvature of the lumbar spine as viewed from the side.

lymph node - A small oval structure located along lymphatic vessels.

lymphatic - Often pertains to the system of vessels involved with drainage of bodily fluids (lymph).

medial - Toward the midline.

muscle - An organ composed of one of three types of muscle tissue (skeletal cardiac or visceral), specialized for contraction.

muscle contracture - An increase of tension in the muscle caused by activation of the contractile mechanism of the muscle (proprioceptors). This often leads to a fibrous condition of the muscle.

myofascial - Skeletal muscles ensheathed by fibrous connective tissue.

origin - The more stationary attachment site of a muscle to a bone. The opposite end is the insertion.

palpable - To touch or touchable.

palpate - To examine or explore by touching (an organ or area of the body), usually as a diagnostic aid.

paravertebrals - Alongside or near the vertebral column.

pelvic girdle - The two hip bones.

pelvic tilt - An anterior (forward), a posterior (backward), or a lateral (sideways) tilt of the pelvis from neutral position.

pelvis - Composed of the two hip bones, sacrum and coccyx.

periosteum - The fibrous connective tissue which surrounds the surface of bones.

posterior - Toward the back or dorsal surface.

posterior tilt - Pelvic tilt in which the vertical plane through the anterior-superior spines is posterior to the vertical plane through the symphysis pubis.

proximal - Nearer to the center or median line, or to the thorax.

range of motion - The range, usually expressed in degrees, through which a joint can move or be moved.

range of motion, active - The free movement across any joint of moving levers that is produced by contracting muscles.

range of motion, passive - The free movement that is produced by external forces across any joint or moving levers.

retinaculum - A network, usually pertaining to a band of connective tissue.

sagittal axis - A horizontal line extending from front to back, about which movements of abduction and adduction take place.

soft tissue - Usually referring to myofascial tissues, or any tissues which do not contain minerals (such as bone).

surface anatomy - The study of structures that can be identified from the outside of the body.

tactile - Pertaining to touch.

tendon - A fibrous tissue connecting skeletal muscle to bone.

thigh - The portion of the lower extremity between the knee and hip joints.

thorax - The region between the neck and abdomen.

tightness - Shortness; denotes a slight to moderate decrease in muscle length; movement in the direction of lengthening the muscle is limited.

trunk - The part of the body to which the upper and lower extremities attach.

unilateral - Pertaining to one side.

Pronunciations and Translations

abdomen	ab-do-men	L. belly
abduct		L. led away
acetabulum	as-e-**tab**-u-lum	L. a little saucer for vinegar
acromion	a-cro-me-on	Gr. *akron*, extremity + *omos*, shoulder
adduct		L. brought together
adipose		L. *adeps*, fat
anconeus	an-**ko**-ni-us	Gr. elbow
annulus	an-u-lus	L. ring
aponeurosis	ap-o-nu-ro-sis	Gr. *apo*, from + *neuron*, nerve, tendon
arrector pili	a-**rek**-tor **pee**-li	L. *pilosus*, hairy
artery	ar-ter-e	Gr. windpipe
basilic		Arabic *basilik*, inner
biceps	bi-**seps**	L. *bis*, twice + *caput*, head
brachial	bray-ke-al	L. relating to the arm
brachialis	bra-**key-al**-is	
brachii	bra-**key**-i	
brachioradialis		
	bra-key-o-ra-de-a-lis	
brevis		L. short
bursa	**bur**-sa	Gr. a leather sack
calcaneus	kal-**ka**-ne-us	L. heel
capitate	**kap**-i-tate	L. head
capitis	**kap**-i-tis	L. referring to the head
capitulum	ka-**pit**-u-lum	L. small head
carotid	ka-**rot**-id	Gr. deep sleep
carpal	**kar**-pul	Gr. pertaining to the wrist
carpi	**kar**-pi	

cartilage	**kar**-ti-lij	L. gristle
cephalic	se-**fa**-lic	Arabic *alkifaL*. outer
cervical	**ser**-vi-kal	L. referring to the neck
chest		AS, box
clavicle	**klav**-i-k'l	L. little key
coccyx	**kok**-siks	Gr. cuckoo
collateral	ko-**lat**-er-al	L. together, pertaining to the side of the body
condyle	kon-**dil**	Gr. knuckle
conoid	**ko**-noid	Gr. cone-shaped
coracobrachialis		
	kor-a-ko-**bra-ke-al**-is	
coracoid	**kor**-a-koyd	Gr. raven's beak
costal	**kos**-tal	L. rib
cranio-		L. skull
cremaster	kre-**mas**-ter	L. to suspend
cricoid	**kri**-koyd	Gr. ring-shaped
cruciate	**kru**-she-ate	L. crossed
cuboid	**ku**-boyd	Gr. cube-shaped
cuneiform	ku-**ne**-i-form	L. wedge-shaped
deltoid	**del**-toid	Gr. *delta*, letter d in Greek alphabet
diaphragm	di-a-**fram**	Gr. a partition
digastric	di-**gas**-trik	Gr. *dis*, twice *gaster*, belly
digit	**di**-jit	L. finger
digitigrade	**di**-ji-tah-grade	
		L. toe-walking
dorsi	**dor**-si	L. back
epi-		Gr. upon
facet	**fac**-et	Fr. small face
facial	**fa**-shal	L. pertaining to the face

fascia	fash-e-a	L. a band
fascicle	fas-i-kl	L. little bundle
femur	fe-mur	L. thigh
fibula	fib-u-la	L. pin
flex		L. to bend
foot		AS, fot
foramen	for-a-men	L. a passage or opening
fossa	fos-a	L. a shallow depression
furcula	fur-ku-la	L. a little fork
gastrocnemius	gas-trok-ne-me-us	
		Gr. *gaster*, stomach + *kneme*, leg
gemelli	jem-el-i	L. twins
genu valgum	je-noo val-gum	
genu varum	je-noo va-rum	
gland		L. acorn
glenoid		Gr. a socket
glossus		Gr. tongue
gluteus	gloo-te-al	Gr. *gloutos*, buttocks
gracilis	gra-ci-lis	L. slender, graceful
hallux	hal-uks	L. big toe
ham		AS. haunch
hamate	ham-ate	L. hooked
hamulus	ham-u-lus	L. a small hook
humerus	hu-mer-us	L. upper arm
hyoid	hi-oyd	Gr. "U" shaped
iliocostalis	il-e-o-kos-ta-lis	
ilium	il-ee-um	L. groin, flank
illiacus	i-li-a-cus	
inferior	in-fe-ree-or	L. below
infraspinatus	in-fra-spi-na-tus	
ischiocavernosus	ish-she-o-ka-ver-no-sus	
ischium	ish-ee-um	Gr. hip
jaw		ME, iawe
joint		L. a joining
jugular	jug-u-lar	L. throat
kyphosis	ki-fo-sis	Gr. humpback
lamina	lam-i-na	L. thin plate
latae	la-ta	L. broad
lateral	lat-er-al	L. side
latissimus dorsi	la-tis-i-mus dor-si	L. widest, back
levator	le-va-tor	L. lifter
levator scapula	le-va-tor skap-u-la	
longissimus	lon-jis-i-mus	
ligament	lig-a-ment	L. a band
linea aspera	lin-e-a as-per-a	L. rough line
longus colli	long-us ko-li	L. to align
lumbar	lum-bar	L. loin
lunate	lu-nate	L. moon shaped
lymph	limf	L. *lympha*, water
magnus	mag-nus	L. large
malleolus	mal-e-o-lus	L. little hammer
mandible	man-di-bl	L. lower jawbone
manubrium	ma-nu-bre-um	L. handle
masseter	mas-se-ter	Gr. chewer
mastoid	mas-toyd	Gr. breast
maxilla	max-il-a	L. jawbone
medial	me-de-ul	L. middle
meniscus	men-is-kus	Gr. crescent-shaped
mental		L. chin
meta-	met-a	Gr. after or beyond
metacarpal	met-a-kar-pul	
metatarsal	met-a-tar-sal	
minimi		L. smaller
multifidi	mul-tif-i-di	L. *fidi*, to split
muscle	mus-el	Gr. to enclose

mylohyoid	my-lo-hi-oyd	Gr. *myle*, milL. *hyoid*, hyoid bone
nasal	na-zl	L. nose
navicular	na-**vik**-u-lar	L. boat-shaped
neck		AS. nape
nuchae	nu-kay	L. nape of neck
oblique	o-bleek	L. diagonal
obturator	ob-tu-**ra**-tor	L. occluded
occipitofrontalis	ok-**sip**-i-to-fron-**ta**-lis	
occiput	ok-si-put	L. the back of skull
olecranon	o-**lek**-ran-on	Gr. elbow
omohyoid	o-mo-**hi**-oyd	Gr. *omo*, shoulder
opponens	o-po-nens	L. hand, palm
parietal	puh-**ri**'e-tul	L. wall
parotid	pa-rot-id	Gr. beside the ear
patella	pa-**tel**-a	L. a small pan
pectinius	pek-tin-e-us	L. comb
pectoralis	**pek**-to-**ra**-lis	L. breast or chest
pedicle	**ped**-i-k'l	
pelvis	**pel**-vis	L. basin
peroneus	per-**o**-ne-us	Gr. to pin
pes anserinus	pes **an**-ser-i-nus	L. *pedes*, foot, L.*anserine*, goose-like
phalanx	fal-anks	Gr. closely knit row
piriformis	pir-i-**form**-is	L. pear-shaped
pisiform	pi-si-form	L. pea-shaped
plantar	**plan**-tar	L. sole of the foot
plantaris	plan-tar-is	
plantigrade		L. sole-walking
platysma	pla-**tiz**-ma	Gr. plate
plexus		L. interwoven
pollex	**pol**-eks	L. strong
popliteus	pop-**lit**-e-us	L. ham of the knee
process	**pros**-es	L. going before

profundus	pro-**fun**-dus	L. located deeper than its reference point
pronate	pro-**nate**	L. prone
psoas	**so**-as	Gr. muscle of loin
pterygoid	**ter**-i-goyd	Gr. wing-shaped
pubis	**pu**-bis	NL. bone of groin
quadratus	**kwod**-rait-us	L. squared
quadriceps	**kwod**-ri-seps	
quadruped		Gr. four-footed
radius	**ra**-de-us	L. ray
ramus	**ray**-mus	L. branch
rectus	**rek**-tus	L. straight
retinaculum	ret-i-**nak**-u-lum	L. halter
rhomboid	**rom**-boyd	
rotatores	ro-ta-to-**rez**	
sacrum	**sa**-krum	L. sacred, from its use in Roman animal sacrifice
saphenous	**sa**-fen-us	Gr. *saphen* clearly visible
sartorius	sar-**to**-re-us	L. tailor
scalene	**skay**-leen	Gr. uneven
scaphoid	**skaf**-oyd	Gr. skiff-shaped
scapula	**skap**-u-la	to dig, like the business end of a shovel
sciatic	si-**at**-ik	Gr. *ischion*, hip joint
semimembranosus	**sem**-e-**mem**-bra-no-sus	
semispinalis	**sem**-e-spi-**na**-lis	
semitendinosus	**sem**-e-ten-di-no-sus	
serratus	ser-**a**-tus	L. a notching
sesamoid	**ses**-a-moyd	L. resembling a sesame seed
skeleton		G. dried up

skull		ME. bow
soleus	so-lay-us	L. sole-shaped
sphenoid	sfe-noyd	G. wedge-shaped
spinalis capitis	spi-**na**-lis **kap**-i-tis	
spinalis cervicis	spi-**na**-lis **ser**-vi-sis	
spine		L. thorn
splenius	sple-ne-us	Gr. bandage
splenius capitis	sple-ne-us **kap**-i-tis	
sternocleidomastoid	ster-no-**kli**-do-**mas**-toyd	
sternohyoid	ster-no-**hi**-oyd	
sternothyroid	ster-no-**thi**-royd	
sternum	ster-num	Gr. chest
stylohyoid	sti-lo-**hi**-oyd	
styloid	sti-loyd	Gr. a pillar
subclavius	sub-**kla**-ve-us	
subscapularis	sub-skap-u-**la**-ris	
superficialis	soo-per-fish-e-**a**-lis	
supinate	su-pi-nate	L. bent backward
supraspinatus	soo-pra-spi-**na**-tus	
sustentaculum	sus-ten-**tak**-u-lum	L. to support
suture	su-chur	L. a seam
symphasis	sim-fi-sis	Gr. growing together
synovial	si-**no**-vi-al	egg white
talus	ta-lus	L. ankle
tarsal	tar-sul	Gr. wickerbasket
temporalis	tem-po-**ra**-lis	
tendon	ten-dun	Gr. to stretch
tensor	ten-sor	L. a stretcher
teres	te-reez	L. round
tertius	ter-she-us	L. third
thoracic	tho-**ras**-ik	Gr. chest
thorax	tho-raks	

thyrohyoid	thi-ro-**hi**-oyd	
thyroid	thi-royd	Gr. shield
tibia	tib-e-a	L. shinbone
trachea	tray-ke-a	Gr. rough artery
trapezium	tra-**pee**-ze-um	
trapezius	tra-**pee**-ze-us	Gr. a little table
trapezoid	trap-e-zoid	
triquetrum	tri-**kwe**-trum	L. three-cornered
trochanter	tro-**kan**-ter	Gr. to run
trochlea	trok-le-a	Gr. pulley
tubercle	tu-ber-kl	L. a little swelling
tuberosity	tu-ber-**os**-i-te	
ulna	ul-na	L. elbow
umbilicus	um-**bil**-i-kus	L. a pit
uvula	u-vu-la	L. a little grape
vastus	vas-tus	L. vast
vertebra	ver-ta-bra	
xiphoid	zif-oyd	Gr. sword-shaped
zygomatic	zi-go-**mat**-ik	Gr. cheekbone

Index

Order Form

Telephone orders:
Toll Free: (800) 775-9227
Have your VISA or MasterCard ready.

Postal orders:
Books of Discovery, PO Box 6107, Boulder, CO, 80306
(Make checks payable to: Books of Discovery)

Please send _____ copies of *Trail Guide to the Body* ($42.95 each).
You may return any book for a full refund, no questions asked.

Company Name: _____

Name: _____

Address: _____

City: _____ State: _____ Zip: _____

Telephone: (_____) _____

Sales Tax:
Please add 3.8% for books shipped to Colorado addresses.

Shipping:
$5.00 for the first book and $1.50 for each additional book. Shipping may take two to three weeks. School/Clinic bulk order discounts available.

Payment:

Check ☐

Credit Card: ☐ VISA ☐ MasterCard

Card number: _____

Name on card: _____ Exp. date: _____ / _____

Call toll free and order now!

Order Form

Telephone orders:
Toll Free: (800) 775-9227
Have your VISA or MasterCard ready.

Postal orders:
Books of Discovery, PO Box 6107, Boulder, CO, 80306
(Make checks payable to: Books of Discovery)

Please send _____ copies of *Trail Guide to the Body* ($42.95 each).
You may return any book for a full refund, no questions asked.

Company Name: _____

Name: _____

Address: _____

City: _____ State: _____ Zip: _____

Telephone: (_____) _____

Sales Tax:
Please add 3.8% for books shipped to Colorado addresses.

Shipping:
$5.00 for the first book and $1.50 for each additional book. Shipping may
take two to three weeks. School/Clinic bulk order discounts available.

Payment:

    Check ☐

    Credit Card: ☐ VISA ☐ MasterCard

Card number: _____

Name on card: _____ Exp. date: _____ / _____

Call toll free and order now!